Dictionary of
Dangerous Pollutants,
Ecology,
and Environment

Dictionary of Dangerous Pollutants, Ecology, and Environment

DAVID F. TVER

INDUSTRIAL PRESS INC.
200 Madison Avenue
New York, New York 10157

Library of Congress Cataloging in Publication Data

Tver, David F.
   Dictionary of dangerous pollutants, ecology, and environment.

   1. Pollution—Dictionaries. 2. Factory and
trade waste—Dictionaries. 3. Ecology—Dictionaries. 4. Power resources—
Dictionaries.
   I. Title.
TD173.T83   304.2′03′21   81-1881
ISBN 0-8311-1060-0   AACR2

First Printing
DICTIONARY OF DANGEROUS POLLUTANTS, ECOLOGY, AND ENVIRONMENT

\

Introduction

We realize today that we live in a constantly changing environment. In the past, however, we took for granted that the air we breathed and the water we drank would remain clear and clean. We believed that there would be enough oil and gas to last for at least several more generations. We increased our consumption of fossil fuels to such an extent that we created serious pollution problems. We were not particularly interested in what happened to industrial wastes that contaminated rivers, lakes and streams, and underground water supplies.

Suddenly, a new generation wondered whether their children would have a world fit to live in. This new generation has become involved in attacking the problem head-on. They have made it a point to become better informed as to the dangers of pollution, from whatever source, and as to the best methods of attacking this problem, which all of us face everyday.

The air you breathe can sometimes make you sick, create permanent injury, or even kill you. Scientific and medical studies have shown positive correlations between air pollution and increases in respiratory ailments, heart disease, and cancer. Just what role pollutants play in making people sick is not yet fully understood; no single disease, but a mixture of ailments are involved, and pollutants usually occur in varying combinations, rather than one at a time.

The atmosphere is a dynamic system. It steadily absorbs a range of solids, liquids, and gases from both natural and man-made sources. These substances may travel through the air, disperse, and react among themselves and with other substances, both chemically and physically. Eventually, whether or not in their original form, they may reach a sink, such as the ocean, or a receptor, such as a man. Some, such as helium, escape from the earth's atmosphere. Others, such as carbon dioxide, may enter the atmosphere faster than they return to their sinks and thus gradually accumulate in the air.

Clean air contains 78.09 percent nitrogen by volume and 20.94 percent oxygen. The remaining 0.97 percent of the gaseous constituents of dry air includes small amounts of carbon dioxide, helium, argon, krypton, and xenon, as well as very small amounts of other inorganic and organic gases whose concentrations may differ with time and place. Water vapor is normally present in air in concentrations of 1–3 percent. The air also contains aerosols—dispersed solid or liquid particles—which range in size from clusters of a few molecules to a diameter of a few tens of microns.

The greatest long-term need for deeper understanding of the air environment lies in the most crucial area: the modes of action and effect of pollutants on man, animals, plants, and inanimate objects. The acute toxicological effects of most air contaminants are reasonably well understood, but the effects of exposure to heterogeneous mixtures of gases and particles at very low concentrations are only just beginning to be comprehended.

A body of water, like an air mass, is a dynamic system, steadily absorbing a range of solids, liquids, and gases, both natural and man-made. Natural waters, moreover, normally team with living organisms, which can powerfully affect the course of events in a given water system. All of these substances may flow, disperse, and interact chemically and physically before they disperse, such as in the ocean or in a receptor such as a fish. En route from source to dispersion they may assume a variety of chemical and physical forms.

Determining the relative significance of man-made sources of water pollution is complicated by the fact that contaminants often enter water in complex mixtures of many substances whose specific chemical identification is largely unknown. For practical purposes, however, waste streams can be described in terms of certain collective characteristics. One of these characteristics is bio-chemical oxygen demand (BOD), which is a measure of the weight of dissolved oxygen consumed in the biological processes that degrade organic matter which enter natural waters. Another collective characteristic is the weight of suspended solids, only part of which is settleable in the waste stream.

The leading source of controlled man-made water pollution or pollutants in the United States is manufacturing and the second leading source is domestic wastes. Other sources include agricultural and urban runoff, acid mine drainage, watercraft, and livestock feedlots.

The main inorganic constituents of most wastes include ions such as sodium, potassium, ammonium, calcium, magnesium, chloride, nitrate, nitrite, bicarbonate, sulfate, and phosphate. The specific organic compounds in waterborne wastes are less well known.

The processes that degrade and convert substances in water to other chemical and physical forms are extremely complicated because of the effects of aquatic life, mainly microorganisms. Microorganisms may control the soluble concentrations of an element in water, notable examples being carbon, nitrogen and phosphorus, the major elements in cells. Microorganisms may convert organic compounds in the water partly into carbon dioxide, or convert dissolved carbon dioxide into organic compounds.

The human body is a complicated organism. Individuals vary widely in their reactions to bodily stress. They also vary widely in occupations and habits that help determine the amounts of pollution they are exposed to.

Our advanced industrial economy relies on thousands of substances which did not exist until the last four decades. These substances were developed because of their extraordinary properties of strength, durability, reactivity, conductivity, or toxicity. Everyone is exposed in one way or another to these substances with every breath of air, drink of water, and swallow of food.

The word "pesticide" is a general term that covers fungicides, herbicides, insecticides, fumigants, and rodenticides. The synthetic organic chemicals are the most important of these compounds, in terms of both rate of growth and potential for contaminating the environment. More than 300 organic pesticidal chemicals are in use in the United States in more than 10,000 formulations, but relatively few of them are used in large enough amounts and are sufficiently long-lived to offer potential environmental hazard.

Both among chemical classes and within classes, the properties of these pesticides can differ widely; their potential as environmental contaminants can vary accordingly. Among the properties that are important in the latter respect are the pesticide's tendency to vaporize, its tendency to dissolve in water and other solvents, and its degree of resistance to various degradation processes.

Pesticide residues in soils may pose several problems to agriculture. Conceivable effects of pesticide residues in soil are injury to crops grown in later years, production of illegal residues in crops that absorb them, or harmful effects on living organisms in the soil. The largest amounts of residues usually result from contamination after crops are sprayed or from applying pesticides directly to the soil. The most common pesticide residues in soils are those of the chlorinated hydrocarbon insecticides. Most other pesticides, such as the organophosphorus and carbamate types, decompose in the soil rapidly enough so that their residues disappear between crop seasons. Most herbicides decompose rapidly in soils, chiefly through the action of microorganisms. The persistence of pesticides in the soil depends on many factors such as soil type, soil moisture, soil temperature, wind or air movement, cover crops, soil cultivation, method of application to the soil, formulation, and soil microorganisms.

Pesticides may contaminate surface and groundwater because of aerial spraying, runoff from treated areas, percolation through soil to groundwaters, waste discharge by pesticide producers, misuse, and other means. Pesticide residues in water may reach humans through drinking water; however, there is no evidence at present to suggest that long-term consumption of such water would produce harmful effects.

To control toxic substances, we must predict their environmental behavior and effects. Every research tool available—epidemiology, toxicology, and chemical analysis—should be used to predict accurately the harmful biological effects of any chemical through any exposure route.

Since the supplies of oil and natural gas are being rapidly depleted, it is necessary to develop other possible sources of energy, e.g., solar, nuclear, coal, and geothermal. We have to learn to conserve the resources we have and to find ways to use waste through biomass conversion, recycling, and other methods.

Coal is this nation's most abundant energy resource, and its chemical conversion into gas offers several significant benefits. The gasification of coal to produce SNG is nearly twice as efficient as the burning of coal to produce electric energy. Reduced pollution is an advantage in that during the production of a given amount of energy, the emission of pollutants is significantly lower from coal gasification than from the combustion of coal. Coal gasification offers promise in helping to alleviate the national shortage of clean energy fuels, including natural gas.

Geothermal energy at present represents a very small fraction of the world total power production, and although geothermal power is expected to rise fairly rapidly during the next decade, the total power contribution from geothermal energy will remain a very small fraction of the total. Nevertheless, for certain developing countries geothermal power may represent an impressive contribution. Similar considerations apply to the use of geothermal energy for industrial and other purposes. The long-term prospects of geothermal energy are prodigious, for within the interior regions of the earth is a store of energy so vast that other energy sources which are now being used pale into insignificance when compared to it. The problem is to tap it, since Nature allows only a fraction of it to leak through near the surface.

Radioactivity in the atmosphere comes from both natural sources and human activities. Natural radioactivity is either of terrestrial origin, consisting mainly of radiative gases, such as radon and thoron, which are released from soils and rocks, or it is produced by the interaction of cosmic radiation with atmospheric constituents. Man-made radioactive contamination of the atmosphere results from the reactor fuel cycle; the use of nuclear energy as a source of propulsive power; the use of radioisotopes in industry, agriculture, medicine, and scientific research; and nuclear weapons testing.

Certain radioactive pollutants are highly persistent and when forced into the stratosphere during a nuclear weapons test, may fall out slowly. They cannot be made harmless by any means now known. Some of the pollutants can cause damage to various parts of the body and produce irreversible changes in genetic material.

It is difficult to assess all the possible effects of radioactive pollution on humans. The effects may be masked since they are the kind that might be produced by pollution or other factors. Because of this we presently must depend on statistical data. It might be several generations before scientists will have the evidence necessary to say that pollution probably does or does not produce certain effects.

Solar energy, the energy received by the earth from the sun, has provided directly and indirectly almost all the sources of energy for the earth since its creation. The human race and all animal and plant life upon the earth have always depended for existence on the sun's energy. The amount of energy reaching the earth's surface is so vast as to be almost incomprehensible. The human race has used solar energy for thousands of years for heating and a wide variety of applications; however, it is only now that we are making any real effort to develop the sun's full potential as an energy source. In the total energy picture, solar energy is one of the major alternative sources and offers promise of making significant and long-range contributions to the solution of our energy problems.

If we are to live in this world and participate in the decisions which effect our lives and the lives of our children, we should be better informed in the areas of ecology, environment, pollution, and energy. This environmental and ecology dictionary was developed for those working in industry, for the student who needs a comprehensive manual, and for the public who would like a better understanding of what is going on and the meaning of what is being discussed. To accomplish this, the *Dictionary of Dangerous Pollutants, Ecology, and Environment* covers the following areas: noise pollution, air pollution, water pollution, nuclear energy, geothermal energy, solar energy, solar design and construction, coal, coal gasification, waste control, biomass conversion, recycling, ecology, meteorology, and climatology.

This dictionary is possibly one of the most comprehensive volumes on these subjects and should be of interest to everyone who desires to see that this world becomes a better, safer, and healthier place to live.

List of Abbreviations and Acronyms

ABS	acrylonitrile butadiene styrene
ABS	alkylbenzene sulfonate
AHH	arylhydrocarbon hydroxylase
API	American Petroleum Institute
ASA	American Standards Association
BATEA	best available technology economically available
BNA	β-naphthylamine
BOD	biochemical oxygen demand
Btu	British thermal unit
CAG	Carcinogen Assessment Group
CAMP	continuous air monitoring program
cfm	cubic feet per minute
COD	chemical oxygen demand
COH	coefficient of haze
COP	coefficient of performance
DDM	diaminodiphenyl methane
DECC	diethycarbamoyl chloride
DMCC	dimethylcarbamoyl chloride
DO	dissolved oxygen
DOE	U.S. Department of Energy
EDB	ethylene dibromide
EIA	environmental impact assessment
EIS	environmental impact statement
EPA	Environmental Protection Agency
GAO	General Accounting Office
HEW	U.S. Department of Health, Education & Welfare
HMPA	hexamethylphosphoric triamide
Hz	hertz; one cycle per second
LC	lethal concentration
LD	lethal dose
LPG	liquefied petroleum gas
MAC	maximum allowable concentration
MHD	magnetohydrodynamics
NAPC	National Air Pollution Control Administration
NASN	National Air Sampling Network

NIOSH National Institute For Occupational Safety & Health
NTA nitrilotriacetic acid
ORD Office of Research & Development
(for the U.S. Environmental Protection Agency)
OSHA Occupational Safety & Health Administration
OTEC ocean thermal energy conversion

PAN peroxyacetyl nitrate
PBB polybrominated biphenyls
PBNA phenyl-β-naphthylamine
PCB polychlorinated biphenyls
ppb parts per billion
ppm parts per million
PSI Pollution Standard Index
psi pounds per square inch

RCRA Resource Conservation and Recovery Act

SIL speech interference level
SLP sound pressure level
SS suspended solids
STP standard temperature and pressure

TCE trichloroethylene
TLV threshold limit value
TOC total organic carbon
TOSCA Toxic Substances Control Act of 1976

A

ab. A prefix attached to the names of the practical electric units to indicate the corresponding unit in the cgs electromagnetic system, e.g., abampere, abvolt.

abatement. A measure taken to reduce or eliminate air or noise pollution, which may involve legislative proceedings and technological applications.

abcoulomb. The electromagnetic unit of charge. The charge which passes a given surface in one second if a steady current of one abampere flows across the surface. Its dimensions are therefore $cm^{1/2}$ g $m^{1/2}$, which differ from the dimensions of the statcoulomb by a factor which has the dimensions of speed. This relationship is connected with the fact that the ratio $2K_e/K_m$ must have the value of the square of the speed of light in any consistent system of units. It further follows that one abcoulomb = $2.997\ 93 \times 10^{10}$ statcoulomb, the speed of light *in vacuo* being $(2.997\ 93 \pm 0.000\ 003) \times 10^{10}$ cm/s.

Abegg's rule. In a helical periodic system, if the maximum positive valence exhibited by an element is numerically added to its maximum negative valence, the sum will tend to equal eight. This tendency is exhibited especially by the elements of the fourth, fifth, sixth, and seventh groups.

absolute humidity. The weight of the water vapor found in a unit volume of air. In the metric system it is the number of grams of water vapor in a cubic meter of air.

absolute pressure. The total pressure exerted by a gas or liquid. The absolute pressure of a vacuum is 0 lb/in.², and that of the atmosphere, about 14.7 lb/in.².

absolute temperature scale. A temperature scale based upon the value of zero as the lowest possible value. Thus all obtainable temperatures are positive. The Kelvin and Rankine scales are absolute scales.

absolute unit. A unit based on the smallest possible number of independent units. Specifically, units of force, work, energy, and power not derived from or dependent on gravitation.

absolute viscosity. The force that will move 1 cm² of plane surface with a speed of 1 cm/s relative to another parallel plane surface from which it is separated by a layer of liquid 1 cm thick. This viscosity is expressed in dynes per square centimeter, its units being the poise, which is equal to 1 dyn s/cm².

absolute zero. The temperature at which a gas would show no pressure if the general law for gases held for all temperatures. It is equal to −273.16°C or −459.69°F.

absorber. (1) A black material which absorbs heat from sunlight. (2) An apparatus used in the process in which one material is employed to retain another. Absorbers are used to selectively remove a gaseous or liquid material from another gas or liquid. Usually the process is performed in cylindrical towers packed with an absorbing material. Devices used for sampling by absorption include the following: scrubber, impinger, packed column. This equipment includes spray chambers, mechanical contactors, bubble cap or sieve plate contactors, or packed towers. A spray chamber is an empty chamber through which a gas stream is passed through curtains of liquid spray. Bubble cap or sieve plate equipment requires that the gas be passed upward through a series of plates on which pools of absorbent are located. Bubble cap trays are used in bubble cap equipment, and porous or perforated plates in sieve plate equipment to support the liquid layers. Packed towers allow liquid absorbent to flow by gravity downward through a bed of packing material while the gas stream moves either concurrently or countercurrently through the tower. In each case, intimate gas–liquid contact is promoted over larger interfacial areas.

absorber plate. A black-painted, flat piece of metal which absorbs sunlight, transforms it to heat by absorption, and in turn radiates heat into the surrounding air and the cover. The absorber plate in a fluid-medium collector is usually constructed differently because it must contain the conduit or piping-circuit medium in the absorber. There are many different designs, with copper and aluminum being the two most commonly used metals.

absorptance. The ratio of the radiant flux absorbed by a body to that which is incident upon it.

absorption. (1) A process in which one material (the absorbent) takes up and retains another (the absorbate) with the formation of a homogeneous solution. The process may involve the physical solution of a gas, liquid, or solid in a liquid, or the chemical reaction of a gas or liquid with a liquid or solid. Absorption may also refer to the process by which molecules of a gas, vapor, liquid, or dissolved substance become attached to a solid surface by physical forces. (2) A process by which a soluble gas is transferred from a gas stream into a liquid. The gas may become physically dissolved in the liquid or may react with a dissolved constituent in the liquid. Absorption serves as a method both for recovering gaseous components and for purifying gas streams. In many air pollution control operations, soluble gaseous components to be removed constitute levels of 1% or less of the main gas. Gas absorption is a diffusional operation which depends upon the rate of molecular and eddy diffusion. Ultimately, the transfer must take place across a liquid–gas interface. The interface may be formed by the use of liquid films, gas bubbles, and liquid droplets.

absorption band. A band of wavelengths (or frequencies) in the electromagnetic spectrum within which radiant energy is absorbed by a substance.

absorption chiller. A refrigeration system used in the solar cooling of buildings, where sun-heated water, usually above 190°F, provides the operating energy.

absorption coefficient. (1) The ratio of the sound energy absorbed by a surface of a medium (or material) exposed to a sound field (or to sound energy incident on the surface). The stated values of this ratio hold for an infinite area of the surface. The conditions under which measurements of absorption coefficients are made are to be stated explicitly. (2) A measure of the amount of normally incident radiant energy absorbed through a unit distance or by a mass of absorbing medium; also, the maximum volume of gas that can be dissolved in a unit volume of water. The absorption coefficient of gases generally decreases with increasing temperature and salinity. (3) The fractional decrease in the intensity of an x-ray or gamma-ray beam per unit thickness (linear absorption coefficient), per unit mass (mass absorption coefficient), or per atom (atomic absorption coefficient) of absorber, caused by the disposition of energy in the absorber. The total absorber coefficient is based on the sum of the individual energies.

absorption coefficient, atomic. The linear absorption coefficient of a nuclide. It is equivalent to the nuclide's total cross section for the given radiation.

absorption coefficient, Compton. The fractional decrease in the energy of an x-ray or gamma-ray beam, caused by the deposition of the energy to electrons produced by the Compton effect in an absorber.

absorption coefficient, mass. The linear absorption coefficient per centimeter divided by the density of the absorber in grams per cubic centimeter. It is frequently expressed as u/p, where u is the linear absorption coefficient and p is the absorber density.

absorption equipment, floating bed scrubber. Essentially a floating-bed–sieve-tray arrangement. Low-density spherical packing is placed on the sieve tray so that the gas bubbles rise through the wetter sphere. This system is reported to be advantageous for use with gas streams containing particulate matter, as the fluidized character of the bed of spheres prevents the buildup of a particulate sludge.

absorption equipment, packed towers. A packed tower is usually a vertical, cylindrical shell packed with solid objects which maximize the gas–liquid interfacial area at low resistance to gas flow. Packing such as Raschig rings, Berl saddles, Pall rings, Intaloz saddles, and Tellerettes may be used. The liquid enters at the top of the tower and is distributed over the packing with spray nozzle, weirs, or perforated plates. The gas enters at the bottom of the tower.

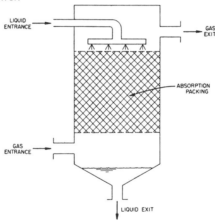

Countercurrent packed absorption tower.

absorption equipment, plate or tray towers. Plate or tray towers contain horizontal plates or trays, usually in a cylindrical shell. Sieve plates and bubble cap plates are commonly used in plate towers. The gases flow upward and bubble through the perforations in the bubble cap or sieve plate. The liquid flows downward through an overflow plate. The gas is prevented from flowing up the liquid downflow by placing the pipe exit below the level of the pool of liquid on the plate. The bubbles provide the gas–liquid interfacial area through which the mass transfer occurs.

absorption equipment, spray towers. Spray towers have a much smaller gas–liquid interfacial area than packed or plate towers. The liquid is introduced through spray nozzles and flows countercurrent, crosscurrent, or co-current with the gas, depending on the design. The pressure drop through a spray tower is substantially less than through either packed or plate towers.

absorption equipment, Venturi scrubbers. A Venturi scrubber is a co-current process with the liquid entering in or near the Venturi throat and flowing with the gas into an entrainment separator. Because of the large pressure drops incurred (10–100 in. of water) and the co-current flow arrangement, Venturi scrubbers do not appear to be best suited for the removal of pollutant gases.

absorption factor. The ratio of the intensity loss by absorption to the total original intensity of radiation. If I_0 represents the original intensity, I_r represents the intensity of reflected radiation, and I_t represents the intensity of the transmitted radiation, the absorption factor is given by the expression $[I_0 - (I_r + I_t)]/I_0$.

absorption line. A minute range of wavelengths (or frequencies) in the electromagnetic spectrum within which radiant energy is absorbed by the medium through which it passes. Each line is associated with a particular mode of electronic excitation induced into the absorbing atoms by the incident radiation.

absorption rate. For surfaces, the energy absorbed per unit area per unit time, expressed in the same units as emissive power. It depends on the incident radiation and the surface characteristics of the body. An absorption rate may be considered for a band, a single wavelength, or the entire spectrum.

absorption ratio differential. The ratio of the concentration of an isotope in a given organ or tissue to the concentration that would be obtained if the same administered quantity of this isotope were uniformly distributed throughout the body.

absorption spectrum. The spectrum obtained by the examination of light from a source, itself giving a continuous spectrum, after this light has passed through an absorbing medium in the gaseous state. The absorption spectrum will consist of dark lines or bands, being the reverse of the emission spectrum of the absorbing substance. When the absorbing medium is in the solid or liquid state, the spectrum of the transmitted light shows broad dark regions which are not resolvable into lines and have no sharp or distinct edges.

absorptivity. The fractional part of the incident radiation that is absorbed by the surface in question. It also varies with the wavelength of the incident radiation and with the temperature of the body. Like the emissivity, it is a dimensionless number between zero and one and can be used as a coefficient of absorption. Also, the ratio of the radiation absorbed by any substance to that absorbed under the same conditions by a black body.

abvolt. The cgs electromagnetic unit of potential difference and electromotive force. It is the potential difference that must exist between two points in order that one erg of work be done when one abcoulomb of charge is moved from one point to the other. One abvolt is 10^{-8} V.

acceleration. (1) The time rate of change of velocity in either speed or direction (cgs unit, one centimeter per second per second). (2) A vector that specifies the time rate of change of velocity. Various self-explanatory modifiers such as peak, average, and rms are often used. The time interval over which the average is taken must be indicated.

acceleration due to gravity. The acceleration of free-falling body in a vacuum. The International Committee on Weights and Measures has adopted 980.665 cm/s² or 32.174 ft/s² as a standard or accepted value.

accelerator. (1) A machine for speeding up subatomic particles to energies running into the millions of electron volts. In an accelerator, basically, particles are accelerated by being attracted to an electrode of opposite charge. If protons are placed at the entrance of a hollow tube, with zero voltage at the entrance and five million volts at the exit, the protons are attracted to the negative end and acquire five million volts of kinetic energy in moving from the entrance to the exit of the tube. In cyclotrons and synchrotrons the particles are made to go in circles by imposing a magnetic field, which causes charged particles to bend their path. (2) In nuclear physics, a device for speeding up charged subatomic particles to energies high enough to smash the nuclei of target atoms; often called an atom smasher. Accelerators are used routinely to produce radioisotopes.

accelerator, linear. A device for accelerating charged particles employing alternate electrodes and gaps arranged proportionally in a straight line, so that when their potentials have the proper amplitudes and frequencies, particles passing through them receive successive increments of energy.

acclimation. The physiological and behavioral adjustments of an organism to changes in the environment.

acclimatization. The adaptation over several generations of a species to a marked change in the environment.

accretion. A process that increases the size of particles by external additions; a form of agglomeration.

accumulator. A small tank installed in the return pipe from the circulating pump in a liquid-type solar heating system. Allows for expansion and contraction of water with temperature changes, provides a convenient place for air to be purged from the system, and makes sure the pump will always be primed for starting.

acenaphthene. $C_{10}H_6(CH_2)_2$, a compound that occurs in the form of white, elongated crystals. It is irritating to the skin and mucous membranes and ingestion of large quantities can cause acute nausea and vomiting. Acenaphthene is an experimental neoplasm and one variation, 5-nitroacenaphthene, has been placed on the list of candidates for review by OSHA as a substance that may cause an increased incidence of benign or malignant tumors among workers exposed to it.

acetamide. CH_3CONH_2, a compound characterized by colorless crystals and a mousy odor. Exposure to the compound can result in irritation of the tissues in contact with the substance. The LD_{50} for experimental animals has been reported as 360 mg/kg. Acetamide has been identified as a carcinogen, and it has been included in a list of suspected occupational sources of cancer to be evaluated by OSHA.

acetanilide. C_8H_9NO, a compound of the aniline category, occurring as white, shining crystalline scales. Exposure to acetanilide can cause contact dermatitis, while skin effects such as eczematous eruptions can also occur through exposure by ingestion or inhalation. The oral LD_{50} has been established as 800 mg/kg in the rat and 1200 mg/kg in the mouse. Acute poisoning in humans has occurred from the ingestion of a few grams of the chemical. Cyanosis, anemia, and damage to blood-forming organs have been reported in patients using medications containing acetanilide. Alterations in red blood cells, with the hemoglobin being converted to methemoglobin or sulfhemoglobin, have been reported in the clinical literature. The 4-phenyl form of acetanilide has been included in the list of suspected occupational carcinogens to be reviewed by OSHA.

acetone. C_3H_6O, a volatile ketone used as a solvent and in the synthesis of other organic compounds. It occurs in small amounts in the urine of normal humans and in large amounts in the urine of diabetic patients. Industrial production of acetone averages over 2000 million pounds per year in the United States, with much of the output employed in the manufacture of celluloid, gunpowder, varnishes and chloroform. Acetone has been found to have a relatively low toxic effect through dermal exposure, although skin irritation is a common result because of the defatting action of the ketone which can damage or destroy subcutaneous tissues. Considerable discomfort can also result from exposure through oral or respiratory routes, with nonfatal, but possibly irreversible, changes to tissue surfaces that come in contact with the substance. The lowest reported toxic concentration for humans is 500 ppm, with conjunctival effects. LD_{50}'s for laboratory animals range from approximately 1300 mg/kg for a mouse via intraperitoneal exposure to 20,000 mg/kg for a rabbit administered acetone dermally. Threshold limit values established by humans for exposure to acetone are 1000 ppm for a time-weighted average concentration and 1250 ppm for short-term exposure.

achromatic. A term applied to lenses signifying their more or less complete correction for chromatic aberration.

acid. For general purposes, a hydrogen-containing substance which dissociates in water solution to produce one or more hydrogen ions. More generally, however, acids are defined according to other concepts. The Bronsted concept states that an acid is any compound which can furnish a proton. Thus NH_4^+ is an acid, since it can give up a proton: $NH_4^+ \rightleftharpoons NH_3 + H^+$. NH_3 is a base, since it accepts a proton. A still more general concept is that of G. N. Lewis, who defines an acid as anything which can attach itself to something with an unshared pair of electrons. Thus in the reaction

$$H^+ + :N \underset{\displaystyle H}{\overset{\displaystyle H}{{<}\!{H}}} \rightleftharpoons NH_4^+$$

the NH_3 is a base because it possesses an unshared pair of electrons. This latter concept explains many phenomena, such as the effect of certain substances other than hydrogen ions in the changing of the color of indicators. It also explains acids and bases in nonaqueous systems such as liquid NH_3 and SO_2.

acid dew point. The dew point of flue gases containing little or no SO_3 is known as the water dew point and is usually in the region of 120°F (49°C); if SO_3 is present in any significant quantity, the dew point is raised considerably, to about 300°F (149°C). This is known as the acid dew point. Acid condensation begins at this temperature, heavy condensation occurring when the surface temperatures fall below 250°F (121°C).

acid number. The number of milligrams of potassium hydroxide required to neutralize the total acidity in one gram of fat, oil, wax, free fatty acids, etc.

acid precipitation (acid rain). A term applied to an increased acidity of wet or dry precipitation, generated by the release of sulfates and nitrates in the atmosphere. With switches from coal to oil and natural gas, and the reduction of particulates that neutralize the sulfates and nitrates released by combustion, there has been an increase in acid rain throughout the world. Not all acid rain is man-made. About half is from natural sources, such as volcanoes and plants (photosynthesis products). There is some belief that acid rain may be responsible for substantial adverse effects on public welfare. These effects may include acidification of lakes, rivers, and groundwaters, with resultant damage to fish and other components of the aquatic ecosystem. According to *Science News*, February 2, 1979, Adirondack lakes were becoming fishless because of acid rain and a 1978–1979 survey of 85 lakes in the Boundary Waters Canoe Area along the Minnesota–Ontario border showed that two-thirds of them were near the brink of acidity where fish-life could not be supported. It was also reported that the nearby Great Lakes region was receiving precipitation that was 5–40 times more acidic than that of normal rain. Acid rain may cause acidification and demineralization of the soil, reduce forest productivity, damage crops, and deteriorate man-made materials. These effects could result from years of exposure or from short heat-acidity episodes.

acid-washed activated carbon. Carbon which has been in contact with an acid solution, with the purpose of dissolving ash in the activated carbon.

acidity. The capacity of wastewater to neutralize a base. It is normally associated with the pressure of carbon dioxide, mineral and organic acids, and salts of strong acids or weak bases.

acoustic. Containing, producing, arising from, actuated by, related to, or associated with sound. Acoustic is used when the term being qualified designates something that has the properties, dimensions, or physical characteristics associated with sound waves; acoustical is used when the term being qualified does not designate explicitly something that has properties, dimensions, or physical characteristics associated with sound waves.

acre foot. (1) The quantity of water required to cover one acre to a depth of one foot, or 43,560 ft³. (2) A term used in sewage treatment in measuring the volume of material in a trickling filter.

acrylonitrile. $CH_2=CHCN$, an explosive, flammable liquid produced by the reaction of propylene with ammonia and oxygen in the presence of a catalyst. About 1,500,000 lbs of acrylonitrile are manufactured each year in the U.S. for use in acrylic and modacrylic fibers employed in the production of clothing, carpeting, blankets, draperies, and upholstery. Other major uses include the manufacture of styrene-type resins, other plastic products, and latices; it is also used as a fumigant. The Food and Drug Administration has banned the application of acrylonitrile materials in the manufacture of beverage containers. NIOSH has estimated that at least 125,000 persons are employed in jobs that may involve contact with acrylonitrile. The chemical structure of acrylonitrile is similar to that of vinyl chloride, a substance that has been identified as a cause of human cancer; animal studies with acrylonitrile indicate that it may be a tumor-inducing agent, including carcinomas. A series of preliminary epidemiological studies by E. I. du Pont de Nemours and Co. indicated an excess risk of lung and colon cancer among workers with potential exposure to acrylonitrile and a cancer mortality rate that was double the expected rate for a population group of similar size. OSHA has recommended a standard for occupational exposure to acrylonitrile of 20 ppm for an eight-hour time-weighted period. However, NIOSH has advised that it "would be prudent to handle acrylonitrile in the workplace as if it were a human carcinogen."

acrylonitrile butadiene styrene (ABS). A black plastic material used to make pipes suitable for carrying sun-heated water in solar space heating and cooling, hot-water, and swimming pool installations.

actinide series. Elements of atomic numbers 89 through 103 analogous to the lanthanide series of the so-called rare earths.

activated carbon. Any form of carbon characterized by high adsorptive capacity for gases, vapors, and colloidal solids. Carbon must usually be activated to develop adsorptive power, achieved by heating to 800–900°C with steam or CO_2, which produces a porous particle structure. This material can be used for clarifying liquids and the purification of solutions (electroplating). The activity of (activated) carbon is the maximum amount of vapor which can be absorbed by a given weight of carbon under specified conditions of temperature, concentration of water vapor, and concentration of other vapors.

activated sludge. A flocculent suspension of aerobic microorganisms capable of effecting removal of polluting matter when aerated with sewage or similar liquids.

active layer. The layer of ground above the permafrost which thaws in the summer and freezes again in the winter.

active mass. The number of gram-molecular weights per liter in solution or in gaseous form.

active solar system. An assembly of collectors, thermal storage device(s), and transfer fluid, which converts solar energy into thermal energy, and in which energy in addition to solar energy is used to accomplish the transfer of thermal energy.

activity coefficient. A factor which, when multiplied by the molecular concentration, yields the active mass. The activity coefficient is evaluated by thermodynamic calculations, usually from data on the emf of certain cells, or the lowering of the freezing point of certain solutions. It is a correction factor which makes the thermodynamic calculations exact.

acute toxicity. Any poisonous effect produced by a single short-term exposure resulting in severe biological harm or death.

adaptation. A characteristic of an organism that improves its chance of surviving in its environment, e.g. the breeding behavior of the grunion and the water-storing capacity of a cactus.

additive. In the context of chemical heat-storage systems, a substance which may be added to other chemicals, resulting in a raising or lowering of the temperature at which a change of state occurs.

adhesion. The force of attraction between molecules of different kinds. When two different substances are in contact, both cohesion and adhesion exist; the characteristics of the particles of the substances determine the resultant force which controls the action at the contact surfaces of the substances.

adiabat. A line on a thermodynamic diagram representing a constant potential temperature.

adiabatic. Involving expansion or compression without gain or loss of heat. The line on the pressure–volume diagram representing the above change is called an adiabatic line.

adiabatic change. A change in the temperature of the air or other gas due specifically to a change in the pressure on the air or gas.

adiabatic efficiency. The degree to which a change in the state of a gas approaches an adiabatic process.

adiabatic lapse rate. The special-process lapse rate of a parcel of dry air as it moves upward in a hydrostatically stable environment and expands slowly to lower environmental pressure without exchange of heat; it is also the rate of increase in temperature for a descending parcel. The lapse rate is g/cpd, where g is the acceleration of gravity and cpd is the specific heat of dry air at constant temperature, equal to 9.767°C/km, or about 5.4°F/1000 ft.

adiabatic process. A thermodynamic change of state of a system in which there is no transfer of heat or mass across the boundaries of the system. In an adiabatic process, compression always results in warming, expansion, or cooling.

adiabatic rate. In a body of air moving upward or downward, the change in temperature that occurs without gain or loss of temperature to the air through which it moves.

adiabatic vaporization. The vaporization of a liquid with practically no heat exchange between the liquid and its surroundings. Normally, the vaporization of a liquid is accompanied by absorption of heat by that liquid.

adiabatic wall temperature. The temperature assumed by a wall in a moving fluid stream when there is no heat transfer between the wall and the stream.

adsorbate. A material adsorbed on the surface of another.

adsorber. An apparatus used for a process in which the molecules of either a gas or liquid are captured by solid material. Usually the capturing solid is activated carbon, which has a large surface area of capillary form (the carbon can be regenerated by steam, which drives off the adsorbed material in the form of a vapor that can be condensed along with the steam).

adsorption. The condensation of gases, liquids, or dissolved substances on the surfaces of solids; a process by which molecules from a gas or liquid stream attach themselves on the surface of a solid in extremely thin layers. The process is based on the attractive force between the solid surface and the adsorbed molecules. These surfaces vary considerably with the nature of the surface and molecules, and the intensity of the force is the basis of subdivision between physical and chemical (chemisorption) adsorption.

adsorption isotherm, activated carbon. A measurement of adsorption determined at a constant temperature by varying the amount of carbon used or the concentration of the impurity in contact with the carbon.

adulterant. A chemical impurity or substance that by law does not belong in a food, plant, animal, or pesticide formulation.

advanced waste treatment. Any treatment method or process employed following biological treatment to (1) increase the removal of pollution load, (2) remove substances that may be deleterious to receiving waters or the environment, (3) produce a high-quality effluent suitable for reuse in any specific manner or for discharge under critical conditions. The term "tertiary treatment" is commonly used to denote advanced waste treatment methods.

advection. The transportation of an atmospheric property solely by the mass motion (velocity field) of the atmosphere; the term is more particularly applied to the transfer of heat by horizontal motion of air. Fog drifts from one place to another by advection. The transfer of heat from high to low altitudes is the most obvious example of advection. Cold air moves from polar regions southward. Large-scale north–south advection is more prominent in the Northern Hemisphere than in the Southern Hemisphere, but west–east advection is prominent on both sides of the equator.

aeration. The intimate contact between air and a liquid by such methods as spraying the liquid in the air, bubbling air through the liquid, or agitating the liquid to promote surface absorption of air. In general, any process whereby a substance becomes permeated with air or another gas.

aerify. To change a liquid into a gaseous or vapor form by the introduction of air; to infuse or force air into.

aerobe. An organism which can live and grow only in the presence of oxygen. An organism which employs aerobic respiration.

aerobic. Depending on the presence of oxygen.

aerobic biological oxidation. Any waste treatment or process utilizing aerobic organisms, in the presence of air or oxygen, as the agent for reducing the pollution load or oxygen demand of organic substances in waste. The term is used in reference to secondary treatment of waste.

aerology. The study of the atmosphere, including the upper air as well as the more general studies understood by world meteorology; frequently used as limiting the study to the upper air. Also, the study of the free atmosphere throughout its vertical extent, as distinguished from studies confined to the layer of atmosphere adjacent to the earth's surface.

aeropause. A region of intermediate limits in the upper atmosphere considered as a boundary or transition region between the denser portion of the atmosphere and space.

aerosol. A suspension of fine solid or liquid particles in air or gas as mist, fog, or smoke; a colloidal system in which gas, usually air, is the continuous medium and particles of solid or liquid are dispersed in it. There is no clear-cut upper limit to the partical size of the dispersed phase in an aerosol, but as in all other colloidal systems, it is commonly set at one micron. Haze, most smokes, and some fogs and clouds may be regarded as aerosols.

afterburner. In incinerator technology, a burner located so that the combustion gases are made to pass through its flame in order to remove smoke and odors. It may be either attached or separated from the incinerator proper.

Agent Orange. A mixture of organochlorine herbicides consisting mainly of 2,4,5-trichlorophenoxyacetic acid (2,4,5-T) and 2,4-dichlorophenoxyacetic acid (2,4-D). The Agent Orange used by U.S. military forces in Vietnam reportedly included supplies that were contaminated by dioxin. In the more than 11,000,000 gal. of Agent Orange sprayed over Vietnamese jungles and rice fields during the 1960s was an estimated total of 100 kg of dioxin. It is believed that thousands of U.S. troops in Vietnam have been exposed to the defoliant Agent Orange. A report by the U.S. General Accounting Office (GAO) stated that some 6000 Marines were within a third of a mile of sprayed areas during and just after defoliation missions were flown. Effects of exposure to Agent Orange on humans began to appear around 1977, when U.S. soldiers who had been stationed in Vietnam began to show signs of liver and gastrointestinal disorders along with skin rashes, extreme fatigue, joint pains resembling arthritis, and numbness in the hands and feet. Many Vietnam veterans exposed to the substance claimed they suffered such effects as cancer, impotence, and personality changes. The U.S. Defense Department, Veterans Administration (VA), and Dow Chemical Co., which manufactured most stocks of the chemical, denied there was any firm evidence linking Agent Orange to any known disease. The GAO reported on November 24, 1979, that as of September 30, 1979, about 4800 former Vietnam servicemen had asked the VA for treatment of what the veterans believed were health problems related to contact with Agent Orange. Of those contacting the VA, some 750 submitted compensation claims. The VA stated that the 4800 or so veterans would be examined once a year for five years to determine whether they were suffering from symptoms linked to Agent Orange. The main components of Agent Orange, 2,4,5-T and 2,4-D, have been used extensively in the U.S. as herbicides to kill weeds along highways and utility lines, to eradicate undesirable plants in forests, pastures, and rangeland, and to control weeds in crops of sugarcane, fruit, wheat, and corn. Because the herbicides are rapidly excreted from the human body and are relatively biodegradable, it was long believed that they were safer to use than alternative herbicides. However, in 1970, laboratory tests indicated that 2,4,5-T could be a cause of teratogenic effects in animals and later evidence suggested that use of the herbicide could be a cause of a higher than average incidence of miscarriages among pregnant women living in forest areas that had been treated with the chemical. Clinical reports have shown that some persons exposed to large doses of 2,4-D experience muscular weakness, tremors, and, in severe cases, convulsions, paralysis, and even death.

agglomeration. The process by which precipitation particles grow larger by collision or contact with cloud particles or other precipitation particles.

agglutination. The process of uniting solid particles coated with a thin layer of adhesive material or of arresting solid particles by impact on a surface coated with an adhesive.

agricultural pollution. The liquid and solid wastes from farming, including runoff from pesticides, fertilizers, and feedlots; erosion and dust from plowing; animal manure and carcasses, crop residues, and debris.

air basin. A geographic region with physical features such as hills, mountains, and bodies of water which determine a common atmospheric interaction for that region. The United States was divided into air basins by a 1971 directive of the EPA. Prior to the EPA action, control of air quality was determind at the level of local governments, mainly by county regulations. The air basin approach to the control of air pollution was advanced as a more realistic method of dealing with the problem, since atmospheric activity transcends local political boundaries. Regional, subregional, and local air-quality standards may be estab-lished within an air basin because of influences of total emission of primary and secondary pollutants, as well as factors such as prevailing winds that may transport pollutants from an industrial site to a neighboring community. In some instances the downwind concen-tration of pollutants can be greater than in the area of a large industrial source because of meteorological effects and physical features of the terrain. High hills and mountains in some geographic regions limit the movement of air pollutants by presenting a barrier to the prevailing wind of the air basin.

air chemistry model. A laboratory representation of the conditions involved in the evolution of secondary pollutants from precursor chemicals in the presence of simulated meteorological influences. The data may be based on or combined with results of studies conducted through aircraft and ground-level monitoring programs. Tags or tracers are usually released to help assess the direction and rate of movement of the pollutants and their chemical precursors. Complicating the accuracy of the representation are the possible presence of substances not included in the laboratory model, residues in the atmosphere from previous days, and unpredictable dynamics of chemical activity as secondary pollutants undergo formation and meteorological factors tend to disperse the pollutants formed. The real atmosphere over a typical urban industrial area may contain more than 100 pollutant substances undergoing a variety of photochemical and other reactions which may or may not be detected by available instruments. As a consequence, most air chemistry models depend upon "barometer" pollutants or reactions, in which the presence of one leads to the assumption that certain others are present in the atmosphere. A basic air chemistry model begins with the effect of sunlight on nitrogen dioxide leading to ozone and nitric oxide, foilowed by the addition of various hydrocar-bons.

air classifier. A waste disposal machine consisting of a stream of air that forces garbage into a storage container. Heavy waste materials such as glass and metal fall out of a hole in the bottom and are recycled separately, other waste materials are forced into an incinerator. The hot gases that escape from these burning wastes power turbines which drive electrical generators.

air conduction. The process by which sound is conducted to the inner ear through the air in the outer ear canal.

air contaminant. Any particulate matter, gas, or combination thereof, other than water vapor or natural air.

air curtain. A method of containing oil spills: Air bubbling through a perforated pipe causes an upward water flow that slows the spread of oil. It can also be used to stop fish from entering polluted water.

air density. The ratio of the mass of air to its volume, expressed as weight per unit of volume.

air ion. A small particle (a molecule or a microscopic dust particle) in the atmosphere which has an induced electrical charge acquired through loss or gain of an electron, or through the adsorption of a molecule which has lost or gained an electron. Sizes are estimated as small as 0.001–0.005 μm; intermediate, 0.005–0.015 μm; large, 0.015–0.1 μm.

air mass. An enormous body of air, often hundreds of thousands of square miles, in which conditions of temperature and moisture are much the same at all points in a horizontal direction. Four fundamental types of air masses are the major weather-makers in the Northern Hemisphere. These are cold–dry, cold–moist, warm–dry, and warm–moist, which, respectively, are referred to as continental polar, maritime polar, continental tropical, and maritime tropical. In the northern half of the Northern Hemisphere, Alaska, Canada, and Siberia are the main source regions for continental polar air masses. The polar regions of the Atlantic and Pacific Oceans are the source regions for cold–moist air masses, the maritime polar variety. Tropical air masses have their origin in the torrid zone, approximately between the Tropic of Cancer and the Tropic of Capricorn. Air masses obtain their characteristics by radiation fluxes and by heat- and water-vapor fluxes through the air–earth boundary layers. The latitude and nature of the underlying surface determine the relative importance of the various processes.

air monitoring. See **monitoring.**

air pollutant. A substance added to the air in sufficient concentrations to produce a measurable effect on man or other animals, vegetation, or material. Pollutants may thus include almost any natural or artificial composition of matter capable of being airborne. They may occur as solid particles, liquid droplets, gases, or in various admixtures of these forms. It is convenient to consider two general groups: those emitted directly from identifiable sources and those produced in the air by interaction among two or more primary pollutants or by reaction with normal atmospheric constituents, with or without photoactivation. Excluding pollen, fog, and dust, which are of natural origin, about 100 contaminants have been identified, and they fall into the following categories: solids, sulfur compounds, organic compounds, nitrogen compounds, oxygen compounds, halogen compounds, radioactive compounds, odors. Solids include carbon fly ash, ZnO, and $PbCl$; gaseous and other airborne compounds include SO_2, SO_3, H_2S, mercaptans, aldehydes, hydrocarbons, tars, NO, NO_2, NH_3, O_3, CO, CO_2, HF, and HCl; radioactive compounds include radioactive gases and aerosols. Pollutants in industry come from many different origins. They can consist of a fume, dust, mist, or just an undesirable gas. Particles formed by mechanical disintegration may generally be classed in the micron-size group. These may be formed by pulverization, crushing, or grinding. Particles created by coarse disintegration, such as sawing, jaw crushing, or tumbling, may be above the 15-μm range. Particles created by a physical change of state, such as sublimation or chemical reformation, or from an intense heating or melting operation, are generally on the order of a micron. Furnace operations, in both steel and nonferrous industries, usually produce exhaust or combustion gases containing metallurgical submicron fumes. The generation of finer fumes is directly related to heat intensity. Pollutants having a composite combination of particles may be found in air which has been swept through the inside of a plant having various processes, or which has been trapped in the hooding around the operating equipment.

air pollution. The presence of contaminant substances in the air that do not disperse properly and interfere with human health.

air pollution control. Measures taken to maintain a degree of purity of air resources consistent with the promotion of public health and welfare, the protection of plant and animal life, the protection of property and other sources, the visibility requirements for safe ground and air transportation, and continued economic development and growth.

air pollution control (mists, dusts, and gaseous constituents). Determining gas density involves questions of whether moisture is present or whether the gas molecular weight is other than that of standard air. The basic technique for sampling aerosol mists can be done with an apparatus such as shown below. The apparatus is required to collect SO_3, P_2O_5, and monoethanolamine (MEA). Filter funnels, three in series, are filled with a 1-in. layer of glass fiber. The gas should have a pressure drop of approximately 10 in. Hg across the funnels for good mist collection. In the presence of water vapor the funnels are mounted in a wooden box fitted with light bulbs to prevent vapor condensation. From the box, impingers remove gaseous constituents such as SO_2. Gases moved by means of a vacuum pump or an ejector are then metered and their pressure and temperature recorded.

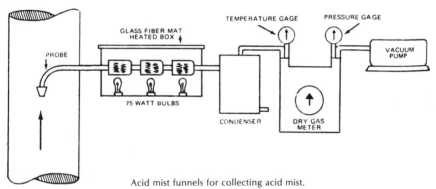

Acid mist funnels for collecting acid mist.

air pollution control (Smith–Greenberg impinger). Similar to a thimble holder, but used if other contaminants such as SO_2 are present. Since condensation causes an increase in the liquid value in the impinger, this amount is measured, calculated back to gaseous volume, and added to the metered air volume. The corrected volume is then used in the calculation of grains per cubic foot or milligrams per cubic foot of contaminants.

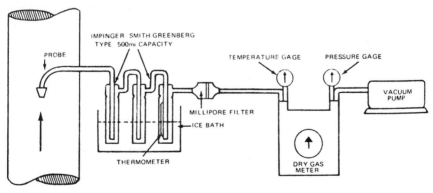

Smith–Greenberg impinger for collecting combined dusts, mists, and gaseous constituents.

air pollution control (thimble method). Utilizes a holder equipped with disks or cylinders made from Whattman paper or a lundum (aluminum oxide). Although the standard thimble holder is ideal for collecting samples of 1–20 g in weight, minute samples necessitate the use of a disk in a retainer ring. Normally, a stripheader or wraparound heating element is installed to keep the temperature above the dew point. After this, the gas enters a condenser or ice bath where water vapor is trapped out, proceeding then through a meter which takes pressure and temperature readings.

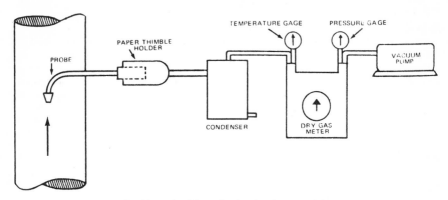

Thimble method for collecting dry dust material.

air pollution episode. A period of abnormally high concentration of air pollutants, often due to low winds and temperature inversion, that can cause illness and death.

air pollution index (API). An arbitrarily derived mathematical combination of air pollutants which gives a single number attempting to describe the ambient air quality. The formula API = 20 [SO_2 + 1 (CO) + 2 (smoke shade)] has no scientifically derived basis. Experience indicates that the average value of this index is 12.0. An index reading of 50.0 is considered adverse and cause for alarm.

air pollution sources (agricultural activities). *Categories:* crop spraying and dusting, field burning, frost damage control, pest and weed control, stubble and slash burning, and smudge pots. *Pollutants:* organic phosphates, chlorinated hydrocarbons, arsenic, lead, smoke, fly ash, and soot.

air pollution sources (combustion). *Categories:* fuel burning, motor vehicles, refuse burning, home heating units and power plants, automobiles, buses, trucks, community and apartment house incinerators, and open burning dumps. *Pollutants:* oxides of sulfur, oxides of nitrogen, carbon monoxide, smoke, fly ash, organic vapors, metal oxides, odors, and particles.

air pollution sources (dust producing processes). *Categories:* crushing, grinding, screening, demolition, milling in road mix plants, and grain elevators. *Pollutants:* mineral and organic particulates.

air pollution sources (manufacturing). *Categories:* metallurgical plants, chemical plants, and waste recovery such as smelters, steel mills, aluminum refineries, petroleum refineries, pulp mills, fertilizer plants, cement mills, metal scrap yards, auto body burning, and rendering plants. *Pollutants:* metal fumes (lead, arsenic, and zinc), fluorides and oxides of sulfur, hydrogen sulfides, fluorides, organic vapors, particles, odors, smoke, and soot.

air pollution sources (nuclear energy activities). *Categories:* ore preparation, fuel fabrication, nuclear fission, spent fuel processing, nuclear device testing, crushing, grinding, screening, gaseous diffusion, nuclear reactors, chemical separation, and atmospheric explosives. *Pollutants:* uranium and beryllium dust, fluoride, ^{41}Ar, ^{131}I, and radioactive fallout (^{90}Sr, ^{137}Cs, and ^{14}C).

air pollution sources (solvents). *Categories:* spray painting, inks, solvent cleaning, automobile assembly, furniture and appliance finishing, photogravure and printing, dry cleaning, and degreasing. *Pollutants:* hydrocarbons and other vapors.

air pressure. The force per unit area exerted by the air, $p = F/A$. It is not necessary to specify the direction of the force or the orientation of the unit area of surface, because the pressure is exerted equally in all directions. If the acceleration of gravity g is the only acceleration present and if M is the total mass of the overlying atmosphere, then $F = Mg$, which is the weight of the atmosphere. Under these conditions the atmosphere is said to be in static equilibrium, and the pressure is defined as the weight per unit area or the weight of a column of air of unit cross section (A) to the top of the atmosphere, i.e., $p = Mg/A$.

air quality control region. As the federal government uses the term, an area where two or more communities, either in the same or different states, share a common air pollution problem. Designated by the Secretary of Health, Education, and Welfare, these regions are required to set and enforce consistent air quality standards.

air quality criteria. As the federal government uses the term, the varying amounts of pollution and lengths of exposure at which specific adverse effects to health and welfare take place.

air quality standard. As the federal government uses the term, the prescribed level of a pollutant in the outside air that cannot be legally exceeded during a specified time in a specified geographical area.

air sampling, freeze-out. An effective but complex method for analyzing atmospheric hydrocarbon vapors. Low temperature condensation using a freeze-out trap essentially involves U-tubes packed with glass wool that are immersed in a Dewar flask containing a refrigerant. Common liquids employed and their corresponding temperatures are as follows:

Refrigerant	Approximate temperature (°F)
Ice	0
Acetone and dry ice	−79
Liquefied gases	
air	−147
oxygen	−183
nitrogen	−196

Sampling rates of a few liters per minute for one or two hours are generally adequate. After collection, the freeze-out trap is brought to the laboratory and the collected contaminant is analyzed by spectrometric means, gas chromatography, or mass spectrometry.

air sampling, grab or instantaneous. An atmospheric analysis method involving manual collection of air in evacuated containers such as glass flasks and stainless-steel bottles. The containers are pumped down in the laboratory and then carried to the field, where the sample is instantaneously collected by merely opening a valve. The vessels containing the gaseous samples are then brought back to the laboratory for analysis.

air sampling, manual. An atmospheric analysis method where a sample of contaminated air is drawn into a collector over a period of time ranging from a few minutes to one or two hours. The sampling system includes a collector, a flow-metering device, and a vacuum source. The collector may contain an appropriate adsorber such as activated carbon, silica gel, or a molecular sieve which is specially selected for retaining the desired pollutant at satisfactory efficiencies. The metering device can be a rotameter or orifice-type flow meter, while the vacuum source is generally a vane, rotary pump, diaphragm pump, or turbine pump. The collected sample is brought back to the laboratory, where the gaseous material is desorbed by means of solvents, steam stripping, or vacuum.

air, static stability of. Air is classified as statically stable if it is nonbuoyant and resists vertical displacement. Stability affects the dispersion of material that is released into the air, as well as cloud formation. For example, a low-lying layer of stable air will cause rising smoke or clouds to spread out horizontally. This layer is created by a temperature inversion. The adiabatic rate, the rate at which dry air cools as it rises adiabatically, is used as a criterion for stability. The rate is equal to $-0.986°C/100$ m. If the vertical gradient of temperature exceeds the adiabatic lapse rate, then the atmosphere is classified as statically unstable. If the gradient is adiabatic, the atmosphere is said to be neutral.

air temperature thermostat. An automatic starting and stopping device that operates on the temperature level of the surrounding air.

airway resistance. The narrowing of the air passage of the respiratory system in response to the presence of irritating substances.

Aitken dust counter. An instrument developed by John Aitken for determining the dust content of the atmosphere. A sample of air is mixed in an expandable chamber with a larger volume of dust-free air containing water vapor. Upon sudden expansion, the chamber adiabatically cools below its dew point and droplets form, with the dust particles as nuclei. Some of these droplets settle on a ruled plate in the instrument and are counted with the aid of a microscope.

Aitken nuclei. The microscopic particles in the atmosphere which serve as condensation nuclei for droplet growth during the adiabatic expansion produced by an Aitken dust counter. These nuclei are both solid and liquid particles, whose diameters are on the order of a tenth of a micron, or even smaller.

albedo. The ratio of the amount of electromagnetic radiation reflected by a body to the amount incident upon it, commonly expressed as a percentage. The albedo is to be distinguished from reflectivity (monochromatic radiation).

alcohol. A compound containing a hydrocarbon group and one or more —OH, hydroxyl, groups. Alcohol can be derived from solar energy in a few months, as compared to the millions of years required by other fuels. It is made by fermentation of vegetable matter, often waste such as molasses, surplus grains, potatoes, or wood, but which are sometimes grown for that purpose. Alternatively, it can be synthesized from ethylene or acetylene, which can be made with electric power from coal and limestone, or similar raw materials. The chemical formula of ethyl alcohol is C_2H_5OH; its composition by weight is: carbon, 54 percent; oxygen, 32 percent; hydrogen, 12 percent.

aldehyde. Any one of a class of compounds of which acetaldehyde is a type. The aldehydes are intermediates between the alcohols and organic acids and differ from the alcohols in having fewer hydrogen atoms in the molecule.

aldrin. $C_{12}H_8Cl_6$, a polycyclic chlorinated insecticide, synthetic and fat-soluble, but water-insoluble. Aldrin is stable for up to three weeks after it has been applied. The toxicity of aldrin in rodents is much higher than other insecticides. The LD_{50} for rats exposed to aldrin is only 10 mg/kg, as compared to 65 mg/kg for Kepone and 300 mg/kg for mirex. Ingestion, inhalation, or absorption through the skin can result in irritability, convulsions, and depression within a few hours after exposure. Chronic exposure may result in liver damage. Aldrin is an experimental carcinogen in animals and has been listed by the EPA's Carcinogen Assessment Group and accepted by OSHA for evaluation as a potential occupational carcinogen.

algae. Any member of numerous groups of simple, unicellular or filamentous plants that are often fast growing and live in fresh or seawater, usually classed as a subdivision of the Thallophyta. Algae contain chlorophyll and are capable of photosynthesis. They may appear blue-green, red, or brown and are classified accordingly. Forms capable of secreting calcium carbonate are important rock builders.

algal bloom. A population explosion of one or more species of algae. When a nutrient normally in short supply, such as phosphorus, is added to a system, it may trigger a rapid growth of organisms in that system. Microorganisms are often held in check by a lack of phosphorus, but when it is made available they use it immediately, and with rapid reproduction rates they can quite suddenly become extremely abundant. Most freshwater ecologists or limnologists call an aquatic population a bloom when there are 500 individuals of a species per milliliter of water. Up to a certain point bloom organisms increase the useful productivity of streams and lakes, making available a more abundant food supply for higher links of the food chain. But in very large numbers these blooms cause problems. Since many bloom organisms are photosynthetic, the oxygen level in the water during the daylight hours is sharply augmented as oxygen is released in the photosynthetic process. But during the night, when there is no photosynthesis to balance respiration which consumes oxygen, the oxygen content of the water may fall below the level necessary for respiration of higher forms of animal life; fish kills often result. This is a particular problem in the still water of lakes. In addition, windrows of algal mats washed up on lake shores decompose, producing hydrogen sulfide and other byproducts and often making bodies of water unsuitable for use of any kind.

aliphatic compound. A member of a major class of organic compounds in which the carbon atoms are lined in long chains.

alkadiene. A straight- or branched-chain hydrocarbon with two double covalent bonds between carbon atoms.

alkali. In chemistry, any substance having marked basic properties. In its restricted sense, the term applies to hydroxides of potassium, sodium, lithium, and ammonium. They are soluble in water; have the power to neutralize acids and form salts; and turn red litmus blue. In a more general sense, the term also applies to hydroxides of the so-called alkaline earth elements—barium, strontium, magnesium, and calcium.

alkali metals. The elements of the first group at the left of the Periodic Table—lithium, sodium, potassium, rubidium, and cesium. They are soft, silvery-white metals with great chemical reactivity. These metals are excellent conductors of electricity and melt at low temperatures. Lithium, sodium, and potassium are lighter than water. The vapors of the alkali metals are mainly monatomic, with a small concentration of diatomic molecules in which the two atoms are held together by a covalent bond. The alkali metals are produced by electrolysis of the molten hydroxides or chlorides. The metals find many uses (especially sodium) in the manufacture of organic chemicals, dyestuffs, and tetraethyl lead. Sodium is used in sodium-vapor lamps and, because of its large heat conductivity, in the stems of valves of airplane engines, to conduct heat away from the valve heads.

alkaline earth metal. One of the group of metals that immediately follow each of the alkali metals in the periodic table; this group includes beryllium, magnesium, calcium, strontium, barium, and radium.

alkalinity. The measure of how alkaline a substance is. Calcium bicarbonate and magnesium bicarbonate, like all bicarbonates and carbonates, are alkaline. They will neutralize acids and liberate carbon dioxide in doing so. The chemist can determine the alkalinity of a water by seeing how much of an acid solution of known strength it takes to neutralize the alkalinity of a measured volume of the water. The addition of acid, or titration, is carried out in the presence of an indicator dye, which changes color when all the alkalinity has been neutralized. The volume of acid required to reach this "end point" measures the alkalinity of the water being tested. Alkalinity in public water supplies should be sufficient to cause effective floc formation, but not high enough to be toxic or to produce a corrosive or encrusting water. Alkalinity is determined by the relative amounts of bicarbonate, carbonate, and hydroxide ions. It is also related to pH and calcium content. In general, alkalinity should not be less than 30 mg/liter, based upon 500 mg/liter dissolved, and have a pH range of 6.0–8.5.

alkaloid. An organic nitrogenous base. Many alkaloids are of great medical importance, such as morphine, strychnine, atropine, cocaine, and quinine. They occur in the animal and vegetable kingdoms, and some have been synthesized.

alkane. Hydrocarbons are compounds composed of hydrogen and carbon alone. An alkane is one of the series of saturated, aliphatic hydrocarbons, with general formula C_nH_{2n+2}. The simplest hydrocarbon is methane, CH_4, a colorless, odorless gas. Natural gas, from oil wells or gas wells, is usually about 85 percent methane. The gas that rises from the bottom of a marsh is methane (plus some carbon dioxide and nitrogen) formed by the anaerobic (air-free) fermentation of vegetable matter. Methane is used as a fuel and for the manufacture of carbon black by combustion with a limited supply of air. Methane is the first member of a series of hydrocarbons called the methane series or paraffin series. Compared with other hydrocarbon series, the alkanes have low chemical reactivity. This stability results from their single covalent bonds. Because they have only single covalent bonds in their molecules, the alkanes are known as saturated hydrocarbons. Saturated bonding is that in which each of the four covalent bonds of each carbon atom in the molecule is a single bond to another atom.

Some physical properties of normal alkanes

Substance	Formula	Melting Point (°C)	Boiling Point (°C)	Density of Liquid (g/ml)
Methane	CH_4	−183	−161	0.54
Ethane	C_2H_6	−172	− 88	0.55
Propane	C_3H_8	−190	− 45	0.58
Butane	C_4H_{10}	−135	− 1	0.60
Pentane	C_5H_{12}	−130	36	0.63
Hexane	C_6H_{14}	− 95	69	0.66
Heptane	C_7H_{16}	− 91	98	0.68
Octane	C_8H_{18}	− 57	126	0.70
Nonane	C_9H_{20}	− 54	151	0.72
Decane	$C_{10}H_{22}$	− 30	174	0.73
Pentadecane	$C_{15}H_{32}$	10	271	0.77
Eicosane	$C_{20}H_{42}$	38		0.78
Triacontane	$C_{30}H_{62}$	70		0.79

alkene. The alkenes, sometimes called the olefin series, are characterized by a double covalent bond between two carbon atoms. The simplest alkene must have two carbon atoms. The names of the alkenes are derived from the names of the alkanes with the same number of carbon atoms by substituting the suffix -ene for the suffix -ane. Since ethane is the alkane with two carbon atoms, the alkene with two carbon atoms is named ethene. The general formula for the alkenes is C_nH_{2n}. Alkenes are made from petroleum by cracking. The cracking of ethane at about 600°C produces ethene and hydrogen.

alkyl halogenide. An alkane is which one or more halogen atoms—fluorine, chlorine, bromine, or iodine—are substituted for a like number of hydrogen atoms. Since R is frequently used to represent an alkyl group and X may be used to represent any halogen, a monosubstituted alkyl halogenide may be represented by RX. Alcohols are alkanes in which the hydroxyl group, $-OH$, has been substituted for hydrogen. An alcohol has the formula ROH. The reaction of an alcohol with a hydrogen halogenide—HCl, HBr, or HI— yields the corresponding alkyl halogenide. Tetrachloromethane, CCl_4 (carbon tetrachloride), is a colorless, volatile solvent used for dry cleaning fabrics, degreasing metals, and extracting oils from seeds. Its vapors are toxic, however, and there must be good ventilation whenever carbon tetrachloride is used. Its most important use is in the preparation of Freon refrigerants and aerosol propellants. Trichloromethane, $CHCl_3$ (chloroform), is a sweet-smelling, colorless liquid used as a solvent and in medicinal preparations. It is manufactured by reducing carbon tetrachloride with moist iron. Dichlorodifluoromethane, CCl_2F_2 (Freon), is used as a refrigerant in mechanical refrigerators and air conditioners. Freon is an odorless, nontoxic, nonflammable, easily liquefied gas. It is prepared from carbon tetrachloride and hydrofluoric acid with antimony compounds as catalysts.

alkylbenzene sulfonate (ABS). A chemical compound sometimes used as a detergent because of its surfactant activity and "sudsing" power. It is ranked as a hard detergent that is not easily biodegradable. After its introduction following World War II, alkylbenzene sulfonate was found to be highly toxic to marine organisms. Its popular use by American households led to the clogging of sewer systems and accumulation of layers of wet foam on streams and other bodies of freshwater. Because of the slow rate of biodegradation of the detergent, it also accumulated in soils. Legislation introduced in the mid-1960s resulted in a general abandonment of the use of ABS in detergents and the substitution of a linear alkyl sulfonate compound that proved to be less toxic to aquatic life and was rapidly degraded by biological organisms. Objections to the petrochemical ABS were partly due to the foaming action, which manufacturers claimed was more of a cosmetic than a toxic effect. The foam eventually appeared from household water taps, on the ground surrounding septic tanks, and in layers 20 ft. thick on some river surfaces before congressional pressure persuaded soap and detergent manufacturers to develop a biodegradable alternative.

alkyne. The alkynes, sometimes called the acetylene series, are characterized by a triple covalent bond between two carbon atoms. The simplest alkyne must have two carbon atoms. The names of the alkynes are derived from the names of the alkanes with the same number of carbon atoms by substituting the suffix -yne for -ane. Therefore, the name of the simplest alkyne is ethyne, a compound more commonly known as acetylene. The general formula for the alkynes is C_nH_{2n-2}.

allergen. A chemically specific substance that sensitizes man and lower animals when exposure of the sensitive host provokes overt illness; also called antigens generically. Although all allergens are recognized as foreign by organisms responding to them, the several forms of hypersensitivity involve allergens of different characteristic sizes and levels of complexity. The molecular species involved in asthma and nasal allergy tend to be nitrogenous, with molecular weights over 5000.

allotropy. The property shown by certain elements of being capable of existence in more than one form, owing to differences in the arrangement of atoms or molecules.

alpha decay. The radioactive transformation of a nuclide by alpha particle emission.

alpha emitter. A radioactive element, whether natural or artificial, which changes into another element by alpha decay.

alpha gain. For transistors, the ratio of the change in collector current to the change in emitter current, with the collector and base voltages held constant.

alpha particle (α particle). The nucleus of a helium atom, two neutrons and two protons, positively charged. The alpha particle is a positively charged fragment of nucleus, consisting of two protons and two neutrons bound so tightly together that they act as a single particle. The alpha particle is also considered to be a helium nucleus stripped of its two electrons. It travels at roughly 10,000 miles/s. Despite such speeds, alpha particles are quickly slowed down by air. Emission of an alpha particle is called alpha decay, causes the element to lose two protons, and reduces its atomic number by two.

alpha ray. One type of ray or stream of alpha particles thrown off by disintegrating atoms. It is actually the control core of a helium atom and consists of two protons and two neutrons packed tightly together.

alveoli. The tiny air spaces at the end of the terminal bronchioles of the lung where the exchange of oxygen and carbon dioxide takes place.

Amagat law. The volume of a mixture of gases is equal to the sum of the volumes of the component gases, each taken at the pressure and the temperature of the mixture. The Amagat law applies exactly only to a mixture of ideal gases.

ambient. The surrounding atmosphere; the environment surrounding a body but undisturbed or unaffected by it. Thus ambient temperature means temperature of the atmosphere at a particular location.

ambient air quality. A general term used to describe the state of the air outside. No qualitative measures are associated with the term. Ambient air quality is usually considered "good" or "bad," depending upon the measurement techniques employed. Some techniques break down and list the actual constituents measured in the air, while others attempt to group all constituents into one arbitrary index number.

Ambient air national quality standards ($\mu g/m^3$, mg/m^3, and ppm)

Pollutant	Average time	Primary standards (health)	Secondary standards (welfare, materials)
Particulates	Annual	75 $\mu g/m^3$	60 $\mu g/m^3$
	24 hr	260 $\mu g/m^3$	150 $\mu g/m^3$
Sulfur dioxide	Annual	80 $\mu g/m^3$ (0.03 ppm)	
	24 hr	365 $\mu g/m^3$ (0.14 ppm)	
	3 hr	—	1300 $\mu g/m^3$ (0.5 ppm)
Carbon monoxide	8 hr	10 mg/m^3 (9 ppm)	Same as primary standards
	1 hr	40 mg/m^3 (35 ppm)	
Hydrocarbons (nonmethane)	3 hr (6–9 AM)	160 $\mu g/m^3$ (0.24 ppm)	Same as primary standards
Nitrogen dioxide	Annual	100 $\mu g/m^3$ (0.05 ppm)	Same as primary standards
Ozone	1 hr	240 $\mu g/m^3$ (0.12 ppm)	Same as primary standards
Lead	3 months	1.5 $\mu g/m^3$ (0.006 ppm)	

ambient noise. The all-encompassing noise associated with a given environment, usually a composite of sounds from many sources, near and far.

ambient temperature. The surrounding or immediate atmosphere either inside or outside of a collector.

amensalism. A relationship between two species whereby one is adversely affected by the second, but the second species is unaffected by the presence of the first.

4-aminodiphenyl. $C_6H_5C_6H_4NH_2$, also known as p-aminodiphenyl and PAB, a regulated carcinogen that occurs as colorless crystals which darken to yellowish brown when exposed to air. Although no longer produced, the chemical has been reported by NIOSH to be found as a contaminant in diphenylamine. Aminodiphenyl induces bladder tumors in humans and causes cancer in rats, mice, and dogs. The effects can result from exposure by ingestion, inhalation, or skin absorption. NIOSH has recommended exposure limits of not more than 0.1 percent by weight or volume.

aminopterin. An anticancer medication. The compound was cited in a 1980 Report to the President by the Toxic Substances Strategy Committee as a substance that can cause birth defects. In addition to fetal death and abnormalities, aminopterin has been found to cause lymphopenia and stomatitis in patients exposed to it.

ammonia. NH_3, a colorless, pungent and strongly alkaline gas: molecular weight, 17.03; specific gravity, 0.817 at $-79°C$; melting point, $-77.7°C$; and boiling point $-33.35°C$ at normal pressure. A byproduct of many industrial processes, it is produced in greatest quantity by the distillation of coal for the production of gas or gas coke, and compounds used in the manufacture of chemicals for textile and chemical process industries. For the most part, ammonia recovery processes are employed in these operations, but often the levels of ammonia in the process effluent rise to 1.0 mg/liter. The waste products are principally free ammonia, cyanide, thiocyanate salts, and a variety of aromatic compounds. The permissible criterion for ammonia levels has been established as 0.5 mg/liter in waters designated as public water supplies. The desirable criterion is less than 0.01 mg/liter.

ammonia plant damage. Plant damage due to ammonia gas, involving a change in the color of the leaves, from a green to a cooked green and then to a brown, or drying. Most toxic gases damage only a certain area of the plant leaf, whereas ammonia attacks the whole leaf, with symptoms occurring within several hours after exposure. Experimental tests to determine the effects of ammonia on some species of plant life revealed the following results. Buckwheat, coleus, and tomato plants showed little or no damage after five hours exposure to 8.3 ppm. Tomato, sunflower, and coleus showed light damage when exposed to 16.6 ppm for four hours. Tomato, sunflower, and coleus plants were damaged when exposed to 40 ppm for one hour. Sometimes light damage was noted at the edge of the plant without any injury to other parts of the plant.

ammonification. The production of ammonia by bacterial action in the decay of nitrogenous organic matter. Subsequent to its incorporation into an organic form in protein and nucleic acid synthesis, inorganic nitrogen (NO_3) is metabolized and returned to the major part of the nitrogen cycle as waste products of that metabolism or as organized protoplasm in dead organisms. Many heterotrophic bacteria, actinomycetes, and fungi occurring in both soil and waste utilize this organic nitrogen-rich substrate; in their metabolism of it they convert it to and release it in an inorganic form: ammonia. This process, ammonification, is also referred to as mineralization.

ampere. A unit of electric current corresponding to the passage of one coulomb of electric charge per second. It is the current which flows when an electromotive force of one volt is applied across a resistance of one ohm.

Ampere's rule. A positive charge moving horizontally is deflected by a force to the right if it is moving in a region where the magnetic field is vertically upward. This may be generalized to currents in wires by recalling that a current in a certain direction is equivalent to the motion of a positive charge in that direction. The force felt by a negative charge is opposite to that felt by a positive charge.

amplitude. The maximum value of the displacement in an oscillatory motion.

amu. The atomic mass unit, a unit of mass equal to one-sixteenth the mass of the atom of oxygen of mass number 16.

anabatic wind. In meteorology, a wind blowing uphill. In general, anabatic refers to winds originating in connection with surface heating, such as a breeze blowing up a valley when the sun warms the ground.

anabolism. A chemical reaction in which simple substances are combined to form more complex substances, resulting in the storage of energy, the production of new cellular materials, and growth.

anacoustic zone. The region above an altitude of about 100 miles where the distance between the air molecules is greater than the wavelength of sound and sound waves can no longer be propagated.

anadromous. Swimming upriver to spawn, like salmon.

anaerobe. An organism for whose life processes a complete or (in some forms) nearly complete absence of oxygen is essential. Facultative anaerobes can utilize free oxygen; obligate anaerobes are poisoned by it.

anaerobic. Able to live in conditions in which oxygen is excluded, and as a result normal life that depends on the presence of oxygen is not possible.

anaerobic biological treatment. Any treatment method or process utilizing anaerobic or facultative organisms, in the absence of air, for the purpose of reducing the organic matter in waste or organic solids settled out of wastes; commonly referred to as anaerobic digestion or sludge digestion when applied to the treatment of sludge solids.

anaerobic digestion. The digestion of materials from biological sources in the absence of air to yield carbon dioxide and methane.

analyzer. An instrument performing frequency analysis. Because of the importance of sound pressure level distribution at various frequency regions in the audible range, frequency analyses are necessary.

analyzer, narrow band. An instrument permitting analysis into discrete frequencies. There are two types of such instruments, both are known as narrow band analyzers and both differ in the sharpness of tuning to individual frequencies. One is a constant-percentage narrow band instrument, the other is a constant-bandwidth narrow band instrument.

anechoic room. A room whose boundaries effectively absorb all the sound incident thereon, thereby affording essentially free-field conditions. Also called a free-field room.

anemometer. A weatherman's instrument for measuring wind velocity. The cup type is the most common, consisting of a set of three or four conical cups fixed to a wheel with a vertical axle. When the wind catches the cups it causes the wheel to rotate at a speed proportional to the speed of the wind. The speed of the rotations is transmitted electrically to an indicator inside the weather station.

aneroid. Literally "not wet," containing no liquid, applied to a kind of barometer which contains no liquid (an aneroid barometer).

aneroid barometer. A weather tool for measuring atmospheric pressure by means of a metal chamber that contains a partial vacuum. The chamber is of thin metal that can expand or contract as the atmospheric pressure changes. The aneroid barometer gets its name from the Greek word for dry, since no liquid is used. Fundamentally, the aneroid barometer consists of an evacuated chamber and a spring. The spring keeps the chamber from collapsing under the pressure of the atmosphere and restores the chamber to a larger shape when the pressure is reduced.

angle. The ratio between the arc and the radius of the arc. Units of angle are the radian, the angle subtended by an arc equal to the radius, and the degree, 1/360 part of the total angle about a point.

angle of azimuth. The angle measured clockwise in a horizontal plane, usually from the north. The north used may be true north, Y north, or magnetic north.

angle of reflection. The angle at which a reflected ray of energy leaves a reflecting surface, measured between the direction of the outgoing ray and a perpendicular to the surface at the point of reflection.

angle of refraction. The angle at which a refracted ray of energy leaves the interface at which the refraction occurred, measured between the direction of the refracted ray and a perpendicular to the interface at the point of refraction.

angstrom (Å). A unit of length used in the measurement of the wavelength of light, x-rays, and other electromagnetic radiation, and in the measurement of molecular and atomic diameters. One angstrom is equal to 10^{-8} cm or 10^{-4} µm.

angular harmonic motion (harmonic motion of rotation). Periodic, oscillatory angular motion in which the restoring torque is proportional to the angular displacement; torsional vibration.

angular momentum (moment of momentum). A quantity of angular motion measured by the product of the angular velocity and the moment of inertia. Its nature is expressed by $g/cm^2/s$. The angular momentum of a mass whose moment of inertia is I, rotating with angular velocity ω, is $I\omega$.

angular momentum quantum number. A number describing the magnitude of the angular momentum of an orbiting electron. Unlike energy, which is scalar quantity, angular momentum is a vector with both magnitude and direction.

angular velocity. The rate of rotation of a body. It is expressed as the angle through which a body turns in unit time; for a particle moving in the arc of a circle, it is the angular displacement in a unit time.

anhydrous ammonia. NH_3, a colorless gas liquefied by compression and commonly used in agriculture as a liquid fertilizer. Toxic exposure is generally via the inhalation route although anhydrous ammonia can also cause adverse health effects such as irritation of the skin, conjunctivitis and irritation of the eyes, and in some cases, corneal ulcers. Inhalation results in irritation of the nose and throat, with respiratory difficulties, coughing, and vomiting. The lowest reported lethal concentration for a human is 10,000 ppm for a period of three hours, but the lowest concentration that produces toxic effects in humans is only 20 ppm. LC_{50} levels for the cat and rabbit are less than 1100 ppm. A spill of anhydrous ammonia from a tank truck in Leland, Illinois, in 1980 required the evacuation of the town and resulted in hospitalization and emergency medical treatment for some of the residents.

aniline. $C_6H_5NH_2$, a hypergolic, colorless, oily liquid fuel used in the manufacture of printing inks, cloth-marking inks, paints, paint removers, and dyes. It is usually oxidized with nitric acid and has molecular weight, 93; melting point, $-6°C$; boiling point, $184°C$; and specific gravity, 1.026. The lowest reported lethal dose in a human is on the order of 350 mg/kg, or somewhat less than the oral LD_{50} of 440 mg/kg established for rats. Human responses to exposure vary considerably with severe toxic effects associated with the ingestion of as little as one gram and less severe reactions occurring from ingestion of as much as an ounce. In addition to toxicity resulting via the oral route, adverse health effects may be associated with exposure by inhalation or skin contact. The most important action of aniline is the formation of methemoglobin through an unknown intermediate step leading to sufficient loss of the oxygen-carrying capacity of the red blood cells, resulting in damage to the cells of the central nervous system through asphyxia. Exposure to aniline in amounts as small as 65 mg has been known to increase the level of methemoglobin in the body by more than 15 percent. Arterial hypotension, cardiac arrhythmia, and jaundice are among other adverse health effects associated with exposure to aniline. Because of evidence that aniline may cause changes in the cells lining the bladder, including the formation of ulcers and papillomas which may become malignant, the compound has been listed as a suspected occupational carcinogen to be evaluated by OSHA.

anion. An ion attracted to the anode of an electrolytic cell; a negative ion.

annealing oven. An oven used in the glass-making process which slowly heats, then slowly cools, glass pieces as a conveyor belt moves them along. Slow heating and cooling helps give the glass an overall strength so that it will not break easily when exposed to high or low temperatures.

annihilate. To make into nothing; specifically, when a particle and its antiparticle (such as an electron and positron) meet and destroy each other, leaving nothing but electromagnetic radiation.

anode. The electrode at which oxidation occurs in a cell. It is also the electrode toward which anions travel because of the electrical potential. In spontaneous cells the anode is considered negative. In nonspontaneous or electrolytic cells the anode is considered positive.

anodized. Coated with protective film on metal (usually aluminum) deposited by chemical or electrolytic methods to prevent corrosion.

antibiosis. The relationship between species in which certain substances produced or excreted by organisms are generally harmful to others.

anticoagulant. A chemical that interferes with blood clotting.

anticyclone. An atmospheric condition of high central pressure, with air current flowing outward; a high pressure area. Circulation is clockwise in the Northern Hemisphere and counterclockwise in the Southern Hemisphere.

anti-degradation clause. A part of air quality and water quality laws that prohibits deterioration where pollution levels are within the legal limit.

antiparticle. A particle having a mass that is the same as, but a charge that is opposite to, that of another particle. The antiparticles are included among the 82 subatomic particles. Although they resemble their normal particles in mass, antiparticles are opposite them either in electric charge or in spin. The antiparticle is the mirror image of its normal particle.

antipritic wind. A wind moving at right angles to isobars (an isobar is a line drawn on a chart connecting places of the earth's surface having equal barometric pressures during a specified time) following a pressure gradient moving from a region of high pressure to a region of low pressure.

antiproton. A particle that has the same mass as the proton but has a negative charge. It exists for brief moments during high-energy nuclear reactions.

anthropogenic source. A source of pollution resulting directly or indirectly from human activity and, in general, producing effects that impact on the health of humans. Examples of industrial anthropogenic pollution factors include petroleum refining activities such as crude oil distillation, catalytic cracking and reforming, polymerization, alkylation, and isomerization; and chemical manufacturing of adipic acid, ammonia, carbon black, hydrochloric and hydrofluoric acids, phosphoric and sulfuric acids, plastics, paints, varnishes, synthetic rubber and synthetic fibers. Other anthropogenic sources are primary aluminum production; copper, lead, and zinc smelting; iron and steel production; cotton ginning; coffee roasting; sugarcane processing; and the manufacturing of asphalt, Portland cement, calcium carbide, ceramic products, glass, fiberglass, and rock wool.

apparent density. The weight per unit volume of activated carbon.

aquastat. A valve automatically controlled by temperature of the fluid, usually water, flowing through it.

aquifer. A subsurface geologic formation which contains water.

Archimedes' principle. A body wholly or partly immersed in a fluid is buoyed up by a force equal to the weight of the fluid displaced. A body of volume V cm³ immersed in a fluid of density ρ g/cm³ is buoyed up by a force in dynes $F = \rho g V$, where g is the acceleration due to gravity. A floating body displaces its own weight in gravity.

area, unit of. The square centimeter; the area of a square whose sides are one centimeter in length. Other units of area are similarly derived.

area source. In air pollution, any small individual fuel combustion source, including vehicles. A more precise legal definition is available in federal regulations.

aromatic compound. A compound derived from benzene with one or more benzene rings of carbon atoms, as distinct from a compound of aliphatic or alicyclic character.

array. A number of individual solar collection devices arrayed in a suitable pattern to collect solar energy effectively.

Arrhenius' theory of electrolytic dissociations. The molecule of an electrolyte can give rise to two or more electrically charged atoms or ions.

arsenic (As). Element 33. (1) Because some inorganic arsenic compounds have a significant vapor pressure, NIOSH has established standards for respiratory protective devices for employee protection under certain defined conditions. They include: full face coverage masks to prevent eye and face irritation (previous standards permitted half-mask respirators); high-efficiency particulate filters for inorganic arsenic compounds that have no significant vapor pressure to enable meeting the recommended OSHA criteria of 2.0 micrograms of arsenic per cubic meter of air; air-purifying respiratory protection with an acid gas canister, as a minimum, for inorganic arsenic compounds with significant vapor pressure—such as arsenic trichloride—to prevent exposure to more than the recommended maximum concentration of arsenic due to the passage of volatile arsenicals through a high-efficiency filter. NIOSH has also recommended that industrial plants using inorganic arsenic install engineering controls so that protective equipment for workers, such as masks, be used only during emergencies, nonroutine operations such as maintenance, or repair activities.

Arsenic pollution is often associated with the manufacture or use of herbicides and pesticides. It may also be a byproduct of mining operations. The permissible criterion for presence of arsenic in public water supplies is 0.05 mg/liter. The desirable criterion is complete absence. Arsenic hydride and trioxide are especially toxic. Arsenate acts by tying up active sites in cellular substituents.

arylhydrocarbon hydroxylase (AHH). A component of diesel engine exhaust. A 1978 EPA study suggested that AHH was associated with the chemical induction of tumors in the lungs, testes, and prostates of four animal species—cats, rats, mice, and guinea pigs—exposed to high levels of diesel exhaust for short periods of time. The EPA reported that levels of AHH increased in the body organs of the animals as a result of the exposure to exhaust gases and particulate portions. However, the EPA study also failed to find tumors in the lungs or other organs in which high levels of arylhydrocarbon hydroxylase had accumulated.

asbestos. A mineral fiber than can pollute air or water and cause cancer if inhaled or ingested.

A-scale sound level. A measurement of sound approximating the sensitivity of the human ear, used to note the intensity or annoyance of sounds.

ash. The mineral content of a product that remains after complete combustion. Ash is the unburnable solid material that is set free when a fuel is burnt. The sparks, which fly upward from a fire, are red-hot particles of ash. In a fire or furnace, where the fuel is coal or coke, much of the ash falls through the fire bars into the ash pit, but an appreciable portion escapes with the flue gases. If coal is finely ground and burnt as pulverized fuel, most of the ash passes into the flues. Most industrial chimneys are less than 200 feet high. In the absence of high wind currents, the largest particles of ash remain in the air for less than five seconds. Because of their short time in the air, large particles of ash do not form a substantial part of the "suspended impurity" in the air, but do contribute appreciably to the pollution deposited on the ground.

assimilation. The ability of a body of water to purify itself of pollutants.

astigmatism. An error of spherical lenses peculiar to the formation of images by oblique pencils. The image of a point when astigmatism is present will consist of two focal lines at right angles to each other, separated by a measurable distance along the axis of the pencil. The error is not eliminated by reduction of aperture as is spherical aberration.

ATGAS. A process for coal gasification. *Process:* Crushed ($< \frac{1}{8}$ in.), dried (4 percent moisture) coal is injected into a molten iron bath through steam lances. Oxygen is introduced through lances located at the iron-bath surface. Coal dissolves in the molten iron where the coal volatiles crack and are converted into carbon monoxide and hydrogen. Fixed carbon of coal reacts with oxygen and steam, producing more carbon monoxide and hydrogen. The sulfur of the coal migrates to a lime slag floating on the molten iron and forms calcium sulfide. The slag-containing ash and sulfur is continuously withdrawn from the gasifier and desulfurized with steam to yield elemental sulfur and desulfurized slag. A portion of the desulfurized slag is recycled to the reactor. Operating conditions of the gasifier are about 2600°F and 5 psig. Raw gas from the gasifier is passed through a heat recovery system. Dust is removed in an electrostatic precipitator. Gas is compressed to about 600 psia and passed through a CO-shift converter. Carbon dioxide

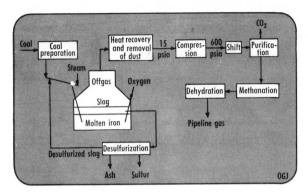

ATGAS process. (Applied Technology Corporation.)

is removed by hot carbonate scrubbing. Purified gas is methanated, dried, and compressed to pipeline quality. This process can be adapted to make low-Btu gas by using air instead of oxygen (Applied Technology Corp.).

atmosphere. The gaseous envelope surrounding the earth, made up of layers of varying pressure and temperature. The term, when used for pressure, is defined as the pressure exerted per square centimeter by a column of mercury 760 mm high at a temperature of 0°C, where the acceleration of gravity is 980.665 cm/s². One atmosphere of pressure equals 1.0133×10^6 dyn/cm².

atmosphere composition.

Constituent	Percent by volume	ppm
Nitrogen	78.084 ± 0.004	
Oxygen	20.946 ± 0.002	
Carbon dioxide	0.033 ± 0.001	
Argon	0.934 ± 0.001	
Neon		18.18 ± 0.04
Helium		5.24 ± 0.004
Krypton		1.14 ± 0.01
Xenon		0.087 ± 0.001
Hydrogen		0.5
Methane (CH_4)		2
Nitrous oxide (N_2O)		0.5 ± 0.1

atmospheric absorption. The absorption by the earth's atmosphere of most frequencies of the electromagnetic spectrum, other than visible light, including x rays and most of the ultraviolet and infrared radiation emitted by the sun. This absorption makes life possible and prevents the earth's surface from getting too hot. The first window ranges from the very softest ultraviolet through the visible spectrum into the infrared and the second covers the shorter radio waves.

atmospheric aerosol pollutant. Fine bits of matter producing undesirable effects in the atmosphere. There is now a better understanding of how the major gaseous pollutants such as carbon monoxide, sulfur oxides, nitrogen oxides, and ozones are formed, and how they affect human health. It is known that these gases react chemically, often to form aerosols. Because of the apparent complexity of the atmospheric chemistry involved, these aerosols cannot be fully characterized. It is known that they are fine bits of matter, often containing solids, liquids, and gases at the same time. It is also known that the gaseous pollutants, especially sulfur and nitrogen oxides, are important chemical precursors to these aerosols. A simplified view of one set of reactions which occur is that the sulfur and nitrogen oxides are oxidized in air and react with water to form acids, which, in turn, react with metallic particulate pollutants to form complex sulfates and nitrates. If the polluted air mass is involved in precipitation, the water-soluble acids may appear in the rain or snow.

atmospheric area.   As the federal government uses the term, a segment of the continental United States in which climate, meteorology, and topography, all of which influence the capacity of the air to dilute and disperse pollutants, are essentially similar.

atmospheric assimilation.   Photochemical reactions whereby chemical substances in the atmosphere are mixed, diffused, and chemically reacted, thereby maintaining reasonably constant concentrations within defined atmospheric regions.

atmospheric condensation and precipitation.   Droplets formed by condensation, usually aggregating to form fog or clouds; precipitation results when droplets or crystals coalesce or grow to such size that they can no longer be held up by currents. The magnitude of the vertical air currents also determines the fall of the condensed or crystallized water. Condensation usually occurs on dust particles or condensation nuclei in the atmosphere. Substances that form condensation nuclei include: sulfur dioxide, which is oxidized to sulfur trioxide and becomes sulfuric acid; sea salt, which contains sodium chloride and a small amount of magnesium chloride; and oxides of nitrogen, particularly nitrogen dioxide. Upon nucleation, the droplet or crystal grows to a visible size in a fraction of a second through diffusion of water vapor, but growth thereafter is slow. Diffusion leads to droplets usually smaller than 10 μm. Most droplets in stratus (nonconvection) clouds without precipitation or fog have diameters less than 10 μm, and an upward air current under 0.5 cm/s is sufficient to keep them from falling. Collision and coalescence of cloud and precipitation elements (accretion) are the methods of growth leading to significant precipitation. Collisions between droplets occur because of the difference in falling speeds due to differences in size.

atmospheric moisture content.   The amount of moisture in the atmosphere. Water vapor has a strong influence on the heat balance in the atmosphere. Evaporation and condensation of water in the free atmosphere requires and releases extremely large amounts of heat. A variety of measures for specifying humidity are used, such as absolute humidity, the weight of water vapor per unit volume of air (used in thermodynamics); vertical temperature, a fictitious temperature that dry air would be at to have the same density as wet air at some actual temperature (used in estimating the stability of air); relative humidity, the ratio of the actual vapor pressure to the saturation vapor pressure at constant temperature (used in weather reports); and wet-bulb temperature the lowest temperature to which air can be cooled by evaporating water into it, depending upon the actual air temperature and the amount of cooling possible by evaporation into that air (used in air conditioning).

atmospheric pressure.   The force exerted by the weight of the atmosphere from the level of measurement to its outer limits.

atmospheric refraction.   The refraction of light from a distant point by the atmosphere, caused by its passing obliquely through various layers of air of different densities.

atmospheric surface turbulence and wind structure.   Large-scale circulation of the atmosphere is complex because of differential heating, the earth's rotation, and friction.

Wind flow at the ground is even more complicated because of the effects of local terrain, vegetation, and buildings. The influence of turbulence on the design of buildings and cooling systems is based on conditions of the boundary layer, which extends from the ground up to about 2000 ft. This layer is important because of the distribution of pollutants that are released into the atmosphere. Major characteristics of the boundary layer are changes in wind direction and speed as a function of distance from the surface of the earth. At some point above the surface, called the gradient level, the wind field is determined by a balance of the pressure gradient and Coriolis and centrifugal forces. At this level, friction due to the ground is negligible. The wind here is called the gradient wind. As the altitude decreases, friction starts to have a greater effect. This decreases the wind speed and turns the wind in the direction of the pressure gradient of the atmosphere. The characteristic turning and reduction when projected on the ground forms what is called the Ekman spiral. The variation of speed with height helps create turbulence. This explains why a smoke plume in the atmosphere does not come out in an ideal conical shape, but is instead distorted by velocity fluctuations in the wind field. Atmospheric turbulence is characterized by a distribution of sizes and eddies. Normally, the scale of turbulence increases above the ground up to a height of about 300–500 ft and then decreases. This increase is because it is difficult for large eddies to form near the earth's surface, since the earth acts as a boundary. Higher up, the eddies get progressively larger. Rough terrain may increase ground-level turbulence by mechanically breaking up the air flow. Turbulence can also increase on a sunny day, since the ground heats up the air more. Vertical turbulence determines how rapidly an elevated plume will be brought to the ground. In a horizontal direction, lateral turbulence affects the spread of the plume, or its width, and therefore the area encompassed by the plume being emitted from its source. Since vertical eddy size is limited as one gets closer to the ground, the vertical meandering of a plume is restricted by the ground and the stability of the atmosphere. The lateral or horizontal meandering of the plume (caused by changes in wind direction lasting ten minutes or longer) spreads pollutants over a wide area.

atom. The smallest unit of a chemical element which retains the chemical characteristics of that element. It consists of an atomic nucleus and electrons. The smallest atom, that of hydrogen, has one proton and one electron, its diameter is about 10^{-10} mm and its mass is 1.67×10^{-24} g. The electrons are distributed around the nucleus in shells or layers. Each shell can accommodate a certain number of electrons, and the chemical character is chiefly determined by the number present in the outer shell. An electron can jump from one orbit into another and back again, with a change in total energy giving rise to emission or absorption of radiation of a particular wavelength. If an electron is detached, the atom becomes ionized. There are four basic forces within the atom which are involved

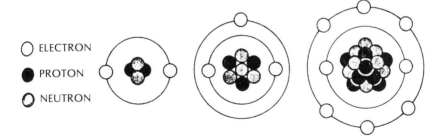

The structure of atoms. Atoms have nuclei of protons and neutrons and, around the nuclei, electrons in orbits. The number and arrangement of the electrons and protons distinguish one element from another. Shown are atoms of helium (left), lithium (center), and oxygen (right).

in any particle interaction in the universe. (a) The strong nuclear force, which occurs very rapidly as pions are exchanged between nucleons (protons and neutrons) billions of times every second. This nuclear force is confined to the short distances encountered in the nucleus. (b) The electromagnetic force, which takes place between charged particles. It involves the exchange of photons. Although it is about 100 times weaker than the strong nuclear force, it extends over a greater distance. (c) The weak force governs some radioactivity and particle decay. It takes about a hundred million times as long to occur as the nuclear force. (d) The gravitational force depends on large masses and has little effect on an atomic level. It is seldom considered in subatomic interactions.

atom, excited. An atom, some or all of whose electrons have been raised to a higher than usual energy state. Excitation may occur by chemical reaction, or absorption of heat or photons.

atom nucleus. The dense core of an atom consisting of neutrons (uncharged) and protons (positively charged).

atom smasher. A machine that causes changes in atoms by bombarding them with parts of other atoms. The bombardment either knocks subatomic particles off the target atom or forces additions to it.

atomic energy. (1) The constitutive internal energy of the atom which was absorbed when it was formed. (2) The energy derived from the mass converted into energy in nuclear transformations; power released when atoms change from one kind into another. Change can occur through splitting (fission), where one heavy atom breaks into two medium-size atoms, or it can occur through joining (fusion), where two light atoms unite into one medium-size atom.

atomic fission. The breaking down of a large, heavy atom with liberation of energy in the form of heat and alpha, beta, or gamma radiation, and creation of new elements from the larger fragments.

atomic half-life. The rate at which a radioactive element decays. Atomic half-life is measured by how long it takes one gram of the element to lose half its mass by radiation and become half a gram. The half-life for radium is 1620 years, indicating that through radiation losses, a gram of radium will be reduced to half a gram in 1620 years. Different radioactive elements have different half-lives, ranging from a fraction of a second to billions of years. ^{236}U, with a half-life of 4.5×10^9 years, has been called a clock for measuring geological time.

atomic heat. The thermal capacity of a gram-atom of an element, obtained by multiplying its specific heat and its atomic weight, expressed in grams.

atomic mass. The mass of a neutral atom of a nuclide. It is usually expressed in terms of the physical scale of atomic masses, that is, in atomic mass units (amu). While the number of protons in the nucleus gives an atom its atomic number, its atomic mass is derived from the total number of protons and neutrons. Often the atomic number of an element is written to the left and slightly below the symbol, i.e., as a subscript, and the atomic mass is written to the left and slightly above it, i.e., as a superscript.

atomic mass unit (amu). A unit measuring the atomic mass of an atom. One amu is equal to 1.660×10^{-27} kg, which is 1/12 of the mass of a ^{12}C atom, the most abundant form of carbon atoms. In atomic mass units, the atomic mass of the most abundant form of hydrogen is 1.007 825, while that of the most abundant form of uranium is 238.0508.

atomic number. The number (Z) of protons within the atomic nucleus. The electrical charge of these protons determines the number and arrangement of the outer electrons of the atom, and therefore the chemical and physical properties of the element. The number of protons in the nucleus of an atom determines the characteristics of that atom and the element to which it belongs, i.e., every atom with 92 protons belongs to the element uranium. The number of protons in the nucleus also determines the element's atomic number, i.e., hydrogen, with one proton in its nucleus, has an atomic number of 1; helium, with two protons, has an atomic number of 2; and uranium, with 92 protons, has an atomic number of 92.

atomic particle.   One of the particles that constitute an atom, as an electron, neutron, or a positively charged nuclear particle.

atomic pile.   A nuclear reactor.

atomic radiation.   The radiation or radioactivity which results from decompositon of a nuclear, fissionable, radioactive material or from the fusion of atomic nuclei.

atomic radius.   The radius of an atom (average distance from the center to the outermost electron of the neutral atom), commonly expressed in angstrom units (10^{-8} cm).

atomic spectrum.   A series of discrete wavelengths of energy emitted by the atoms of an element when they are excited above ground state.

atomic structure.   According to the currently accepted view, the atom consists of a central part, called the nucleus, and a number of electrons (called orbital or planetary electrons) circling about the latter, like planets about the sun. The nucleus has a high specific weight; it contains most of the mass of the entire atom (its mass is considered equal to the atomic mass) and is composed of positively charged particles called protons (the number of which always equals the atomic number Z), and particles of zero charge, called neutrons (the number of which equals the difference between the atomic weight and the atomic number, $A - Z$). The diameter of the nucleus is between 10^{-13} and 10^{-12} cm, and the relatively vast distance in which the orbital electrons circle about it is illustrated by the fact that this nuclear diameter is only 1/100,000 to 1/10,000 of the entire atomic diameter. While the nucleus carries an integral number of positive charges (an integral number of protons) each of 1.6×10^{-19} C, each electron carries one negative charge of 1.6×10^{-19} C, and since the number of orbital electrons is equal to the number of protons in the nucleus, the atom as a whole has a net charge of zero. The electrons are arranged in successive shells (q.v.) around the nucleus; the maximum number of electrons in each shell is determined by natural laws, and the extranuclear electronic structure of the atom is characteristic of the element. The electrons in the inner shells are tightly bound to the nucleus; this inner structure can be altered by high-energy particles, gamma rays of radium, or x rays. The electrons in the outer shells are responsible for the chemical properties of the element.

atomic valence.   The way an atom behaves in either adding or giving away electrons from its outer shell. Atoms combine with other atoms to complete their outermost shell of electrons. They combine by gaining, losing, or sharing their outer electrons with another atom. The valence of an atom shows how many outer electrons the atom will gain, lose, or share when it unites, or makes a chemical bond with another atom. When an atom has a negative valence, it will tend to gain electrons to complete its outer shell. A positive valence indicates that the atom tends to give electrons away to the atom with which it bonds; the atom has some unneeded extra electrons. In chemical bonds where electrons are shared, the valence is written as 1 or 2 and so on, without positive or negative signs. Some atoms exhibit both a negative and a positive valence, since under some conditions they tend to give up their outer electrons, while at other times they tend to gain the electrons needed to close their outer shell.

atomic volume.   The atomic weight of an element or substance divided by its specific gravity.

atomic waste. The radioactive ash produced by the splitting of uranium or other nuclear fuel in a nuclear reactor; it may include products made radioactive in such an apparatus.

atomic weight. The relative weight of an atom on the basis of oxygen's weight being 16. For a pure isotope, the atomic weight rounded off to the nearest integer gives the total number of nucleons (neutrons and protons) making up the atomic nucleus. If these weights are expressed in grams, they are called gram-atomic weights. Numerically, the mass of an atom is approximately equal to the total number of protons and neutrons in the nucleus of an atom. All elements exist in nature as a mixture of their various isotopes. The atomic weight of an element takes into account these varying isotopes and represents an average of them all. Atomic weight is derived by determining the average mass of all the natural isotopes of an element, adjusted according to their abundance in nature. Then this weighted average is compared to a ^{12}C standard. ^{12}C has been assigned an atomic weight of 12 atomic mass units, or amu, 1 amu being the approximate mass of one proton or one neutron. Both atomic mass and atomic weight are measured in amu.

atomization. The dividing of a liquid into extremely minute particles either by impact with a jet of steam or compressed air, or by passage through some mechanical device.

atomize. To divide a liquid into extremely minute particles, either by impact with a jet of steam or compressed air, or by passage through some mechanical device.

atomizer. A nozzle through which fuel is sprayed, its function being to break up the fuel into a fine mist to ensure good dispersion and combustion.

atomized suspension. A pyrolysis process in which waste material is atomized into a vertical stainless-steel tube which is externally heated to 1500°F or more. The solids are dried in this manner, and the resulting volatiles are vented to afterburning if necessary.

atopic disease. A disease associated with allergies. The association of bronchial asthma, nasal allergy (allergic rhinitis), and certain immediate reactions to food, along with a chronic skin disorder—atopic dermatitis (infantile eczema)—occur together in families and in the cumulative health experience of individuals. To distinguish these conditions from other forms of allergy, the term "atopic diseases" was applied. The most fundamental trait of atopic persons is their tendency to produce skin-sensitizing antibodies (atopic reagins) in response to common allergens encountered by inhalation and ingestion. When definite symptoms occur, atopic persons may present three cardinal disease types, singly or in combination: nasal allergy, bronchial asthma related to inhalant or ingestant allergies, and atopic dermatitis.

Atterberg grade scale. A decimal grade scale for particle size, with 2 mm as the reference point, and involving the fixed ratio 10. Subdivisions are the geometric means of the grade limits: 0.2, 0.6, and 2.0.

attractant. A chemical or agent that lures insects or other pests by stimulating their sense of smell.

attrition. The wearing or grinding down of a substance by friction, a contributing factor in air pollution, as with dust.

audible window. The range of vibrations recognized as music, between about 30 and 30,000 vibrations per second. Above are ultrasonic waves inaudible to the human ear, and below are individual pulses and large earthquake waves with frequencies of down to about 0.06 vibrations per second, or a wavelength of 16,384 ft.

audio frequency. The range of frequencies of sound waves from about 25 to 15,000 Hz, being audible to persons of normal hearing.

audiometer. An instrument that measures hearing sensitivity.

auramine. $C_{17}H_{21}N_3$, a common name for one or more of the 4,4'-(imidocarbonyl)bis(N,N-dimethyl) anilines. Auramine may occur in the form of yellow needles. It is an experimental carcinogen in animals and an established cause of bladder cancer in humans. Auramine occurs in the manufacture and use of dyestuffs. Rubber workers, textile workers, and persons employed in the manufacture of paints are among those who may be exposed to this chemical.

aurora. The light radiated by atoms and ions in the ionosphere; it is induced by the influx of particles from the sun and is seen mostly in the polar regions.

autecology. The study of the ecology of individual organisms of a species.

authigenic. A term applied to products of chemical and biochemical action which originate in sediments at the time of or after deposition, and before burial and consolidation, such as calcium carbonate or manganese oxide deposition.

automatic damper. A device which cuts off the flow of hot or cold air to or from a room when the thermostat indicates that the room is sufficiently warm or cool.

automatic data plotter. An instrument that plots contours on weather maps. The automatic data plotter gets its working data from a magnetic tape fed into a converter and changes the information from the digital type to the analog type. The analog data are transformed into electronic impulses which instruct a mechanical hand on the plotting board how the contours on the weather map should be drawn. The mechanical hand guides an inked stylus that does the actual drawing.

automobile air pollution. The air pollution from cars, three major sources being the evaporation from gas tanks and carburetors, crankcase blow-by, and tailpipe emission. Although the gas caps normally keep gasoline in the tank, the seal does not prevent the escape of vapors. Evaporation also takes place from the carburetor. When the car is decelerated or idles, more gas flows into the carburetor than is burned; much of this excess gas evaporates. About 20 percent of the hydrocarbon emitted by an automobile is the result of evaporation. Because the sealing of pistons in the cylinders is not airtight, some combustion products slip past the pistons after the air–gas mixture is ignited by the spark plugs. This exhaust gets into the crankcase, where it is released into the air by a breather tube. This source of hydrocarbons is called crankcase blow-by and amounts to 25–35 percent of the exhaust products of the engine. Most of the combustion products are vented to the atmosphere through the tailpipe. Gasoline burns most efficiently at a 15:1 air–fuel ratio, but a high-compression engine requires a rich mixture of fuel with air, 13:1 or 12:1. This means that the hydrocarbons in the fuel are not completely burned. The chief emissions are carbon monoxide, nitrogen oxides, unburned gasoline, carbon dioxide, water, and lead.

automobile engine. In the typical automobile engine with no air pollution controls, a mixture of fuel and air is fed into a cylinder by the carburetor, compressed and ignited by a spark from the spark plug. The explosive energy of the burning mixture moves the pistons and the pistons' motions are transmitted to the crankshaft that drives the car. The burnt, spent mixture passes out of the engine and then out through the exhaust tailpipe. A pound of gasoline can burn completely when mixed with about 15 lb of air. For maximum power, however, the proportion of air to fuel must be less. Most driving takes place at less than the 15:1 ratio; combustion is incomplete and substantial amounts of material other than carbon dioxide and water are ejected into the environment. One result of insufficient air is the emission of carbon monoxide instead of carbon dioxide;

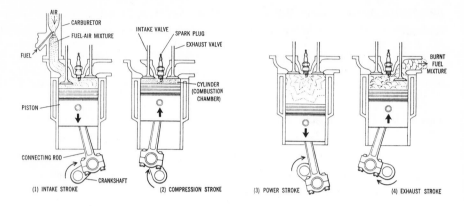

Combustion in an automobile engine. One cylinder of a typical automobile engine shown.

another byproduct is unburned gasoline, and a variety of hydrocarbon products that react easily with other chemicals. There are also nitrogen oxides and oxygen in the extremely hot air of the engine. Together, the hydrocarbons and nitrogen oxides produce photochemical smog.

automobile engine crankcase. The flow of air past a moving vehicle is directed through the crankcase in order to rid it of any gas–air mixture which has blown past the pistons, any evaporated lubricating oil, and any escaped exhaust products. The air is drawn in through a vent and emitted through a tube extending from the crankcase at a rate that depends on the speed of the car. About 20 to 40 percent of the car's total hydrocarbon emissions are sent into the atmosphere from the crankcase. Crankcase emissions are known as blow-by gases. Together with tailpipe exhaust, they account for almost all the pollution from automobiles.

automobile exhaust emission control (Chrysler Clean Air Package). The present clean air system produces satisfactory emission levels mainly by means of leaner fuel–air mixture, adjusted ignition timing, higher engine speed and air flow at idle, and special choke calibration. The basic design is illustrated in the accompanying figure. Other modifications are made peculiar to each engine, including redesigned combustion chambers, improved manifold heat-transfer facility, and new carburetor metering systems to improve carburation and cylinder distribution. Most of the modifications are directed toward the idle, acceleration and deceleration modes in which the buildup of exhaust emission occurs.

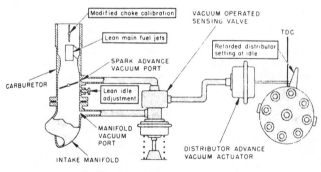

Chrysler cleaner air package schematic.

automobile exhaust emission control (combination split flow ventilation systems). In these systems, crankcase ventilation is achieved through connections to both air-cleaner and intake manifold. These systems incorporate an orifice to restrict flow in the intake manifold connection so that almost all emissions are drawn through this route at speeds up to about 35 mph at road load. Above this value, air-cleaner suction takes effect, ensuring adequate ventilation under all conditions. A check valve may be installed in the air-cleaner connection to prevent bypassing of the filter element at low engine speeds. In some designs, an extra filter element is incorporated for this purpose, but a restriction may still be required in the air-cleaner connection to act as a flame arrester.

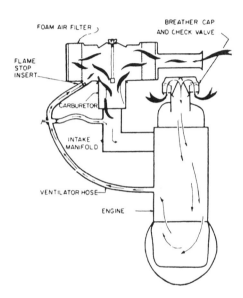

Split-flow crankcase ventilation system.

automobile exhaust emission control system [General Motors Air Injection Reaction (AIR) System]. Basically, an air injection system combined with engine modifications to increase its effectiveness. While this air distribution system is common to General Motors vehicles, necessary engine modifications vary to some extent for particular engine–carburetor–transmission combinations. Modifications of the standard carburetor flow characteristics is necessary to achieve close control of mixture ratio in the low-flow region. Retarded ignition timing at idle effectively contributes to the reduction of emissions without reducing fuel economy or performance. An antibackfire gulp valve is incorporated, which permits a controlled amount of air to enter the intake manifold on throttle closure and thus prevents buildup of unburnt fuel in the exhaust. Suitable pressure relief and check valves are also incorporated for protection of the air injection system.

automobile exhaust emission control system (Volvo Dual Manifold Emission Control System). The emission control system relies on lean carburetion, retarded spark timing at idle, increased idle speed, and throttle bypass valve for deceleration. The modified induction system incorporates a heat transfer and turbulence chamber. The carburetor throttle and manifold throttle are linked at the opening point of the manifold throttle, corresponding to a cruising speed around 75 mph and a static spark timing of 5° before top dead center is used, this relatively moderate retard being chosen to avoid "running on" and overload of the cooling system. The deceleration valve is a simple vacuum-operated poppet valve which claims to show only a moderate limitation of the engine braking effect.

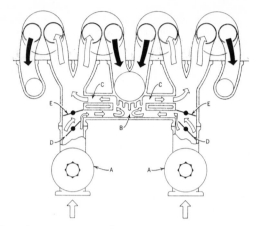

Induction system for Volvo emission control engine: *A*, carburetor; *B* and *C*, cross flow pipes; *D*, carburetor throttle; *E*, manifold throttle.

automobile exhaust emission control system (Zenith Duplex Induction System). This system incorporates, in parallel with the induction manifold, a primary feed of smaller cross section, whereby the mixture is taken through an exhaust-heat chamber. Flow in the primary feed pipe is controlled by a separate throttle (C), the main larger section being closed during part-throttle running by the "secondary" throttle valve (D). Carburetor features are of primary importance in mixture control and a special thermostatic valve is incorporated to offset the rechening effect as the engine warms up. In addition, a diaphragm bypass valve is incorporated, which short-circuits the primary throttle at a selected vacuum, thus lowering the manifold depression and supplying extra mixture to support combustion. Ignition is retarded at idle and deceleration by means of a small valve, operated by lost motion in the acceleration linkage, which feeds vacuum to a retard capsule on the distributor.

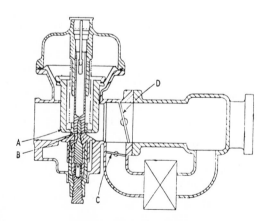

Zenith duplex induction system.

autotrophic nutrition. That process by which an organism manufactures its own food from inorganic compounds.

auxiliary energy subsystem. Equipment utilizing energy other than solar, both to supplement the output provided by the solar energy system and to provide full energy backup during periods when the solar domestic hot-water systems are not operating.

auxin. An organic compound normally produced by the leaves of a plant and which, in the proper concentrations, serves to keep the leaves attached to the stem or stalk. In the fall, auxin levels normally drop and a layer of large, thin-walled cells forms where the leaf connects to the plant. When these cells rupture, the leaf falls. By applying a compound which lowers the auxin content of a leaf, it is possible to defoliate a plant prematurely at will. This is commonly done before harvesting cotton. Conversely, by adding auxin at the right concentration, leaf fall and fruit drop can be inhibited, thereby decreasing loss from preharvest drop of fruit. If an excess of auxin is applied, however, plants respond by increasing their respiratory activities considerably beyond their ability to produce food, causing the plant to literally grow themselves to death.

Avogadro's law. Equal volumes of different gases at the same pressure and temperature contain the same number of molecules.

Avogadro's number. The number of molecules in one mole or gram-molecular weight of a substance. A number of values of the Avogadro number, which is usually denoted by $\mathcal{N}$, have been found by various methods. $\mathcal{N}$ generally lies within a range of 1 percent about the value $(6.024\ 86 \pm 0.000\ 16) \times 10^{23}$ per gram-mole (physical), $(6.023\ 22 \pm 0.000\ 16) \times 10^{23}$ per gram-mole (chemical).

azeotrope. A liquid mixture which shows a maximum or a minimum boiling point. It implies the ability of a liquid to be mixed with one or several liquids to form a mixture which boils at a constant temperature, either higher or lower than those of the components.

B

Babo's law. The addition of a nonvolatile solid to a liquid in which it is soluble lowers the vapor pressure of the solvent in proportion to the amount of substance dissolved.

backfill. The material used to refill an excavation, or the process of doing so.

background level. In air pollution, the level of pollutants from natural sources present in ambient air.

background noise. The total of all sources of interference in a system used for the production, detection, measurement, or recording of a signal, independent of the presence of the signal.

background scatter. The scattering of radiant energy into the hemisphere of space bounded by a plane normal to the direction of the incident radiation and lying on the same side as the incident ray; the opposite of forward scatter.

backsiphonage. The flowing back of contaminated or polluted water from a plumbing fixture or cross connection into a water supply line, due to a lowering of the pressure in such a line.

backwashing. The process of cleaning a rapid sand or mechanical filter by reversing the flow of water.

bacteria. Single-celled microorganisms that lack chlorophyll. Some cause diseases, others aid in pollution control by breaking down organic matter in air and water.

baffle. A deflector vane, guide, grid, grating, or a similar device constructed or placed in flowing water or sewage to check or effect a more uniform distribution of velocity; absorb energy; divert, guide, or agitate the liquids; and check eddy currents.

baffle chamber. In incinerator design, a chamber designed to promote the settling of fly ash and coarse particulate matter by changing the direction and/or reducing the velocity of the gases produced by the combustion of the refuse.

bag filter. A fabric filter used in industrial operations to recover valuable matter, as well as to control atmospheric pollution at the source. Such a filter is commonly of a tubular form or similar to an envelope and can be slipped over a wire frame. Applications of this type of filter are used by carbon black, cement, clay, and pharmaceutical plants, as well as by power stations, and where operations involve abrasives and irritating chemicals.

Particle-laden air is introduced inside the tube, usually to allow the larger particles to be projected into the dust hopper before air enters the tubes.

baghouse filtering.   One of the many processes which may be used to eliminate intermediate and large (greater than 20 μm in diameter) particles in a "bag filter." This device operates in a way similar to the bag of an electric vacuum cleaner, passing the air and smaller particulate matter, while entrapping the larger particulates.

balanced amplifier.   An amplifier circuit in which there are two identical signal branches connected so as to operate in phase opposition and with input and output connections each balanced to ground.

baling.   A recycling process where waste materials, such as used newspapers, are compacted into bundles held together with wire.

ballistic separator.   A machine that sorts organic from inorganic matter for composting.

Balmer series.   A family of lines in the visible and near-ultraviolet atomic spectrum of hydrogen.

band application.   In pesticides, the spreading of chemicals over or next to each row of plants in a field.

band-pass filter.   A wave filter that has a single transmission band extending from a lower cutoff frequency greater than zero to a finite upper cutoff frequency.

band pressure level.   For a specified frequency band, the sound pressure level for the sound contained within the restricted band. The reference pressure must be specified.

band spectrum.   An emission or absorption spectrum consisting of a number of bands, each band being a very large number of closely spaced lines which overlap on low-dispersion spectrograms.

bandwidth.   The range of frequencies above and/or below the frequency of the carrier wave, required for modulation of a signal. The bandwidth for medium waves is 9 kHz and for television, 3 MHz.

bank, sludge.   An accumulation of deposits of solids of sewage or industrial waste origin on the bed of a waterway.

bar.   A unit of pressure equal to 10^6 dyn/cm^2 (10^6 barye), 1000 mbar, or 29.53 in. Hg.

bar screen.   In wastewater treatment, a device that removes large solids.

barium (Ba).   A silvery-white, lustrous, malleable metallic element which forms salts used in pesticides, depilatories, and, when combined with the sulfate ion, as a radiopaque contrast medium. The barium ion apparently affects the polarity or permeability of tissue cell membranes so that muscle cells are stimulated indiscriminately. The fatal dose of barium in a human is approximately one gram. The sulfate salt of barium, used in human x-ray examinations, is insoluble but may occasionally be contaminated with a soluble salt such as barium carbonate or barium hydroxide. Radiologists are advised to test barium sulfate preparations by adding a small amount of magnesium sulfate or sodium sulfate solution to a sample portion; the appearance of a precipitate indicates the presence of a soluble barium salt. Sulfide, oxide, and carbonate salts produce irritation of the eyes, nose, and throat. The salts have been shown to produce paralysis in laboratory animals. Accidental human poisoning by the oxide salt has resulted in severe abdominal pain with vomiting, difficult breathing, rapid pulse, paralysis of the limbs, cyanosis, and death. In 1980, New Orleans public health officials advised against using local tap water for preparing baby formulas or in the operation of dialysis equipment for kidney disease patients because of contamination of the water supplies by barium from an illegal spill. The permissible criterion for barium in public waters is 1.0 mg/liter. Barium may exert its detrimental physiological effect by forming a barium sulfate precipitate, thus effectively diminishing sulfate ion concentration below body requirements.

barium oxide. BaO, a chemical compound used as a water vapor absorbent.

barn. A unit of area for measuring a nuclear cross section. One barn equals 10^{-24} cm^2.

baroclinity. The state of stratification in a fluid in which surfaces of constant pressure (isobaric) intersect surfaces of constant density (isosteric). The number per unit area of isobaric–isosteric solenoids intersecting a given surface is a measure of the baroclinity.

barograph. A self-recording aneroid barometer usually so arranged that changes in pressure are recorded on a rotating paper-covered drum.

barotrophy. The state of a fluid in which surfaces of constant density (or temperature) are coincident with surfaces of constant pressure; it is the state of zero baroclinity.

barotropic disturbance. (1) An atmospheric wave in a two-dimensional nondivergent flow, for which the driving mechanism lies in the variation of vorticity of the basic current and/or in the variation of the vorticity of the earth about the local vertical. When the basic current is uniform, the wave is a Rossby wave, also called a barotropic wave. (2) An atmospheric wave of cyclonic scale in which troughs and ridges are approximately vertical.

baryon. A joint name for nucleons and hyperons. The baryon family has the greatest mass of all subatomic particles. Baryons include the basic proton and neutron plus the heavier brother xi, sigma, and lambda baryons. The last three baryons are called hyperons because they are bigger than either the proton or the neutron or the neutron members of their family. Hyperons can be created in the laboratory by adding energy to a proton. The proton absorbing the energy gains mass and becomes the larger hyperon.

basal application. In pesticides, the spreading of a chemical on stems or trunks just above the soil line.

base. A proton acceptor; a substance which reacts with an acid to form a salt. For many purposes it is sufficient to say that a base is a substance which dissociates on solution in water to produce one or more hydroxyl ions. More generally, bases are defined according to other concepts. The Bronsted concept states that a base is any compound which can accept a proton. Thus NH_3 is a base, since it can accept a proton to form ammonium ions: $NH_3 + H^+ \rightleftharpoons NH_4^+$. A still more general concept is that of G.N. Lewis, who defines a base as anything which has an unshared pair of electrons. Thus in the reaction

$$H^+ + :N \underset{\displaystyle H}{\overset{\displaystyle H}{\lessgtr}} H \rightleftharpoons NH_4^+$$

the NH_3 is a base because it possesses an unshared pair of electrons. The latter concept explains many phenomena, such as the effect of certain substances other than hydrogen ions in the changing of the color of indicators. It also explains acids and bases in nonaqueous systems such as liquid NH_3 and SO_2.

BATEA. An acronym for Best Available Technology Economically Available, terminology applied by the EPA in guideline regulations for controlling 65 pollutants that may be present in wastewater effluents from a group of 21 major industries. The pollutants are also referred to as consent-decree chemicals because they are covered in a 1976 EPA consent decree. The technologies required for the control of priority toxic pollutants include such techniques as carbon adsorption devices. The industrial categories include pesticides, textiles, rubbers, plastics, pharmaceuticals, wood-preserving, and refineries.

batt. A sheet of insulating material such as fiberglass.

beat. A periodic variation in amplitude or a superposition of disturbances having different frequencies. When the variations (oscillations) occur in an electric circuit, the phenomenon is known as heterodyning. Two vibrations of slightly different frequencies f_1 and f_2 which, when added together, produce, in a detector sensitive to both these frequencies, a regularly varying response which rises and falls at the "beat" frequency $f_b = |f_1 - f_2|$. It is important to note that a resonator which is sharply tuned to f_b alone will not resound at all in the presence of these two beating frequencies.

beat frequency. The beat of two different frequencies of signals on a nonlinear circuit when they combine or beat together. It has a frequency equal to the difference of the two applied frequencies.

Beaufort Wind Scale For Land Use (U.S. Weather Bureau).

Beaufort number	Miles per hour	Knots	Wind effects observed on hand	Terms used in U.S.W.B. forecasts
0	Less than 1	Less than 1	Calm; smoke rises vertically	Light
1	1–3	1–3	Direction of wind shown by smoke drift; but not by wind vanes	Light
2	4–7	4–6	Wind felt on face; leaves rustle; ordinary vane moved by wind	Light
3	8–12	7–10	Leaves and small twigs in constant motion; wind extends light flag	Gentle
4	13–18	11–16	Raises dust, loose paper; small branches are moved	Moderate
5	19–24	17–21	Small trees in leaf begin to sway; crested wavelets form on inland waters	Fresh
6	25–31	22–27	Large branches in motion; whistling heard in telegraph wires; umbrellas used with difficulty	Strong
7	32–38	28–33	Whole trees in motion; inconvenience felt walking against wind	Strong
8	39–46	34–40	Breaks twigs off trees; generally impedes progress	Gale
9	47–54	41–47	Slight structural damage occurs (chimney pots, slates removed)	Gale
10	55–63	48–55	Seldom experienced inland; trees uprooted; considerable structural damage occurs	Whole gale
11	64–72	56–63	Very rarely experienced; accompanied by widespread damage	Whole gale
12 or more	73 or more	64 or more	Very rarely experienced; accompanied by widespread damage	Hurricane

bed depth. The amount of carbon, expressed in length units, which is parallel to the flow of the stream and through which the stream must pass.

Beer's law. If two solutions of the same colored compound be made in the same solvent, one of which is, say, twice the concentration of the other, the absorption due to a given thickness of the first solution should be equal to that of twice the thickness of the second. Mathematically, this may be expressed $I_1c_1 = I_2c_2$ when the intensity of light passing through the two solutions is a constant and if the intensity and wavelength of light incident upon each solution are the same.

bel. A unit of level when the base of the logarithm is 10. A unit used for the logarithmic expression of the ratios of power, voltage, or current in wire or radio communications; also in comparing the ratio of noise intensities.

benthic region. The bottom layer of a body of water.

benthos. The plants and animals that inhabit the bottom of a water body.

benzene. C_6H_6, a clear, colorless liquid with a molecular weight of 78.11 and a boiling point of 80.1°C; it is used as a solvent and as an intermediate in the manufacture of organic chemicals, such as phenol. In Europe, benzene is one of the constituents of motor fuel. Benzene enters the body most commonly via the inhalation route, but may also be absorbed through the skin. Benzene has a cumulative effect; while a single exposure to a relatively high concentration is not too serious to the person experiencing it, prolonged or repeated exposures result in chronic benzene poisoning, which leads to symptoms ranging from fatigue and nausea to anemia, leukopenia, thrombocytopenia, reticulocytosis, and prolonged bleeding times.

benzene exposure limitations. The Department of Labor issued rules on February 2, 1978, to limit worker exposure to benzene, a widely used chemical suspected of causing leukemia and other blood ailments. Benzene was used in the production of other chemicals that were used in the manufacture of rubber, plastics, resins, disinfectants, and pharmaceuticals. It was also used as a component of motor fuels and in the production of detergents, pesticides, solvents, and paint removers. In 1978 about 11×10^9 lbs of benzene were produced on an annual basis. Because "the available scientific evidence established that employee exposure to benzene presents a cancer danger," the new rules stated "the standard limits employee exposure to the lowest feasible level." The rules would require employers to limit work exposure to the chemical to a maximum of 1 ppm of air over an eight-hour period or 5 ppm for any 15-minute exposure period. The new standard was a 90 percent reduction from the department's permanent standard for an eight-hour period.

benzidine. $NH_2C_6H_4C_6H_4NH_2$, an aromatic amine that occurs as white to reddish crystals, powder, or flakes. It may be used in the production of sulfur, azo, and aniline dyes; clinical detection of blood; printing; and quantitative analysis. Benzidine, also known as benzidine base, is a recognized carcinogen in rats and mice, and it has been found to induce bladder tumors in humans, primarily by absorption through the skin. It also can cause damage to blood cells through hemolysis and bone marrow damage. Ingestion results in signs and symptoms ranging from nausea and vomiting to liver and kidney damage. When heated, the chemical may decompose to produce highly toxic fumes. The exposure limit to benzidine established by NIOSH is 0.1 percent by weight or volume. The oral LD_{50} in rats has been reported as 300 mg/kg.

Bernoulli's theorem. At any point in a tube through which a liquid is flowing, the sum of the pressure energy, potential energy, and kinetic energy is constant. If p is pressure, h is the height above a reference plane, d is the density of the liquid, and v is the velocity of flow, $p + hdg + \frac{1}{2}dv^2 = $ const.

Berthelot principle of maximum work. Of all possible chemical processes which can proceed without the aid of external energy, that process always takes place which is accompanied by the greatest evolution of heat. This law holds for low temperatures only and does not account for endothermic reactions.

beryllium (Be). A metallic element resembling magnesium in appearance and chemical properties and hard enough to scratch glass. Atomic weight, 9.0; specific gravity, 1.80; melting point, 1280°C; room temperature modulus of 43×10^6 lb/in.^{2}. The ratio of modulus to density is more than six times greater than that of aluminum, magnesium, steel, or titanium.

beta decay. One of the processes by which certain radioactive atoms disintegrate. In beta decay, the atomic core loses a neutron and gains a proton, at the same time an electron is expelled from the atom. Beta decay is involved in the conversion of ordinary uranium into atomic fuel plutonium.

beta emitter. A radioactive element, whether natural or artificial, which changes into another element by beta decay.

beta gain. For transistors, the ratio of the change in collector current with the collector plus emitter voltage held constant.

beta particle (β particle). One of the particles which can be emitted by a radioactive atomic nucleus. It has a mass about $\frac{1}{1837}$ that of the proton. The negatively charged beta particle is identical with the ordinary electron, while the positively charged type (positron) differs from the electron in having equal but opposite electrical properties. The emission of an electron entails the change of a neutron into a proton inside the nucleus. The emission of a positron is similarly associated with the change of a proton into a neutron. Beta particles have no independent existence inside the nucleus, but are created at the instant of emission.

beta radiation. The emission of either an electron or a positron by an atomic nucleus.

beta ray. A stream of electrons. The name beta ray was applied before scientists knew what it was. Still commonly used to describe an electron expelled from a disintegrating atom.

betatron. A device used to accelerate beta particles (electrons). It uses the principle of the transformer, in which the primary is a huge electromagnet and the secondary is the stream of electrons being accelerated. In a betatron an electric gun shoots electrons into a circular tube called a "doughnut." This tube is placed between the poles of a powerful electromagnet. Changes in the magnetic field make the electrons gain energy and cause them to be pushed along at steadily increasing speeds. The electrons whirl around in the doughnut as the magnetic field holds them in their path. The betatron keeps electrons in the same circular path, instead of causing them to spiral outward as do particles in the cyclotron.

BeV. A billion electron volts. An electron possessing this much energy travels at a speed close to that of light, 186,000 mile/s.

bevatron. A six or more billion electron volt accelerator of protons and other atomic particles; makes use of a Cockcroft–Walton transformer cascade accelerator and a linear as well as an electromagnetic field in the buildup.

BIGAS. *Process:* Coal is crushed, dried, and pulverized (70 percent through 200 mesh). Coal and steam are fed to the upper reactor by four upward-directed, concentric injecting nozzles which are spaced around the shell so that streams do not strike the refractory, but impinge in the gas space. Here the coal is rapidly heated to 1700°F by hot gas from the lower reactor. Devolatilization of coal occurs in producing methane and char. The residence time of the coal is a few seconds. Entrained char, separated from the raw gas in a cyclone, is fed into the gasifier. Preheated oxygen (1200°F) and steam are also fed to the gasifier, which operates at a temperature of about 2700°F, causing slagging of the ash. Molten slag from the gasifier drains into quench water. Raw gas from the upper reactor is quenched by hot condensate to 660°F, filtered through sand, and shifted. It is then purified to remove CO_2 and H_2S, methanated, and dried to produce pipeline-quality gas. The gasifier may also be operated on air, rather than oxygen, at moderate system pressures, producing a low-Btu raw gas (Bituminous Coal Research Inc.).

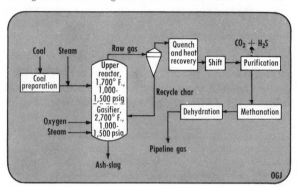

BIGAS process. (Bituminous Coal Research Inc.)

binding energy of a nucleus. The energy which would be generated if that nucleus was built up from its nucleons, and which would have to be supplied to decompose that nucleus into its nucleons. It can be obtained quite accurately by comparing the mass of that nucleus with that of all its protons and neutrons taken separately; the binding energy is the Einstein equivalent of that mass difference and is calculated by dividing it by the square of the speed of light. The mass of one proton is equivalent to 931 MeV. By the binding energy of a proton or neutron in a nucleus, we mean the energy needed to remove it from that nucleus (usually 6–8 MeV).

bioaccumulation. The process by which certain plants and animals retain chemical pollutants in their cell tissues. Tissues of fish concentrate low levels of chemicals such as DDT and PCB, which are less of a threat to the fish themselves than to those who eat the fish. Bioaccumulation is often related to the amount of lipids, or fats, in an animal and the fat solubility of the chemical. In the example of fish, chemicals that are only slightly soluble in water, but highly soluble in lipids, tend to become concentrated in the fatty tissues of the fish whose habitat is contaminated by the toxic substance. In general, the amount of chemical found in a fish is directly proportional to the amount of chemical in the water. The ratio of the amount of chemical in the fish tissues to the amount of chemical in the water is known as the bioconcentration factor.

bioassay. Using living organisms to measure the effect of a substance, factor, or condition.

biochemical oxygen demand (BOD). (1) A test which measures the quantity of oxygen utilized in the biochemical oxidation of organic matter in a specified time and at a specified temperature (usually 5 days at 20°C). (2) The amount of oxygen required for the biological oxidation of the organic matter in a liquid.

biocide. A poison or other substance used to kill such pests as injurious insects, but which also kills useful insects and other living things.

bioconversion. The conversion of organic wastes into methane (natural gas) through the action of microorganisms.

biodegradable. Capable of being broken down by bacteria into basic elements. Most wastes which are or were once alive, such as food remains and paper, are biodegradable.

biofouling and corrosion. The determination of methods to prevent, alleviate, and control the formation of biological slime, corrosion, and other coatings on ocean thermal energy conversion (OTEC) heat-exchanger surfaces.

biogenesis. The origin of living organisms from other living organisms.

biogeochemical cycle. The cyclical series of transformations of a chemical element through the organisms in a biotic community and their physical environment. The more or less circular paths of chemical elements passing back and forth between organisms and environment are known as biogeochemical cycles. *Bio* refers to living organisms, and *geo* to the rocks, soil, air, and water of the earth. Biogeochemistry is the study of the exchange (back and forth movement) of materials between living and nonliving components of the biosphere.

biological control. The suppression of reproduction of a pest organism by utilizing behavioral traits or other organisms rather than chemical means.

biological half-life. The time required for the body to eliminate one-half of an administered dose of any substance by regular processes of elimination. This time is approximately the same for both stable and radioactive isotopes of a particular element.

biological magnification. The concentration of certain substances up a food chain, a very important mechanism in concentrating pesticides and heavy metals in organisms such as fish.

biological oxidation. The way that bacteria and microorganisms feed on and decompose complex organic materials. Used in self-purification of water bodies and activated sludge wastewater treatment.

biological oxygen demand (BOD). The demand for oxygen by bacteria breaking down organic pollutants. The most common pollutants are organic and are attacked by bacteria and broken down into simpler compounds. To do this, bacteria require oxygen. The greater the supply of organic food, the larger the population of bacteria that can be supported, and the greater the demand on the oxygen supply in the water. The BOD is a useful index of pollution, especially that related to the organic load of the water. Since all stream animals are dependent upon the oxygen supply in the water, the BOD is of particular importance in determining which forms of life a polluted stream is capable of supporting.

bioluminescence. The production of light without sensible heat by living organisms as a result of a chemical reaction either within certain cells or organs, or extracellularly in some form of secretion.

biomass. The total quantity of living material (expressed in dry weight) present at a given time for a particular area. Although biomass is usually expressed in dry weight per unit surface of a certain habitat, in studying aquatic areas, soil strata, and certain microhabitats, such as logs, it is sometimes convenient and meaningful to express biomass as the volume or weight per unit volume of habitat. In biomass, no allowance is made for nonnutritious parts of organisms such as shells, skeletons, and other parts that cannot be digested or taken as food.

biome. A climatically controlled area including a number of different communities in various stages of succession. The entire region is dominated by a typical climax type (climatic climax), but will include a number of diverse climax types (disclimax, edaphic climax, etc.) in accordance with existing ecological conditions.

biomonitoring. The use of living organisms to test water quality at a discharge site or downstream.

biosphere. The transition zone between earth and atmosphere within which most forms of terrestrial life are commonly found; the outer portion of the geosphere and inner or lower portion of the atmosphere.

biostabilizer. A machine that converts solid waste into compost by grinding and aeration.

biota. The fauna and flora of a given region.

biotic. Pertaining to life.

biotic potential. The growth rate inherently possible in a population under ideal conditions. This maximum growth rate is equivalent to the maximum natality minus the minimum mortality, with no restraints such as lack of food, predation, parasitism, or competition for space.

biotic succession. The natural replacement of one or more groups of marine organisms growing in a specific habitat by other groups, the preceding groups in some way preparing or favorably modifying the habitat for the succeeding groups.

biotron. A test chamber used for biological research within which the environmental conditions can be completely controlled, thus allowing observations of the effect of variations in environment on living organisms.

Bishop's ring. A whitish ring, centered on the sun or moon, with a slight bluish tinge on the inside and reddish brown on the outside.

bitumen. A tarry hydrocarbon mixture soluble in carbon dioxide.

bivane. A sensitive wind vane used in turbulence studies to obtain a record of the horizontal and vertical components of the wind.

black body. That part of a collector which accepts short-wave radiant energy and converts it to long-wave radiant energy. In the process, the black body temperature is raised by the radiant heating of the insolation. The black body acts as a transfer or conversion device, which changes insolation to heat that is ready to be carried away by the media fluid stream. It acts as a heat exchanger.

black body radiation. The amount of energy radiated from a black body, depending largely upon the temperature of the body. At a given temperature, there is an upper limit to the amount of radiant energy that can be emitted in a given time by a unit surface of a body. This maximum amount of radiation for a given temperature is called the black body radiation. A body that radiates for every wavelength the maximum intensity of radiation possible at a given temperature is known as a black body. At a given temperature, the maximum is the same for every black body, regardless of its structure or composition. All black bodies emit a continuous spectrum. Gases, however, have a discontinuous spectrum, showing only emission and absorption in various parts of the spectrum, called lines. These lines are characteristic for each substance and serve as a means of identification.

black lung disease. A type of pneumoconiosis that affects coal miners, causing their lungs to become black from coal dust and producing a disabling inflammation. The disorder, also known as anthracosilicosis, begins with very small deposits of coal dust less than one-fourth of an inch in area, which collect on the walls of the alveoli and terminal bronchioles. The surrounding lung tissue develops localized areas of emphysema. The lesions may not progress further or they may evolve into fibrous nodules that merge into larger fibrotic masses of tissue. It has been suggested that whether a case of black lung disease is self-limiting or progresses to a more serious form depends upon other particles in the coal dust, such as silica, which can aggravate the condition. Some coal miners exhibit no disease symptoms. But a typical case begins with shortness of breath and coughing spells, followed by bronchospasms and emphysema. The miner often becomes unable to work and may eventually die of heart failure involving the right ventricle of the heart, the chamber responsible for pumping blood through the lungs. Clinical studies indicate that once black lung disease has become established, the disorder may be expected to become progressively more serious, even if the coal miner is removed from a coal dust environment. More than 135,000 disabled coal miners have filed claims for compensation under the black lung legislation passed in the early 1970s.

blanket. The layer of an element, such as ^{238}U or Th, in a breeder or converter reactor, in which fissionable material is produced by neutron capture. From ^{232}Th and ^{238}U, ^{233}U and ^{239}Pu result eventually, after two beta particles are emitted in each case.

blast furnace. Furnaces where the substance and fuel are in contact with each other. The function of the fuel is not only to heat the ore, but to produce gases which react chemically with it. Blast furnaces perform two functions: They reduce iron ore to iron, and separate the iron from the mixtures of minerals in which it then lies by melting. The gases which pass out of the top of a blast furnace still contain carbon monoxide. In the more modern installations, the blast furnace gas is dedusted and usually employed as a fuel in hot blast stoves for preheating the air of the blast furnace, or in coke ovens.

blast furnace gas. A residual gas which passes out of the top of a blast furnace. In a blast furnace the ore is reduced from oxide to metal by freshly made producer gas, only a part of which is used in the chemical reaction. Blast furnace gas is similar in constitution to producer gas, but it contains a higher proportion of carbon dioxide. It retains about 60 percent of the heat of the original coke, and its calorific value is 70–80 percent of that of producer gas.

bloom. A population explosion of microorganisms caused by the sudden availability of an essential substance. Phosphorus is a trigger factor in most aquatic systems. Microorganisms are often held in check by a lack of phosphorus, but when it is made available they use it immediately, and with their rapid reproduction rates they can quite suddenly become extremely abundant. This population explosion, usually of one or more species of algae, is called a bloom. Normally, ecologists call an aquatic population a bloom when there are 500 individuals of a species per milliliter of water.

blowdown. Hydrocarbons purged during refinery shutdowns and startups which can be manifolded to blowdown systems for recovery, safe venting, or flaring.

blowoff. A controlled outlet on a pipeline, tank, or conduit which is used to discharge water or an accumulation of materials carried by the water.

BOD$_5$. The amount of dissolved oxygen consumed in five days by biological processes breaking down organic matter in an effluent.

bog. A swamp or tract of wet land commonly covered with peat.

Bohr radius. The smallest possible radius of an electron orbit in the Bohr model of the atom, $5.291\,67 \times 10^{-9}$ cm.

Bohr's atomic theory. The theory that atoms can exist for a duration solely in certain states, characterized by definite electronic orbits, i.e., by definite energy levels of their extranuclear electrons, and in these stationary states they do not emit radiation; the jump of an electron from one orbit to another of a smaller radius is accompanied by monochromatic radiation.

boiling. (1) The rapid vaporization which disturbs a liquid, occurring when the vapor pressure of the liquid equals the pressure on its surface. When water is so hot that its vapor pressure equals the pressure on the liquid surface, bubbles of vapor reach the surface freely. Vaporization then occurs at such a rapid rate throughout the water that the water becomes agitated. If the pressure on the liquid surface is greater than one atmosphere, boiling occurs at a temperature higher than the normal boiling point; if the pressure on the liquid surface is less than an atmosphere, boiling occurs at a temperature lower than the normal boiling point. (2) Rapid evaporation from within and from the surface of a liquid, occurring at a specific temperature called the boiling point, which is characteristic for each liquid.

boiling point. The temperature at which the equilibrium vapor pressure between a liquid and its vapor is equal to the external pressure on the liquid.

boiling point of water. The temperature at which the equilibrium vapor pressure between water and its vapor is equal to the external pressure. Since the vapor pressure of water is 760 mm Hg, or one atmosphere, at 100°C this is the normal boiling point of water. If the air pressure is reduced to 525.8 mm Hg, water boils at 90°C, because at this temperature the vapor pressure of water is 525.8 mm Hg. In order to make water boil at 50°C, the pressure must be reduced to 92.5 mm Hg. If the pressure is increased to 787.5 mm Hg, water will not boil until its temperature reaches 101°C.

boiling water reactor. A reactor in which water is allowed to boil directly inside the reactor itself, thus producing steam without the usual intermediate step of transferring the heat from a coolant to a boiler in which steam is made.

bolometer. An instrument for measuring radiant energy, either thermal or radar. In the former case, the change of electrical resistance of an exposed thin strip of platinum foil, coated with finely divided "black" platinum, is measured, while carefully screened from all radiation except that which is being measured.

Boltzmann's constant. The ratio of the universal gas constant to Avogadro's number; equal to $1.380\ 54 \times 10^{-16}$ erg/°K.

bond, chemical. The mechanism that joins atoms together to form molecules.

bond, covalent. A chemical bond in which atoms are held together by electron pairs shared more or less equally.

bonding. The joining together of atoms or molecules by means of electron-sharing ion formation, or other processes.

bone conduction. The process by which sound is conducted to the inner ear through the cranial bones.

boom. A floating device used to contain oil on a body of water.

bora. A cold downslope wind. The temperature is so low at its source that even though it is impressed and warmed as it pours down the mountainside, when it reaches the lower areas it is still colder than the air it displaces.

boron (B). A yellow, monoclinic crystal or brown amorphous powder; atomic weight, 10.82; specific gravity, 2.3; melting point, 2030°C; boiling point, 2550°C. Industrial boron pollution occurs as a result of the manufacture or use of synthetic boranes. Boron may be present naturally in concentrations as high as 15 mg/liter. The permissible criterion for public water supply is 1.0 mg/liter. Boron is toxic for many organisms in concentrations as low as 1 mg/liter, while other organisms tolerate levels above 15 mg/liter. Boron is an essential nutrient in concentrations up to 0.5 mg/liter.

botanical pesticides. A plant-produced chemical used to control pests; for example, nicotine or strychnine.

box model. A mathematical model employed to estimate the effects of air pollutants entering a hypothetical "box" through which air is being moved by a wind of constant velocity. The box model is perhaps the most simple of theoretical devices used to study the possibilities of gases and particulates being mixed with and diluted by a moving air mass. Wind velocity is assumed constant and is measured in meters per second. The pollutant is estimated to enter the atmosphere at a rate calculated in grams per second. The dimensions of the imaginary box are measured in meters. The pollutant is assumed to be uniformly distributed within the dimensions of the hypothetical box and the concentration is assumed to be constant with respect to time. The formula for estimating the effect is $C = Q/uWD$, where Q is the emission rate of pollutant, u is the wind velocity, and W and D represent the width and depth of the model space, respectively. More sophisticated models for estimating industrial air pollution are available, including laboratory devices that permit control over the various factors entering the model and the resultant effects of dilution, mixing, and interaction.

Boyle's law for gases. At a constant temperature, the volume of a given quantity of any gas varies inversely as the pressure to which the gas is subjected. For a perfect gas, changing from pressure p and volume v to pressure p' and volume v' without change of temperature, $pv = p'v'$.

brackish water. A mixture of fresh and saltwater.

breathing. (1) In humans, a cyclic inspiration and expiration, repeated 15–18 times per minute when at rest. Air enters the lung in response to a partial vacuum created by enlarging the chest cavity. This is accomplished by contractions of the diaphragm, a sheetlike muscle separating the thoracic and abdominal cavities, and the rib muscles. The diaphragm is pulled downward and the rib muscles elevate the rib cage with the concerted effect of expanding the thoracic volume. Expiration is primarily due to the elasticity of the lungs, but is assisted by the weight of the chest wall. (2) The net flow of oxygen to the cell of the body and the return flow of carbon dioxide is in response to concentration gradients resulting from the process of internal respiration. These gradients apply to each gas and their order of magnitude individually. The blood passes through the lung and tissue capillaries too quickly for the gases to be completely equilibrated with those in the alveoli and cells of the body. A man at rest circulates blood at the rate of about 5 liter/min. The difference between the rate required and that actually observed is due to the action of a chemical compound called hemoglobin, which acts as a carrier for some 98 percent of the oxygen transported by the blood. Hemoglobin combines with oxygen to form oxyhemoglobin and oxygen is then reliberated in response to the concentration gradient of this gas, and to a lesser extent, a response to the amount of carbon dioxide present. Thus hemoglobin combines with oxygen at high O_2 tensions and dissociates at low tensions. As the CO_2 of the blood is lowered, the dissociation of hemoglobin and oxygen is accelerated. In this way, a man is supplied with the 250 ml/min (at rest) or more of oxygen needed to bring about the liberation of energy from organic foodstuffs. Carbon dioxide is sparingly soluble in plasma and depends upon hemoglobin for its export from the body tissues.

breeching. A passage or conduit to conduct products of combustion to the stack or chimney.

breeder reactor. A reactor in which fissionable material, such as ^{239}Pu, is produced at a greater rate than the fuel, ^{235}U, is consumed. Conventional nuclear reactors use up the initial loading of fuel, which must be replaced with new fuel. In a breeder reactor, however, the initial fuel generates and also produces new fuel in the form of plutonium or other fissionable material. The rate that a breeder reactor "breeds" this additional fuel is its "doubling time"; this is the time it takes a reactor to make an amount of new fuel equal to the amount put in originally. The doubling time ranges from 5 to 20 years.

breeze. Very fine particles of coke.

Brewster's law. The tangent of the polarizing angle for a substance is equal to the index of refraction. The polarizing angle is that angle of incidence for which the reflected polarized ray is at right angles to the refracted ray. If n is the index of refraction and θ is the polarizing angle, $n = \tan\theta$.

bright-line spectrum. A spectrum consisting of one or more discrete wavelengths of light (which appear as bright lines) separated by regions of darkness where there is no radiation. A bright-line spectrum is emitted by incandescent bases.

brightness. The quality or condition of being bright, measured by the flux emitted per unit of emissive area as projected onto a plane normal to the line of sight. The unit of brightness is that of a perfectly diffusing surface giving out one lumen per square centimeter of projected surface and is called the lambert. The millilambert (0.001 L) is a more convenient unit. One candle per square centimeter is the brightness of a surface which has in the direction considered, a luminous intensity of one candle per square centimeter.

British thermal unit (Btu). The quantity of heat required to raise the temperature of one pound of water one degree Fahrenheit at or near its point of maximum density (39.1°F). The Btu is equivalent to 0.252 kcal.

broadcast application. In pesticides, the spreading of a chemical over an entire area.

brown fumes (brown smoke). The smoke emitted by superoxygenated air-blown, or oxygen-blown, iron or steel converters.

Brownian movement. A continuous agitation of particles in a colloidal solution caused by unbalanced impacts with molecules of the surrounding medium. The motion may be observed with a microscope when a strong beam of light is caused to traverse the solution across the line of sight.

bubble chamber. A device used for the detection and study of elementary particles and nuclear reactions. In a bubble chamber, a container is filled with a liquid and the tracks of particles become visible as a series of bubbles when the liquid boils. One substance used in bubble chambers is liquid oxygen. The advantage of the bubble chamber over the cloud chamber is that more nuclear collisions take place in the dense liquid of the bubble chamber. This difference makes it possible to study particles and rays having speeds that are much higher than those which produce recordable events in a cloud chamber.

bubble concept. An alternative emission-reduction plan offered in state implementation programs for air pollution control. The plan allows an industrial facility to reduce controls in a part of its plant where pollution control costs are high, in exchange for a comparable increase in pollution control in another part of the same plant where abatement is less expensive. Since the EPA-sponsored proposal treats an industrial facility in terms of its total emissions—as if it had a bubble over it—the program is known as the bubble concept. The cost-effective mix of pollution controls is an alternative to a standard plan in which regulations are applied to each individual process of the industrial plant.

bubbler. An absorption apparatus used in gas sampling, usually U-tube absorbers filled with a specific amount of reagent and fitted with a glass partition, so that the air or gas led into them passes through the reagent solution in the form of very fine bubbles. The sampling rate is usually about 100–150 liter/hr of gas stream.

buffer. A substance in a solution which tends to lessen the change in hydrogen ion concentration (pH), which would otherwise be produced by adding acids or bases.

buffer strip. A strip of grass or other erosion-resisting vegetation between or below cultivated strips or fields.

Bugwatcher. A rapid assay system used by the EPA to measure the ability of small animals to respond physically to the presence of toxic substances in their environments. The system consists of a closed-circuit television device that monitors and records a test animal's movements, which in turn are analyzed by a computer. The computer traces the paths traveled by the test animals after exposure to a toxic substance and calculates the various characteristics of their altered movements. The system is capable of handling 20 control and 20 experimental animals at the same time. The entire experiment requires only a few hours, compared to a standard 96-hour short-term toxicity assay. Effects observed in such tests include retarded growth in aquatic animals, associated with a sharp increase in swimming speed, the loss of ability of an animal to sense the presence of food, or abnormal responses by an animal to changes in light and darkness.

bulkhead line. A legally set limit on commercial filling things under ideal conditions.

bulking, sludge. A phenomenon that occurs in activated sludge plants whereby the sludge occupies excessive volumes and will not concentrate readily.

burial ground (graveyard). A disposal site for unwanted radioactive materials that uses earth or water for a shield.

burst. In cosmic ray studies, an exceptionally large electric pulse visible in an ionization chamber; it indicates the emission or incidence of several or many ionizing particles simultaneously. The cause of the phenomenon may be cosmic ray showers or a nuclear disintegration known as spallation.

bushing. A removable cylindrical lining for an opening, such as the end of a pipe. These linings limit the size of the opening, resist abrasion, and act as a watertight guide. Thus, when two different sizes of pipe are mated, one or more bushings may be used.

Buys–Ballot's law. If an observer in the Northern Hemisphere stands with his back to the wind, lower pressure will be on his left.

byssinosis. An occupational disease that mainly affects people who work with textiles; also known as brown lung disease. It is primarily a respiratory ailment associated with atmospheric concentrations of dust from cotton, flax, and other fabric materials, and complicated by the presence of foreign materials such as molds and fungi that may collect on fabric materials. Wheezing and shortness of breath are among the characteristic symptoms. Workers are most often affected with respiratory symptoms when they return to the plant after a weekend or holiday absence from work. Prolonged exposure often leads to chronic bronchitis and emphysema, although there is no causal evidence linking byssinosis to the type of pulmonary fibrosis associated with other occupational lung disorders. As with other occupational diseases of the lungs, removing the worker from a source of textile dust is unlikely to result in recovery from the effects once the symptoms have progressed to the point of reducing the ability of the worker to be fully productive. However, a textile worker who develops byssinosis is not sensitized to all other types of atmospheric dust. Because pulmonary function becomes limited, an added strain is placed on the right side of the heart and right ventricular heart failure often is a part of the prognosis.

C

cadmium (Cd). A metallic element widely used by industry in the production of copper, lead, silver, and aluminum alloys. It is also used in metal finishing, ceramic manufacturing, and photography and is a byproduct of nuclear reactor operation. Salts of cadmium are used as insecticides and as antiparasitic agents. The permissible criterion for cadmium in public water is 0.01 mg/liter. Cadmium is dangerous, since it may combine synergically with other toxic substances. Its principal effect upon aquatic life is the inhibition of bivalve shell production in mollusks. Cadmium is an experimental carcinogen, and its inhalation or ingestion results in nausea, salvation, vomiting, diarrhea, abdominal pain, and other gastrointestinal symptoms that resemble food poisoning. High concentrations may result in pulmonary edema and death.

caesium (cesium) (Cs). A silver-white metallic element of atomic weight 132.91; specific gravity, 1.90; melting point 28.6°C; boiling point 713°C. Caesium iodide (CsI) crystals are used for the detection of gamma rays by the production of light.

calciferol. A form of vitamin D in the human body. Studies of industrial air pollution indicate that levels of calciferol are reduced in human populations living in urban areas affected by high sulfate and particulate densities in the atmosphere. Calciferol plays a vital role in the hormonal influence on calcium levels in the blood, which in turn determine the mineral composition of the teeth and bones. Rickets is a disease associated with faulty calcium metabolism. Air pollution reduces the amount of sunlight on the skin. The ultraviolet frequencies of sunlight cause provitamin D on the skin to become converted to the calciferol form of vitamin D. The irradiated product is absorbed through the skin and carried to the liver and other organs of the body, including the small intestine, where the substance enhances the absorption of calcium from food being digested. Children are particularly sensitive to the interference by heavy concentrations of particulates in the atmosphere with ultraviolet light activity on skin surfaces because of their still developing bone structures. Adults are also affected by the air pollution effects on calciferol synthesis and may develop a form of bone softening called osteomalacia.

calcite. A common mineral, $CaCO_3$, calcium carbonate, occurring in a variety of crystalline forms, chalk, limestone, marble, etc., and it is one of the layers which make up the shells of marine invertebrates.

calcium (Ca). An element of the alkaline earth family. A white, crystalline metal which is soluble in water, with the formation of hydroxide, and readily dissolved in acids, with the formation of salts.

calorie. (1) A unit of heat; the amount of heat required to raise one gram of water one degree centigrade (strictly from 14.5 to 15.5°C). A kilocalorie is a unit 1000 times larger, the amount of heat required to raise one kilogram of water one degree centigrade. (2) By international agreement, a specific number of joules, 1 cal = 4.186 05 J. Since the kilocalorie equals 10^3 calories, 1 kcal = 4.186 05 $\times$ 10^3 J. The size of the calorie defined in this way is very nearly the same as the original calorie.

calorific value. The number of heat units obtained by the complete combustion of a unit mass of fuel, usually expressed in British thermal units (Btu) per pound of solid or liquid fuel and per standard cubic foot for gaseous fuels, or in calories in the cgs and mks systems of units.

candela. The unit of luminous intensity in the International System of Units, 1960; equal to 1/60 of the luminous intensity from one square centimeter of a black body at 2046°K (the temperature of solidification for platinum).

candle. One-sixtieth of the intensity of one square centimeter of a black body radiator at the temperature of solidification for platinum (2046°K).

capacitance. The property of a capacitor which determines the amount of charge that can be stored, measured by the charge which must be communicated to a body to raise its potential by one unit. An electrostatic unit of capacitance is that which requires one electrostatic unit of charge to raise the potential by one electrostatic unit. The farad = 9 $\times$ 10^{11} electrostatic units. A capacitance of one farad requires one coulomb of electricity to raise its potential by one volt.

capillary fringe. Water held above the water table, whether it is the main water table or a perched water table, by capillary force. In the capillary fringe the small pores of the soil are filled with water, held in the pores by capillarity opposing the force of gravity. The thickness of this fringe will vary with soil texture; fine textured soils (silts and clays) will have a much thicker capillary fringe than that of a coarser texture (sand). The water table may be relatively close to the soil surface in areas adjacent to ponds and lakes; in this case, trees and other vegetation may use this water by the growth of root systems into regions of the capillary fringe.

capillary water. Water which is retained around soil particles (by cohesive attraction) and in capillary (minute) pores of the soil structure after gravitational water has left the area. It is an important component of soil water, for it is the capillary water that is important in chemical and biological activities within the soil. Capillary water is readily available to plants and animals.

carbide. A compound of carbon with one or more metallic elements.

carbon (C). A nonmetallic element of atomic weight 12; melting point 3500°C; and boiling point, 4200°C. Carbon atoms can link themselves in long chains and rings to form the highly complicated molecules that are the basis of all living matter; no other atoms except silicon possess this property to a comparable degree. Although carbon composes less than one percent of all matter, carbon is found in all living things. A common form of carbon is graphite; diamonds are a crystal form of carbon. Carbon combines with hydrogen to form compounds called hydrocarbons. These compounds, in turn, combine readily with oxygen to produce heat and light energy, such fuels as gasoline, natural gas, and oil, which are hydrocarbon compounds. Carbon combined with oxygen forms the gas carbon dioxide. There are nearly a million known compounds that contain carbon, called organic compounds.

carbon-14 (^{14}C). A radioactive isotope of carbon. ^{14}C, with a half-life of 5600 years, is an excellent atomic clock for determining the age of less ancient objects. ^{14}C is formed when cosmic rays from outer space enter the atmosphere and collide with nitrogen atoms. In the collision, nitrogen atoms are sometimes broken down and lose one proton. When this happens, nitrogen transmutes and becomes carbon. The change is not to ordinary ^{12}C, with six protons and six neutrons, but to a rare radioactive isotope, ^{14}C, with six protons and eight neutrons. There is only one atom of radioactive ^{14}C for every trillion atoms of ordinary ^{12}C.

carbon black. Finely divided forms of carbon made by the incomplete combustion or thermal decomposition of natural gas or liquid hydrocarbons. Principal types, according to the method of production, are channel black, furnace black, and thermal black.

carbon column. A column filled with granular activated carbon whose primary function is the preferential adsorption of a particular type or types of molecules.

carbon cycle. An essentially perfect cycle in which carbon is returned to the environment about as fast as it is removed. It involves a gaseous phase, atmospheric carbon dioxide. The basic movement of carbon is from the atmospheric reservoir to producers to consumers and from both these groups to the decomposers, and then back to the reservoir. (2) A sequence of atomic nuclear reactions and spontaneous radioactive decay which serves to convert matter into energy in the form of radiation and high-speed particles. It is regarded as one of the principal sources of energy for the sun and other similar stars.

carbon dioxide. CO_2, a colorless gas which forms 0.3 percent of normal air and is an essential raw material for photosynthetic activity by green plants. Carbon dioxide combines chemically with water to produce carbonic acid (H_2CO_3), which influences the hydrogen ion concentration (pH) of water. Carbonic acid dissociates to produce hydrogen (H^+) and bicarbonate (HCO_3^-) ions. The bicarbonate radical may undergo further dissociation forming more hydrogen (H^+) and carbonate (CO_3^{2-}). The amount of free carbon dioxide in water is about 0.5 ml of carbon dioxide per liter of water, but much more carbon dioxide is present in ionized form as bicarbonate and carbonate radicals. Free (uncombined) carbon dioxide will exist as carbon dioxide (CO_2) or carbonic acid, which remains in equilibrium with the combined forms. Seawater with a salinity of 35 parts per thousand contains about 46.8 ml of carbon dioxide per liter.

Not normally considered an air pollutant because the uncontaminated atmosphere has a concentration of approximately 300 ppm, it is essential for animal and plant life, and there must be at least 5000 ppm in the air before man's respiration is adversely affected. However, since the middle of the 19th century, worldwide atmospheric concentrations of carbon dioxide have been rising steadily because of the increasing dependence of industry on fossil fuels. Huge quantities of carbon dioxide, the main product of combustion, are emitted each day into city air. The carbon dioxide concentrations over heavily industrialized areas are at times as high as 1000 ppm. By the end of the century, as much as 50×10^9 tons may be released annually into the atmosphere. About 2300×10^9 tons of carbon dioxide are now present in the atmosphere. Ten times as much carbon dioxide is being emitted into the air from coal fires and metallurgical furnaces alone as from natural sources of breathing. The principal undesirable local effect of atmospheric carbon dioxide is the deterioration of building stones, in particular, carbonate rocks such as limestone. In the presence of moisture, carbon dioxide produces carbonic acid, which converts calcium carbonate to the water-soluble bicarbonate that is then leached away. Carbon dioxide is also responsible in part for the atmospheric corrosion of magnesium, and perhaps of other structural metals. It is unlikely that the climate has been seriously affected as yet by the small increase in atmospheric carbon dioxide that has taken place since industrial coal combustion began. Both biological and geochemical processes provide a natural disposal, as well as a replenishment system for carbon dioxide. The potential of plant life in the sea and on land to adjust the world carbon dioxide supply is not known, but it may be considerable. The only other real possibility for disposing of the extra amounts of carbon dioxide from the combustion of fossil fuels is through the oceans, but this could take as long as 10,000 years for all the extra carbon dioxide to be dissolved. There is a legitimate concern that an unchecked increase in the rate of combustion of carbon fuels may eventually extend carbon dioxide levels to meteorological and physical significance and that carbon dioxide concentrations may become great enough to cause climatic changes.

carbon dioxide buildup. The accumulation of carbon dioxide in the atmosphere. Carbon dioxide in the air is efficient in absorbing near-infrared and infrared radiation, since it has a number of absorption bands between 2.1 and 15µm. The band at 14µm is especially effective, since this is the portion of the spectrum where radiation from the earth's surface is most intense. Increasing CO_2 levels in the atmosphere have the potential of increasing the temperature of the atmosphere and producing other climatic changes. Research data indicate that carbon dioxide levels have increased by more than 5 percent in the past 20 years, with a strong indication that the rate of increase from year to year is increasing. The annual oscillation of about 5 ppm, with peaks in the late winter and lows in the late summer, is believed to be the result of the storing of carbon in vegetation during the warm months due to photosynthesis. Until recently, the CO_2 increase had been thought to be primarily the result of additions from the burning of fossil fuels. It is now believed that, in addition, the destruction of forests, principally tropical forests, is as important as the burning of fuels in increasing atmospheric CO_2. Destroyed forests do not take up CO_2 by photosynthesis, and the increased amount of dead vegetation from deforestation increases CO_2 production through microbial decay. The replacement of forests with agricultural areas does not help, because the leaf area per unit area for agricultural operations is nowhere near that of a mature forest. Although there is great uncertainty about the specific nature of the CO_2 balance between atmospheric sources and sinks, it is certain that CO_2 levels are increasing and that man's activities are playing an important role.

carbon disulfide. CS_2, a solvent used mainly in the textile industry in the production of rayon fibers. In a 1980 report by the U.S. Toxic Substances Strategy Committee to the President, carbon disulfide was cited as a chemical whose toxic effects include symptoms of mental illness, as evidenced by an otherwise unexplained high rate of suicides among workers exposed to the substance. Exposure by inhalation and skin absorption also may result in bizarre sensations in the arms and legs, with loss of sensation and increased muscular weakness. Mental health effects include depression, anorexia, insomnia, irritability, and loss of memory. Loss of vision, dizziness, and symptoms of Parkinson's disease are also reported as effects of exposure to carbon disulfide. Clinical evidence of effects include diminished or lost corneal and pupillary reflexes, blanching of the optic disk, and damage to the blood vessels of the retina.

carbon monoxide. CO, a gas that is one of the three most common products of fuel combustion; carbon dioxide and water vapor are the other two. Its molecular weight is 28 and its boiling point is −190°C. Most of the carbon monoxide in the atmosphere results from the incomplete combustion of carbonaceous materials. Automobiles are one of the principal factors in producing this gas. Carbon monoxide is quite stable in the atmosphere. It is probably converted to carbon dioxide, but the rate of conversion is slow. Because it remains unchanged for several days, carbon monoxide has been used to calculate the dispersed volume of other pollutants. Carbon monoxide is an odorless and colorless gas and in this lies the danger. It is a poisonous inhalant and no other toxic gaseous air pollutant is found at such relatively high concentrations in the urban atmosphere. Carbon is dangerous because it has strong affinity for hemoglobin, which carries oxygen to the body tissues. The effect of carbon monoxide is to deprive the tissue of necessary oxygen. At concentrations of slightly more than 1000 ppm, carbon monoxide kills quickly. One hundred parts per million is generally considered the upper limit of safety in industry for healthy persons within certain age ranges, when exposure may continue for an eight hour period. Most people experience dizziness, headache, lassitude, and other symptoms at approximately 100 ppm. Many scientists believe that carbon monoxide is not a cumulative poison; when exposure is discontinued, the carbon monoxide that had combined with hemoglobin is spontaneously released and the blood is cleared of half its carbon monoxide, at least in healthy people, in three to four hours. Carbon monoxide can cause acute poisoning as a result of exposure to high air concentrations of the gas, but chronic poisoning does not occur as a result of long-continued exposure to relatively low concentrations. Some particularly susceptible persons, such as those already afflicted with a disease that involves a decrease in the oxygen capacity of the blood (such as anemia), or persons with cardiorespiratory diseases, may be affected by the carbon monoxide levels reached in city streets. People who already have variable amounts of carbon monoxide in their blood from smoking, or from exposure in their occupations, may be further affected by inhaling an additional amount of the gas from the carbon-monoxide-contaminated air in their communities. At high levels of concentration, carbon monoxide, more than any other air pollutant, has been identified as a participant in synergetic reactions.

carbon monoxide toxicology. The study of the harmful effects of carbon monoxide, which include asphyxiation, with degrees ranging from severe to fatal, with a threshold limit value (industrial) of 50 ppm. CO toxicity is due primarily to its affinity for hemoglobin, the oxygen carrier of the blood. It is nontoxic to insects and other lower forms of life which have no red blood cells. If 500 ppm of CO are inhaled until equilibrium is reached, 50 percent of the hemoglobin will be combined with CO, cutting the oxygen carrying capacity in half, causing confusion and possible fainting.

carbon–nitrogen cycle. A collision between a proton (nucleus of hydrogen atom) and a carbon atom containing twelve protons, yielding nitrogen-13 and a gamma ray. The nitrogen decays spontaneously into carbon-13, a positive electron, and a neutrino. A series of further reactions ultimately reforms carbon-12, together with an alpha particle, and there are three releases of energy in the whole process.

carbon tetrachloride activity. The maximum percentage increase in weight of a bed of activated carbon after air saturated with carbon tetrachloride is passed through it at a given temperature.

carcinogenic. Cancer producing. See **chemical carcinogens.**

Carnot cycle. A sequence of operations forming the working cycle of an ideal heat engine of maximum thermal efficiency. It consists of an isothermal expansion, an adiabatic expansion, an isothermal compression, and an adiabatic compression to the initial state.

carrying capacity. (1) In recreation, the amount of use a recreation area can sustain without deterioration of its quality. (2) In wildlife, the maximum number of animals an area can support during a given period of the year.

carryover (carry through). A term used in designing solar installations, a measure of the number of days without sunshine during which the heat storage system can provide adequate space heating and domestic hot water.

cascade impactor. A sampling device in which air is drawn through a series of jets against a series of slides. Air particles then adhere to the microscopic slides, which are coated with an adsorbing medium. Jet openings are sized to obtain a size distribution of particles.

catabolism. A chemical reaction by which complex substances are converted within living cells into simpler compounds with the release of energy; the breakdown of food compounds or of protoplasm by living cells into simpler compounds.

catalysis. A process in which the speed of a reaction is altered by the presence of an added substance which remains unchanged at the end of the reaction. Catalysts are usually employed to accelerate reactions and are called "positive" catalysts; in some cases "negative" or retarding catalysts are employed, e.g., glycerol, phenol, and mannite retard the oxidation of sodium sulfite.

catalyst. A substance or combination of substances which accelerates a chemical reaction without being used up itself, or being permanently changed by the reaction.

catalytic agent. A substance whose mere presence alters the velocity of a reaction, and may be recovered unaltered in nature or amount at the end of the reaction.

catalytic converter. An air pollution abatement device that removes organic contaminants by oxidizing them into carbon dioxide and water.

catalytic incineration. A disposal process applied to gaseous wastes containing low concentrations of combustible materials and air. Usually noble metals such as platinum and palladium are the catalytic agents. A catalyst is defined as a material which promotes a chemical reaction without taking a part in it. The catalyst does not change nor is it used up. These catalysts must be supported in the hot waste-gas stream in such a manner that they present the greatest surface area to the waste gas so that the conduction reaction can occur on the surface, producing nontoxic effluent gases of carbon dioxide, nitrogen, and water vapor. The advantage of the catalyst is that the reaction temperature in catalytic systems is lower than it is in thermal systems, because the catalyst promotes the reaction at a lower temperature. Catalytic systems have been used widely in the oxidation of paint solvents, odors arising from chemical manufacture, food preparation, wire enameling ovens, lithographing ovens, and similar applications.

cathode. The electrode at which reduction occurs. It is the negative electrode in a cell through which current is being forced, but it is the positive pole of a battery. In a vacuum tube, the cathode is the electrode from which electrons are liberated.

cation. A positively charged ion.

caustic soda. Sodium hydroxide (NaOH), a strong alkaline substance used as the cleaning agent in some detergents.

cavitation. The formation and collapse of vapor-pressure bubbles owing to the movement of a body in a fluid or in the effect of this action.

ceiling. The height of a cloud above the ground is called the ceiling, provided the cloud covers at least half the sky. Ceiling height may be measured from the ground in several ways. One way is to release a small balloon inflated to a known lift rate of rise, and time its ascent to the cloud base—used during daylight. Another way is to shine a searchlight beam upward and observe its intersection with the cloud base with a clinometer—used at night. In the cloud base and top indicator operated by radar, a powerful radar beam of short wavelength is projected vertically. As the beam strikes a layer of clouds a portion of the energy is reflected to the surface, where it is detected and recorded. By elaborate circuitry and the proper choice of radar pulse length and pulse repetition rate, the device is capable of giving a complete cross section of all clouds above it.

ceilometer. An automatic, recording, cloud-height indicator.

cell. In solid waste disposal, a hole where waste is dumped, compacted, and covered with layers of dirt daily.

cell, primary. An electrochemical cell in which the reacting materials must be replaced after a given amount of energy has been supplied to the external circuit.

Celsius scale (centigrade scale). A temperature scale with 100 degrees between the freezing and boiling points of water. The freezing point is zero degrees; the boiling point is 100 degrees.

centigrade scale. The temperature scale on which water boils at 100 degrees at sea level and freezes at zero degrees; hence a centigrade degree is 1/100th of the temperature difference between the boiling and freezing points of water.

centimeter. A unit of distance in the metric system, equal to 1/100 of a meter.

centimeter–gram–second (cgs) system. A system of units based on the centimeter as the unit of length, the gram as the unit of mass, and the second as the unit of time.

centipoise. A unit of viscosity in the metric system. The viscosity of water at 20°C is approximately one centipoise, and one centipoise is 1/100 of a poise, which is the absolute unit of viscosity in the metric system. Water at 20°C has a viscosity of 1.002 cP.

central receiving. A solar–electric system concept: Solar energy is collected by a single large heliostat field and focused on a tower-mounted receiver–boiler.

centrifugal collector. A mechanical system using centrifugal force to remove aerosols from a gas stream or to de-water sludge.

centrifugal pump. A pump for increasing the pressure of a liquid through the agency of centrifugal force.

centripetal force. The force restraining bodies to move in a curved path. According to Newton's first law of motion, a body in motion will continue in the same direction in a straight line and with the same speed unless acted upon by some external force. This means that in order for a body to move in a curved path some force must be continually applied. This force is called the centripetal force, and it is always directed toward the center of rotation.

ceramic. An inorganic compound or mixture requiring heat treatment to fuse it into a homogeneous mass, usually possessing high temperature strength but low ductility.

ceramic filter. A component of a stack sampling system, also known as a ceramic thimble and suitable for high-temperature (1000°F) use.

Cerenkov radiation. The radiation from a charged particle whose velocity is greater than the phase velocity an electromagnetic wave would have if it were propagating in the medium. The particle will continue to lose energy by radiation until its velocity is less than the phase velocity.

cesium (See caesium).

cesspool. An underground pit into which raw household sewage or other untreated liquid waste is discharged and from which the liquid seeps into the surrounding soil or is otherwise removed. Sometimes called a leaching cesspool.

cgs. Centimeter-gram-second, as in cgs system.

cfm. Cubic feet per minute, a measure of air flow in air-type solar heating systems.

cfs. Cubic feet per second, a measure of the amount of water passing a given point.

chain reaction. In general, any self-sustaining process, whether molecular or nuclear, the products of which are instrumental in and directly contribute to the propagation of the process. Specifically, a fission chain reaction, where the energy liberated or particles produced (fission products) by the fission of an atom causes the fission of other nuclei, which in turn propagate the fission reaction in the same manner.

chamber detritus. A detention chamber larger than a grit chamber, usually with provisions for removing sediment without interrupting the flow of liquid. A settling tank or short detention period primarily designed to remove heavy settleable solids.

chamber, flowing-through. The upper compartment of a two-story sedimentation tank.

chamber, grit. A small detention chamber or an enlargement of a sewer designed to reduce the velocity of flow of the liquid to permit the separation of minerals from organic solids by differential sedimentation.

channelization. The straightening and deepening of streams so water will move faster; a flood reduction or marsh drainage tactic that can interfere with waste assimilation capacity and disturb fish habitat.

channelization impact. An effect on water quality and aquatic life resulting from alteration in stream flow by construction of power or other industrial plants on or near sources of freshwater. Sediment often becomes suspended in the water during construction stages, increasing the turbidity of water downstream. After construction, natural pools and riffles may be lost to development of a uniform stream bed and steady water flow through the area, eliminating the preferred feeding and breeding places for fish and other aquatic life. Water temperatures are often altered either by the effluent of the industrial plant or through the loss of vegetation that provides protective shading from the sun. Other effects may include a change in the concentration of dissolved oxygen in the water and, in some cases, an increase in the concentration of potentially harmful chemicals in the public water supplies downstream. Channelization in rural areas often increases the rate of runoff of topsoil and agricultural chemicals from neighboring farms into the downstream water supplies.

chaparral. A biome consisting of dense thickets of stiff- or tough-leafed shrubs, found in areas with a Mediterranean climate.

charcoal. A product of the destructive distillation of wood used for absorbent, filtering, and decolorizing media. Charcoal can absorb various gases and vapors, showing preferential absorption for vapors of liquids that are volatile at ordinary temperatures. One volume of carbon can absorb 170 volumes of NH_3 or 148 volumes of alcohol vapors, as compared with about 17 volumes of O_2 and N_2.

Charles' law or Gay–Lussac's law. The volumes assumed by a given mass of a gas at different temperatures, the pressure remaining constant, are, within moderate ranges of temperature, directly proportional to the corresponding absolute temperature.

check valve. A one-way valve usually used in liquid-type solar heating systems to prevent heated water from going through a pipe in an undesirable direction.

check work. A pattern of multiple openings in a refractory wall to promote turbulent mixing of combustion products.

chemical carcinogens. (See chart on page 61.)

chemical oxidation and reduction. A process involving the addition of an oxidizing or reducing agent under proper chemical (pH) conditions, e.g., destruction of cyanide by the alkaline chlorination method and the reduction of hexavalent chromium to trivalent chromium. It may also be used for the destruction of certain producing bodies, taste- and odor-producing bodies, and phenols.

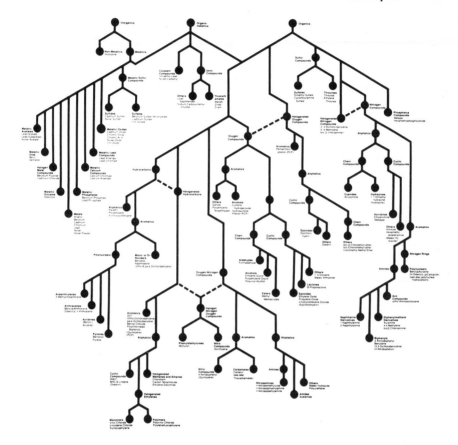

Potential chemical carcinogens.

chemical oxygen demand (COD). (1) A test based on the fact that all organic compounds, with few exceptions, can be oxidized to carbon dioxide and water by the action of strong oxidizing agents under acid conditions. Organic matter is converted to carbon dioxide and water, regardless of the biological assimilability of the substances. One of the chief limitations is the test's inability to differentiate between biologically oxidizable and biologically inert organic matter. The major advantage of this test is the short time required for evaluation (two hours). (2) The amount of oxygen required for the chemical oxidation of organic compounds in a liquid.

chemiluminescence. The emission of light during a chemical reaction.

chemisorption. The binding of a liquid or gas on the surface or in the interior of a solid by chemical bonds or forces; chemical adsorption which depends on chemical bond formation between the adsorbent and adsorbate, but which is distinct from a chemical reaction, in that it takes place only in a monolayer on the surface of the adsorbent.

chemosterilant. A chemical that controls pests by preventing reproduction.

chemosynthesis. The synthesis (by organisms) of organic carbon from CO_2, using energy released from some other chemical reaction. Some bacteria are able to synthesize high-energy organic compounds from inorganic raw materials without utilizing light energy at all. Such bacteria oxidize inorganic compounds and trap the energy thus released. There are many examples of bacteria that oxidize various nitrogen and sulfur compounds, and even molecular hydrogen. These bacteria share an ability to trap the small quantities of energy released by their oxidation of inorganic raw materials and use this energy in the synthesis of carbohydrates. Such organisms are called chemosynthetic.

chemotrophic nutrition. The process by which an organism manufactures its food by using the energy derived from oxidizing organic matter.

chernozem soil. A black topsoil that is found in temperate to cool, humid environments. The soil is characterized by a distinct calcium carbonate (hard pan) stratum in the C horizon. The A horizon is nearly black, high in organic content, and slightly acid, overlying a lighter-colored B horizon. Tall and mixed grasses flourish on this soil group. A narrow zone of chernozem soil extends west of the prairie soils from central Canada to the Gulf Coast in the United States.

chestnut and brown soils. Two soils that occupy extensive areas west of the chernozem types in the United States and South America. These two soil groups are situated in temperate or cool areas with arid or semiarid conditions. These soils, characterized by brown or dark-brown surface soils overlying lighter-colored B horizons, are like the chernozem soils in that there is a zone of hard pan (calcium carbonate accumulation) in the C horizon.

chilling effect. The lowering of the earth's temperature because of increased particles in the air blocking the sun's rays.

chimney effect. A phenomenon consisting of a vertical movement of a localized mass of air or other gases due to temperature differences.

chimney/stack. A stack or vertical steel flue for conducting cooled combustion products of a process to the atmosphere, sometimes movement-activated by an induced draft fan.

chimney superelevation. As used in dispersion calculations, the effective chimney height denoting the maximum height of the center of a plume path above the ground.

chloracne. A form of dermatitis caused by environmental exposure to halogenated aromatic compounds. The disorder is also called environmental halogen acne. In addition to the refractory pustular symptoms of the disease, it is often accompanied by systemic toxicity. Chloracne was first observed in 1897 and until the end of World War II most cases were associated with exposure to the chloronaphthalenes and polychlorinated biphenyls (PCB's). The polychloronaphthalene compounds have been used mainly in the production of nonmagnetic boat hull coatings and in the electronics industry in the manufacture of insulating materials, sealing compounds, and dielectric condensers, products which in recent years have generally been replaced by other materials. The source of most cases of chloracne in recent years has been chemicals involved in the production of pesticides. Presently, six categories of chemicals are identified as producers of chloracne. They are the polyhalogenated naphthalenes, biphenyls, and dibenzofurans, contaminants of polychlorophenol compounds and of 3,4-dichloroaniline types of compounds, and a group of miscellaneous substances including DDT and Dichlobenil. Chloracne usually affects workers involved in the manufacture of halogenated aromatic chemical products, but occasional public health disasters have occurred, as in the western

Japanese community of Kyushu in 1968 when 1200 persons were poisoned by rice oil that had been contaminated by PCB's that had leaked from a heat exchanger. The initial symptoms were those of chloracne, dark pigmentation of the skin accompanied by rash, edema, and acne-like pustules that form about the hair follicles. The symptoms may persist for as long as 15 years after the exposure to the chemicals has ceased. Prevention and control of chloracne requires totally enclosed chemical manufacturing processes to eliminate skin contact or inhalation of the toxic substances.

chlordane. $C_{10}H_6Cl_8$, chlorine-based pesticide manufactured under several different trade names. It is regarded as a highly toxic substance, easily absorbed through the skin, and capable of causing the death of a human in less than an hour after exposure. The LD_{50} for oral doses administered to rats has been reported as 283 mg/kg and the dermal LD_{50} for rats is about 700 mg/kg. Animals exposed to small experimental doses show hyperexcitability, tremors, and convulsions. For those that do not die of acute effects, there is loss of appetite and weight. One human fatality resulting from chlordane involved a man who received an accidental application on the skin of a 25 percent solution of chlordane. He died of respiratory failure before medical help could arrive. Two other victims who died after ingesting doses of two to four grams of chlordane showed severe fatty liver degeneration on post-mortem examination. Ingestion by farm animals of alfalfa or other plants treated with chlordane results in a metabolic conversion of the pesticide to an epoxide, oxychlordane. The epoxide residue has been found in the milk and cheese from dairy cows exposed to chlordane and recent studies of the chronic effects of feeding chlordane to animals has led to evidence that it may be a carcinogen. Chlordane is one of the oldest of the chlorinated hydrocarbon insecticides to be used for general agricultural purposes and it is one of the slowest to be degraded through bioprocesses. It has been estimated by the Soil Science Society of America that chlordane can be expected to remain in the soil for more than 12 years, as compared to the estimated persistence of DDT of more than 10 years.

chlorella. A form of unicellular green algae which have been considered for helping maintain a closed environment (within a space cabin for instance), by producing oxygen and consuming carbon dioxide; also a source for food.

chloride. A compound in which chlorine is combined with another element or radical. Chloride and sulfate wastes are closely associated with totally dissolved solids and the control of the latter is accomplished in much the same way as for chloride waste. It is recommended that chloride concentrations not exceed 1 mg/liter, and that it ideally be less than 25 mg/liter.

chlorinated dibenzofuran. A compound that is associated with polychlorinated biphenyls (PCB's) as a contaminant. The health effects of chlorinated dibenzofuran (CDBF) are similar to those of PCB's, being implicated in acute skin, eye, reproductive, developmental, neurological, and respiratory disorders.

chlorinated hydrocarbons. A class of persistent, broad-spectrum insecticides, notably DDT, that linger in the environment and accumulate in the food chain. Other examples are aldrin, dieldrin, heptachlor, chlordane, lindane, endrin, mirex, benzene, hexachloride, and toxaphene.

chlorination. (1) The process of introducing one or more chlorine atoms into a compound. A common chlorinating agent is gaseous chlorine, using light heat or a catalyst as a reaction promoter. (2) The application of chlorine to water, sewage, or industrial wastes, generally for the purpose of disinfection, but frequently for accomplishing other biological or chemical results. Prechlorination is chlorination prior to treatment, and postchlorination follows treatment.

chlorination breaking point. The application of chlorine to water, sewage, or industrial waste containing free ammonia, to the point where free residual chlorine is available.

chlorination, free residual. The application of chlorine to water, sewage, or industrial wastes to produce directly, or through the destruction of ammonia or certain organic nitrogenous compounds, a free available chlorine residual.

chlorinator. A device that adds chlorine to water in gas or liquid form.

chlorine, available. A term used in rating chlorinated lime and hypochlorites as to their total oxidizing power.

chlorine, combined available residual. The portion of the total residual chlorine remaining in water, sewage, or industrial wastes at the end of a specified contact period, which will react chemically and biologically as chloramines, or organic chloramines.

chlorine-contact chamber. That part of a waste treatment plant where effluent is disinfected by chlorine before being discharged.

chlorine demand. The quantity of chlorine absorbed by wastewater (or water) in a given length of time.

chlorite, high-test hypo. A combination of lime and chlorine, consisting largely of calcium hypochlorite.

chlorite, sodium hypo. A water solution of sodium hydroxide and chlorine, in which sodium hypochlorite is the essential ingredient.

chloroform. $CHCl_3$, a colorless volatile liquid manufactured by the chlorination of methane in a process that may also yield varying proportions of methyl chloride, methylene chloride, and carbon tetrachloride. The annual production of chloroform in the United States averages above 300 million pounds, of which about 80 percent is consumed domestically. Most of the chloroform is produced as a raw material in the preparation of fluorocarbons. It is used in the extraction and purification of antibiotics; in the purification of alkaloids; in the solvent extraction of vitamins and flavors; as a general solvent; as an intermediate in the preparation of dyes, drugs, and pesticides; and as an anesthetic. Chloroform has been used in cough and cold preparations; in dental preparations such as toothache drops, toothpastes, and mouthwashes; and in topical liniments. Its narcotic effects on the central nervous system are well documented; its use as an anesthetic dates back to 1847. Human exposure to chloroform has resulted in toxic hepatitis, cardiac irregularities, local irritation of the skin, lassitude, dry mouth, depression, irritability, and painful urination, according to various clinical studies. However, there have been no published reports of a documented association between chloroform and cancer in humans. In 1976, the National Cancer Institute released its report on the carcinogenic bioassay of chloroform and stated that the chemical, when administered orally, produced kidney tumors in rats and hepatocellular carcinoma in mice. NIOSH estimates that 40,000 workers are occupationally exposed to chloroform. OSHA has recommended a value standard for concentrations of chloroform in workplace air of 50 ppm. NIOSH has advised that no worker be exposed to chloroform in excess of 10 ppm, determined as a time-weighted average for up to 10 hours a day, 40 hours per week.

chlorophyll. A pigment which gives plants their green color and is of paramount importance in transforming radiant energy to chemical energy in the process of photosynthesis.

chloroprene. $CH_2CHCClCH_2$, a colorless liquid used in the production of neoprene rubber products. Chloroprene has been used by American industry since 1931, but it was not until 1974 that NIOSH expressed concern about its potential carcinogenicity as a result of a review of the world's medical literature. During the literature review, it was found that Russian doctors had reported cases of skin and lung cancer from "escaping chloroprene" during the manufacture of neoprene. In addition to the clinical reports of cancer in neoprene workers, the Russians had reported on the results of animal experiments in which chloroprene also caused adverse effects on the development of embryos in rats and mice. The oral LD_{50} for rats ingesting chloroprene has been established as 1600 mg/kg. The lowest published lethal dose for rats exposed via the subcutaneous route is 500 mg/kg. Animal experiments indicate that a concentration of 250 ppm of chloroprene may be toxic and a continued exposure of 75 ppm may also be toxic. Exposure to chloroprene in the atmosphere results in irritation of the respiratory tract, followed by breathing difficulties, and finally asphyxia. The vapor is a central nervous system depressant and it also causes degenerative tissue changes in the liver and kidneys. Corneal necrosis, anemia, and temporary hair loss are other physical effects. In animal tests, chloroprene was found to cause an increase in embryo mortality and a reduction in the fetal weight of offspring of females exposed during pregnancy. The automotive industry is the largest user of neoprene, and because of its resistance to oil and weather, this product of chloroprene is commonly used in cable sheaths, rubber hoses, fabrics, and adhesives.

chlorosis. A discoloration of normally green plant parts that can be caused by disease, lack of nutrients, or various air pollutants.

chopper. A mechanical device that filters a stream of neutrons coming from an atomic furnace so that only neutrons of a given speed reach a detector. It works like a shutter, opening and closing the neutron path in time with the operation of the detector. The chopper is used to study the effect that neutrons have on various materials.

Christiansen effect. When finely powdered substances, such as glass or quartz, are immersed in a liquid of the same index of refraction, complete transparency can only be obtained for monochromatic light. If white light is employed, the transmitted color corresponds to the particular wavelength for which the two substances, solid and liquid, have exactly the same index of refraction. Owing to differences in dispersion, the indices of refraction will match for only a narrow band of the spectrum.

chromatic aberration. A property of lenses that causes light of various wavelengths from the same source to not be focused at the same point. This is due to the difference in the index of refraction for different wavelengths.

chromatographic analysis. A method of separation based on selective adsorption. A solution of the substance or substances desired is allowed to flow slowly through a column of adsorbent. Different substances will pass with different speeds down the column and will eventually be separated into zones. The column core can then be pushed out and the zones of material cut apart.

chromatography. The process of separating constituents of a mixture by permitting a solution of the mixture to flow through a column of adsorbent on which the different substances are selectively separated into distinct bands or spots. (1) Partition chromatography involves selective solution of the desired material between two solvents. The final solvent, usually water, is used to wet the solid material packed in the column, and the first solvent, containing the desired material is poured into the column as above. (2) Paper chromatography is a micro method. A drop of the liquid to be investigated is placed near one end of a strip of paper. This end is immersed in solvent, which travels down the paper and selectively distributes the materials present in the original drop. Comparison with known substances allows identification.

chromatography, gas. An analytical technique for separating mixtures of volatile substances. The procedure consists of introducing the sample mixture into a separating column and washing it down with an inert gas. The column is packed with absorbent materials (20–200 mesh) which selectively retard the components of the sample. The surface is usually coated with a relatively nonvolatile liquid designated as the stationary phase. This gives rise to the term gas–liquid chromatography. If a liquid is not present, the process is gas–solid chromatography. Different components move through the bed of packing at different rates and appear one after another at the effluent end, where they are detected and measured by thermal conductivity changes, density differences, or various types of ionization detectors. Gas chromatography is advantageous for minute-quantity analysis of complete mixtures.

chrome pigment. A NIOSH report in 1976 stated that a study of chrome or chromate pigments had found an excess of respiratory cancer in persons exposed to lead chromate and recommended that lead chromate be regarded as a carcinogen. The study was sponsored by the Dry Color Manufacturers Association and covered lead–molybdenum chromate and zinc chromate pigments in addition to lead chromate pigment exposure. Ninety-day animal feeding tests of rats and dogs were conducted, in addition to a survey of workers at plants in the United States. Of 548 workers in three plants, 53 deaths had occurred prior to the survey, and lung cancer was found to account for 29 percent of the deaths, as compared with an average of about 8 percent of all deaths among industrial workers. In addition, five of the deaths were due to stomach cancer, a figure which was seven times the expected number for a worker population of that size. Prior to the U.S. study, there had been limited reports since 1940 of lung cancer deaths among chrome pigment workers in Germany and Norway. Previous animal studies had also shown a potential for chrome pigments to produce tumors at the site where the pigment was implanted or injected. An OSHA survey found that nearly one-half the atmospheric samples of chrome pigments in the plants checked reached or exceeded the standards at that time. The levels of lead and chromium had been assumed to be lower than levels in the past because of process changes and improved engineering controls and work practices, according to a NIOSH report.

chromite. A ferrochromite of iron and chromium; a hard, brownish black, submetallic mineral with a pitchy luster, occurring as disseminated grains or as granular masses. The iron–chromium ratio varies. Chromate is used as refractory brick and as a source of metallic chromium and chromium chemicals.

chromium (Cr). A multivalent, bluish-white metallic element used mainly in the manufacture of alloys and in the electroplating of other metals. It is estimated to account for about one-fourth of one percent of the total weight of the earth and slightly more than one-half of one percent of the mantle of the earth. Chromium occurs naturally in amounts ranging from less than 1 ppb to more than 10 ppb in lakes and rivers of North America. The U.S. Public Health Service has determined that chromium in the hexavalent form is a toxic substance in drinking water and water containing chromium in concentrations greater than 50 µg/liter should not be used. The toxicity of chromium varies with pH, temperature, the organism exposed, and the valence form. In any given form, toxicity may vary with the presence of other compounds. As a primary toxicant, chromium has been associated with fish and plankton kills, cattle poisoning, and accumulations in fish and shellfish. It has been reported to cause pulmonary disorders in humans exposed to the dust and is considered a potential carcinogen in water discharged from industries involved in metal plating, anodizing, and chrome tanning processes. One study has found that chromium administered to rats offered protection against the toxic effects of lead.

Chromium compounds may be present in industrial wastes from a vast variety of processes, including tanning, electroplating, and cooling tower effluents. Chromium waste lends itself to chromate recovery, as commonly practiced in the metal plating industry. For final disposal, chromium is precipitated as the metal hydroxide, after reduction to the trivalent state.

chromosphere. The thin gaseous layer of the sun's atmosphere, having a pressure of about 10^{-7} atm just outside the photosphere. The gases in the upper atmosphere are predominantly hydrogen and calcium, and they may extend some 30,000 mi. above the photosphere. Its mean temperature is about 600,000°K.

chronic. Long lasting or frequently recurring, as a disease.

cinnabar. Mercury sulfide, HgS, the chief ore of mercury; a soft, heavy (specific gravity, 8), scarlet or brownish-red to brownish-black mineral with a scarlet to reddish-brown streak.

clarification. The process of removing turbidity and suspended solids by settling. An essential part of any process where a contaminant is removed by precipitation, such as phosphate removal. It may also be used for removal of colloidal organic material, including some color bodies, by adsorption on formed chemical flocs. Clarification involves the addition of chemical coagulant acids, and pH adjustment to form a stable, rapidly settling floc which is then separated from the water by sedimentation. Chemical clarification improves settling rates and provides a more consistent effluent quality than is obtainable with straight sedimentation.

clarifier. A sedimentation tank.

clay. A firm, fine-grained earth, plastic when wet. Clays are classified in different ways. The behavior of a clay and its value and suitability for any specific use depends on its chemical and physical properties. In ceramic uses, the important physical properties are plasticity, cohesion, tensile strength, texture, and drying shrinkage. Chemically, clay minerals are complex hydrous silicates of aluminum. Some of them contain small amounts of potassium, sodium, magnesium, and iron. The chief use of clay is in making bricks. A pure, fine-grained white clay is required for the manufacture of the better grades of pottery, chinaware, and porcelain. A type of clay that withstands high temperature is called fire clay. It owes its refractory properties to the fact that it contains only very small amounts of elements which act as fluxes. Its principal uses are in making firebrick and other heat-resisting materials, such as furnace linings required in the iron and steel industry, and in making coke. Fuller's earths are clays which possess the ability to decolorize or bleach oils, fats, and greases, and are used extensively in the petroleum industry for decolorizing and stabilizing lubricants and in the fat and vegetable oil industry. Bentonite, an absorbent clay which swells when soaked with water, is used in the preparation of oil well drilling muds, in oil refining, and as a bonding material in molding sands.

clay industry furnace. There are various types of furnaces in the clay industry, which make (a) refractory bricks of fireclay, silica, and other materials for industrial use; (b) building bricks, tiles, and pipes; and (c) china, earthenware, glazed tiles, porcelain, and other similar articles. The method is much the same in every case. Wet clay is molded to the required shape and heated gradually, at first, to about 120°C to evaporate the free water, then more strongly to 900°C to drive water out of the molecules of clay, and finally to the baking temperature, between 1200° and 1450°C. The down-draught kiln, whether circular or rectangular, is the most efficient and satisfactory of intermittent kilns for firing all kinds of clay products.

clear cut. A forest management technique that involves harvesting all the trees in one area at one time. Under certain soil and slope conditions it can contribute sediment to water pollution.

clear well. A reservoir containing water which has been previously filtered or purified before going into the standpipes or distribution system.

climax community. The last aggregation in the successional series. Provided that the climatic conditions do not change and no catastrophic event alters the area, the community will sustain itself indefinitely. Climax communities are found in virgin forests, or in areas that have not been denuded by man or natural disasters. The climax community can always be predicted for any region of comparable climatic conditions. The species will be similar in these climatic communities.

climograph. A graph over a period of time, generally a twelve-month period, of climatic conditions. It may be used to compare the environmental conditions in several areas, because it enables one to make a very rapid comparison between two localities. Temperature, precipitation, vapor pressure deficit, or some other climatic factor may be plotted along one axis, and time or another factor along the other axis. The hytherograph, a specialized type of climograph, always records temperature and precipitation.

cline. A smooth transition of forms within a species, changing systematically with geographical position.

clipping circuit. A pulse-shaping network which removes that part of a waveform which tends to extend above, or below, a chosen voltage level.

clo. A unit of thermal insulation, usually applied to clothing or bed covers. It is defined as the amount of insulation necessary to maintain comfort and a mean skin temperature of 92°F for a person producing heat at the standard metabolic rate (50 kilogram-calories per square meter of body surface per hour) in an indoor environment characterized by a temperature of 70°F, a relative humidity of less than 50 percent, and an air motion of 20 feet per minute.

clone. An organism that has descended from a common ancestor by asexual propagation only.

closed ecological system. A system that provides for the maintenance of life in an isolated living chamber through complete reutilization of the material available, in particular, by means of a cycle wherein exhaled carbon dioxide, urine, and other waste matter are converted chemically or by photosynthesis into oxygen, water, and food.

closed shell. In the outer layers of an atom, a set of electrons in which all the spaces are filled.

cloth filters. A versatile, highly efficient method for collecting solid particulate matter in a wide range of sizes. Efficiency increases in direct proportion to the amount of cloth in the filter. For efficient design and cost, and to still maximize filter surface per unit volume, designers frequently determine size by selecting the highest filter rate consistent with good operation. The actual fabric of the filter is chosen on the basis of its ability to withstand temperatures and stresses inherent in the process, as well as its compatibility with the pollutants collected. The basic equipment in a filterhouse (baghouse) includes a number of tubular cloth filter bags which are supported top and bottom with gas entering from one end. The dust-bearing gas is moved by either suction (negative pressure) or propulsion (positive pressure) through the filter, so that dust particles are trapped on the approach side of the fabric, while clean gas passes through. One type of reverse-flow cleaning-type cloth filter is illustrated.

cloud. A condensation of invisible vapor suspended above the earth's surface at various heights.

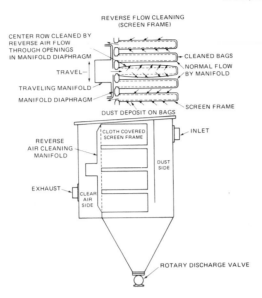

REVERSE FLOW CLEANING
(SCREEN FRAME)

CENTER ROW CLEANED BY
REVERSE AIR FLOW
THROUGH OPENINGS
IN MANIFOLD DIAPHRAGM

CLEANED BAGS

NORMAL FLOW
BY MANIFOLD

TRAVEL—

TRAVELING MANIFOLD

MANIFOLD DIAPHRAGM

SCREEN FRAME

DUST DEPOSIT ON BAGS

CLOTH COVERED
SCREEN FRAME

INLET

REVERSE
AIR CLEANING
MANIFOLD

DUST
SIDE

EXHAUST

CLEAR
AIR
SIDE

ROTARY DISCHARGE VALVE

Reverse flow cleaning (screen frame).

cloud, altocumulus. (1) A white or gray, or both white and gray, patch, sheet, or layer of cloud, generally with shading, composed of laminae, rounded masses, rolls, etc., which is sometimes partly fibrous or diffuse and which may or may not be merged; most of the regularly arranged small elements usually have an apparent width of one to five degrees. (2) A cloud formed chiefly of small water droplets, usually supercooled. If the proper low temperatures exist within such a cloud group, ice crystals may also be present. Seven of the nine basic types that make up the middle-altitude family of clouds are altocumulus-like in character. They are identified as thin altocumulus, altocumulus in patches, altocumulus in parallel bands, altocumulus formed from cumulus, altocumulus in two or more layers, castellated altocumulus, and altocumulus of chaotic sky. For the most part, these clouds assume a checkerboard pattern of white or gray layers.

cloud, altostratus. A cloud formed of a mixture of supercooled water droplets and ice crystals. This formation makes a gray or bluish cloud cover, thin and quite smooth in appearance, that often blankets the entire visible sky. It has an opaque quality about it that is particularly evident when the sun or moon attempts to shine through, but only succeeds in producing a weak light. Altostratus clouds often produce precipitation, light in quantity but of prolonged persistency.

cloud, cirrocumulus. A thin, white patch, sheet, or layer of clouds without shading, composed of very small elements in the form of grains, ripples, etc., merged or separate and more or less regularly arranged; most of the elements have an apparent width of less than one degree. Cirrocumulus clouds are formed either by supercooled water droplets, small ice crystals, or a mixture of both.

cloud, cirrostratus. A cloud formed of ice crystals, looking like a thin, white, uniform veil, smooth and transparent, sometimes covering the entire sky, and sometimes forming huge patches. Cirrostratus clouds are wispy, and often so transparent as to be almost invisible; they occasionally are confused with haze. The sun shines through them with a kind of diffused glow. From time to time they cause halos around the sun and moon because of the myriad of ice crystals they contain.

cloud, cirrus. (1) A detached cloud in the form of a white, delicate filament or white, or mostly white, patch or narrow band. These clouds have a fibrous (hairlike) appearance. (2) A cloud formed by ice particles of different sizes. As a rule, cirrus clouds are shaped in tenuous filaments, grouped in patches or narrow bands. The formation often has considerable vertical depth, since many of the ice particles have just enough weight to fall slowly earthward. Long, feathery cirrus clouds are sometimes referred to as "mare's tails." The variations of cirrus cloud formations include cirrus below 45 degrees, cirrus above 45 degrees, cirrus filaments, dense cirrus, cirrus with hooks or tufts, and dense cirrus in patches.

cloud, cumulonimbus. A cloud formed in the upper portion of the atmosphere, and containing mostly ice crystals and supercooled water droplets. Such a cloud group is the lower section of an immense cloud formation that soars upward from the lower, through the middle, to the high cloud region. The uppermost portions of cumulonimbus clouds are frequently classed with the cirrus variety because of their great height. These high portions almost always assume a distinctive anvil shape, which is caused by strong horizontal winds that flatten the towering, rounded heads.

cloud, cumulus. (1) A detached cloud, generally dense and with sharp outlines developing vertically in the form of rising mounds, domes, or towers, of which the bulging upper part often resembles a cauliflower. The sunlit parts of these clouds are mostly brilliant white; their base is relatively dark and nearly horizontal. (2) A cloud formed of closely packed, small water droplets that, for the most part, are supercooled. Larger drops often develop and fall from the base of such a cloud formation as rain or traces of evaporating rain (called virga). Ice crystals may form in the upper parts of a large cumulus cloud, and sometimes these grow larger by taking water from the water droplets. Cumulus clouds are a type formed by the vertical movement of air currents.

cloud, nimbostratus. (1) A gray cloud layer, often dark, the appearance of which is rendered diffuse by more or less continuously falling rain or snow, which in most cases reaches the ground. It is thick enough throughout to blot out the sun. (2) A cloud formed by suspended water droplets, sometimes supercooled, and by falling raindrops or snowflakes. This type of cloud is dark gray in color and has an immense range across the sky. It can also be identified as a thick altostratus.

cloud, stratocumulus. A cloud formed usually of small water droplets, sometimes large ones, soft hail, and very rarely by snowflakes. These clouds are gray or whitish in appearance and are separated into rolls or rounded shapes.

cloud, stratus. A cloud formed by minute water droplets and sometimes, if the temperature is low enough, of ice crystals. Stratus clouds are gray with a rather uniform, structureless base. Precipitation is not generally associated with this class of cloud, which quite often comes into existence as a result of the dissipation or lifting of the lower layers of a fog bank. Stratus is usually a low-hanging cloud formation.

cloud chamber. An apparatus containing moist air or another gas which on sudden expansion condenses moisture to droplets on dust particles or other nuclei. Thus charged particles or ions in the space become nuclei and their numbers and behavior, when properly illuminated, may be studied.

cloud chamber, Wilson. (1) A device which illuminates normally invisible paths of high-energy subatomic particles. A supersaturated vapor condition is created in a chamber filled with dust-free air by a sudden adiabatic expansion and cooling. In this environment, the small molecular ions formed along the path of a high-energy particle act as effective condensation nuclei. The line of droplets so formed can be used to mark the path. (2) A device for observing the paths of ionizing particles, based on the principle that supersaturated vapor condenses more readily on ions than on neutral molecules.

cloud-height indicators. The general term for instruments which measure the height of cloud bases. They may be classified according to their principle of operation. One class is based upon height determination by means of triangulation. Examples are the ceilometer and the ceiling light. Another class is based upon pulse techniques; the time required for a pulse of energy to travel from a source located on the ground to the cloud base and back to the ground is measured electrically. Examples are the pulsed-light cloud-height indicator and the vertically directed cloud-detection radar.

cloud seeding. The sprinkling of particles of dry ice, silver iodide, etc., into clouds in an attempt to induce rainfall. Seeding certain types of clouds with silver iodide crystals provides water-attracting nuclei for condensation of the tiny, supercooled droplets of water that make up the cloud. As the droplets condense on the silver iodide nuclei, drops of water are formed that may ultimately grow to precipitable size. The most successful seeding attempts have used cumulus clouds or orographic storms, that is, unstable clouds formed by air being forced up over mountains. Efforts to seed large cyclonic storms, that is, those associated with low pressure areas, have not been very successful.

CO_2 acceptor. *Process:* A basic feature of this process is to provide heat for reaction of carbon and steam by reacting the CO_2 formed with calcined dolomite. Removal of CO_2 also enhances the CO-shift and methanation reactions, both being exothermic. The combined effect allows operation of the gasification system without oxygen or an external hydrogen supply. Heat for regenerating spent dolomite (acceptor) is supplied by burning char with air. Crushed and dried coal is fed to the fluid-bed devolatizer, which operates at about 1500°F and 150–300 psi. Devolatized char is conveyed by superheated steam to the gasifier, where additional char is gasified with steam at about 1600°F. Dolomite is used in both reactors to remove the CO_2 and H_2S. Spent dolomite from both reactors is carried by compressed air to the regenerator. Char from the gasifier is transferred to the regenerator by steam. This char is burned with air at about 1900°F to provide heat for the calcination of the spent dolomite. Cyclones separate char, ash, and dolomite fines from the regenerator effluent gas, which is used for power generation and steam production. Raw gas from the devolatilizer passes through a heat recovery section, water quench, and acid-gas removal. Purified gas has an H_2/CO ratio of about 3.2, so that no shift is required before methanation. Dehydration of the methanated gas produces pipeline gas (Consolidation Coal Co.).

coagulation. A process of converting a finely divided or colloidally dispersed suspension of a solid into larger-size particles to cause rapid settling or precipitation. Often accomplished by addition of a di- or trivalent metal. Alum, aluminum sulfate, and ferric sulfate are commonly used to clarify water from suspended particles.

coal. A dark-colored, organic rock, derived from peat, occurring in lenses or continuous beds, interbedded with other sedimentary rocks. Beds of coal range in thickness from a fraction of an inch to many tens of feet. Coal was formed by the accumulation and partial decomposition of vegetation in ancient peat bogs or swamps, buried under a thick cover of other rocks, and conversion of peat to coal by heat and pressure. Peat accumulates in swamps or bogs, whenever the rate of growth of swamp plants exceeds the rate of decay. Where the beds were never deeply buried or crumpled by earth movements, the coal is lignite. Where they were both deeply covered and highly folded, the coal is anthracite. In between come intermediate ranks of coal. The coal series is lignite (brown coal), sub-bituminous, bituminous, and anthracite. Lignite contains 30 to 40 percent water; sub-bituminous coal, 10 to 20 percent; bituminous, 5 to 10 percent; and anthracite, less than 3 percent. The ash content of coal increases from a few percent in clean, low-rank coals to 8 or 10 percent in anthracite. Many coals contain foreign mineral matter, so their ash content is high. With increases in such impurities, coal grades into "bone coal" and black shale.

coal, anthracite. A hard coal.

coal, bituminous. A soft coal containing 80 percent carbon and 10 percent oxygen.

coal combustion, gaseous emission factors (lb/ton of coal burned).

Pollutant	Power plants	Industrial	Domestic and commercial
Aldehydes (HCHO)	0.005	0.005	0.005
Carbon monoxide	0.5	3	50
Hydrocarbons (CH_4)	0.2	1	10
Oxides of nitrogen (NO_2)	20	20	8
Oxides of sulfur (SO_2)	38S[a]	38S[a]	38S[a]

[a]S is percent sulfur in coal.

coal desulfurization. The process of removing sulfur from coal. Some of the sulfur in coal can be removed by mechanical cleaning processes which depend upon the difference of the specific gravity of coal and the unwanted materials. Thus iron pyrites, which account for about one-half of the total sulfur content of coal, can be removed mechanically because the specific gravity of iron pyrites is 5, while that of coal is between 1.25 and 1.45.

coal fuel value. The value of lignite is 7000 to 8000 British thermal units (Btu) per pound (one unit will raise the temperature of one pound of water one degree Fahrenheit). High-grade bituminous and anthracite coal yield 13,000 to 15,000 Btu per pound. Impurities in coal lower the fuel value.

coal gas. A smokeless, ashless coal that can be made sulfur free. About 25 percent of the calorific value of good gas coal is recovered in the form of gas, 50 percent in the form of coke, and 7 percent as tar and other byproducts. The calorific value of coal gas is about 16,500 Btu/lb, or about 500 Btu/ft^3, where 1 Btu/ft^3 = 8.9 cal/m^3.

coal gasification. (1) A process that removes sulfur and particulates before the fuel is burned, trace elements such as heavy metals also being removed in the gas cleanup process. This clean gas is a convenient fuel for both small and large users. The basic coal gasification process consists of heating coal in the presence of steam in a gasifier. This causes some of the hydrogen in the steam to unite with the carbon in coal to form methane, CH_4. Under existing technology, not all the carbon and hydrogen in this steam–coal mix unite to form methane during the gasification process. Besides producing methane, the coal gasifier also generates carbon monoxide and hydrogen. These two gases can be made to react to form more methane in a step called methanation. During the gasification process, some of the carbon is burned in the presence of air to produce the heat that makes the process work. This burning yields carbon dioxide as a waste product. Gasification produces other waste products, ash and sulfur (which can be sold or stored). (2) A process that converts coal into a liquid or gaseous fuel, depending on the processing procedure. The four routes possible are the following. (a) Pyrolysis, which produces a co-product, a gas and a liquid. The volatile or complex organic hydrocarbon matter is thermally decomposed. Depending on the rate of heating, the amount of gaseous and liquid products can be varied. More rapid heating tends to convert more of the coal and leave less char. (b) Solvation, which produces a liquid that can be further upgraded by hydrogenation. By use of a hydrogenation donor solvent, coal molecules can be broken down to a size which gives a product with properties of a liquid. Relatively light hydrogen is required to convert coal to a liquid. (c) Hydrogenation. *Direct process:*

$$\text{coal} + H_2 - \frac{\text{catalytic} \rightarrow \quad \text{liquids}}{\text{methane} \rightarrow \text{destructive}}$$

Liquids, controlled: A catalytic process in which hydrogen, coal, and a catalyst are brought into intermittent contact at elevated temperatures and pressures. *Gaseous, destructive:* Coal + $H_2 \rightarrow CH_4$ + char − noncatalytic. (d) Synthesis gas. *Indirect process:* Coal + $O_2 \rightarrow$ syngas catalyst → liquids/gases. Partial combustion of coal (or char) produces carbon monoxide (CO) and hydrogen (H_2). Pure oxygen usually is used with steam as a temperature moderator and reactant. After cleanup, gases are passed over an appropriate catalyst to produce the end product desired. In the diagram the gasification of coal takes two routes. The first route to clean gas produces only low (100–250 Btu/cf) heating value gas. This is because the gas contains a considerable amount of nitrogen. The nitrogen is introduced when air is used in the system for combustion to furnish the heat required for the gasification reactions. The second route produces clean gas of either medium (250–550 Btu/cf) or high (950–1000 Btu/cf) values. The latter is a supplement to pipeline-quality natural gas (SNG). Production of clean liquids or clean solids from coal is shown to be carried out by three principal routes. In the first route, clean gas containing appropriate proportions of CO and H_2 (synthesis gas) is converted by the Fischer–Tropsch process to hydrocarbon oil. The second route involves heating the coal to drive out the naturally occurring oils in it (pyrolysis) and then treating these oils with hydrogen for desulfurization and quality improvement. Pyrolysis processes produce significant quantities of byproduct and char, which must be disposed of economically. The third route to clean liquid fuel involves dissolving the coal in a solvent and filtering out ashes which include the pyritic sulfur. After removing the solvent, the resulting heavy crude oil (syncrude) is treated with

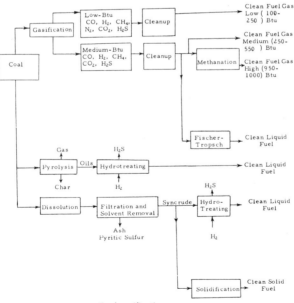

Coal gasification process.

hydrogen to remove sulfur and at the same time improve its quality. In one process (SRC), a solid fuel is produced if the syncrude is allowed to cool before the hydrotreating step.

coal tars (tar oils). In gas works, about 5 percent of the original coal is recovered as crude tar and small amounts are recovered in coke ovens. Tar can be used as a liquid fuel in furnaces if its most volatile constituents are removed by distillation until its flash point rises above 150°F. Tar is not normally burnt in its nearly crude state, for some of its substances are too valuable to burn. Usually, tar is carefully divided into fractions by distillation and the light oils driven off below 170°C (350°F) and redistilled to produce motor benzol. The residual tar is separated into solvent naphtha and heavy naphtha, anthracene oil, cresote, solid aromatic hydrocarbons, pitch, and road asphalt.

coalescence. The act of combining or uniting to form one body, whether through chemical affinity, simple mixture, or concentration.

coastal zone. Ocean waters and adjacent lands that exert an influence on the uses of the sea and its ecology.

cobalt (Co). A silver-gray, magnetic metal with atomic weight, 58.94; specific gravity, 8.9; melting point, 1492°C. Its alloys are important materials for high-temperature applications where resistance to deformation and failure is required in the temperature range 800–1000°C.

COED. *Process:* The char-oil energy development (COED) process produces synthetic crude oil by pyrolysis of coal. Coal is crushed to less than 1/8 in. It is dried and heated to successively higher temperatures in a series of fluidized-bed reactors. In the first stage, coal is heated to 600°F by hot flue gases and devolatized. In the subsequent stages, the coal is subjected to increasing temperatures of 850, 1000, and 1600°F. The pressure of the operation is between 6 and 10 psig. Some of the char is burned with oxygen in the fourth stage to maintain a 1600°F temperature and provide hot gases for heating the second and third stages. Gas from the fourth stage flows countercurrent to the solids through the third to the second stage, from which most of the volatile products are collected. Exit gas from the first stage is quenched with recirculating liquor. Oil is recovered. Part of the gas is used in the coal drier and part in the char cooler. Volatile products in the second-stage gas pass to the product recovery section. Gas is quenched directly with water to condense the oil. The oil is separated from the water and filtered to remove char carryover. Separator gas from the oil recovery section is purified to remove ammonia, CO_2, and H_2S. It is then steam reformed to produce hydrogen. Filtered synthetic crude oil is about minus 4° API. Hydrotreating at about 750°F and 1500–3000 psig removes sulfur, nitrogen, and oxygen from the oil and produces a 25° API synthetic crude oil (FMC Corp.).

coefficient of area expansion. When solids are heated they increase in all dimensions; the change in unit area per degree change in temperature is approximately twice the coefficient of linear expansion.

coefficient of haze (COH). A measurement of visibility interference in the atmosphere.

coefficient of thermal expansion. The ratio of the change of length per unit length (linear), or change of volume per unit volume (voluminal), to the change of temperature.

coffee roasting. The bulk of coffee roasting in the United States is concentrated in and around New York City, Los Angeles, Chicago, and New Orleans. However, a survey of 37 cities by the U.S. Public Health Service found that 18 of the cities have, or have had, problems with this source of air pollution. Some indicated the problem was of minor importance. Many of the coffee roasters were originally located on the outskirts of cities, but they have since become engulfed by burgeoning residential growth. As a result, pollution "receptors" have developed where previously there were none, and a demand for control has followed. Coffee processing produces four kinds of emissions: dust, chaff, odor, and smoke. Dust is generated in the handling of green beans, which usually are bagged in burlap. Chaff consists of the outer covering or skin of the coffee beans, which burst as the bean swells during roasting temperatures generally above 400°F. The odor and smoke are combinations of organic constituents volatilized at roasting temperatures and steam produced when the roast is quenched with water. Further processing to produce powdered, or instant, coffees causes an additional emission in the form of powdered coffee which escapes during the drying process. During decaffeination, odors may be produced by decaffeinating solvents used to extract caffeine from the green beans. The decaffeinating solvents approved by the Food and Drug Administration in 1980 were methylene chloride and trichloroethylene (TCE), although trichloroethylene use was discontinued by American processors after an alert was issued by the National Cancer Institute in 1975 of a possible carcinogenic association between TCE and laboratory animals.

coffin. A thick-walled container (usually lead) used for transporting radioactive materials.

coil, digester. A part of a system of pipes for hot water or steam installed in a sludge digestion tank for the purpose of heating the sludge being treated.

coke. The solid, combustible residue left after the destructive distillation of coal or crude petroleum. If any member of the coal family is strongly heated in a vessel from which air is excluded, it gives off combustible gases and leaves a residue of which a high proportion is carbon. The process is called "carbonization." Wood and peat yield coherent residues which may be used as lump fuel. Lignite and the bituminous coals of high oxygen content leave residues which crumble easily into powder, as does anthracite and carbonaceous coal. A number of bituminous coals of low oxygen content behave quite differently. On heating they first appear to melt; the bubbles of volatile matter escaping from the viscous, semi-molten mass cause it to swell, in some cases, to several times its original volume, and on further heating the mass hardens and takes on the familiar spongy appearance of coke, which it retains when it cools to a normal temperature. These are the "strongly caking" coals which are among the most useful for the manufacture of gas and coke.

coke, metallurgical. The highest quality is made from strongly caking coals of 20 to 30 percent volatile matter; less strongly caking coals of 30 to 35 percent volatile matter are also used. The coal is carbonized at 1000°C or more in coke ovens. The most important use of metallurgical coke is for smelting iron in the blast furnace.

coke oven. An oven mostly employed in the manufacture of metallurgical coke. Coke ovens are built in batteries, and each charge of 15–25 tons of coal per oven is heated for 12–22 hours, its final temperature being about 1000°C. The coke from most coke ovens is quenched by water, requiring about three tons of water per ton of coke.

coke oven emission. Three stages of coke production and use have been identified as sources of toxic emissions. The first stage of emissions occurs during preheating of dry coal when the temperature of the material is raised to approximately 550°F before its introduction into the coke oven. Emissions at this stage may contain some carcinogenic benzene-soluble organic compounds. A second stage involves the coke quenching operation, when coke is cooled rapidly by the application of water. EPA analysis shows the presence of polynuclear hydrocarbon carcinogens in the emissions when clean water is used for quenching, with a tenfold increase in the level of such substances when contaminated quench water is used. The third stage, emissions from coke byproduct industries, may result in the release of benzene, a known carcinogen, and several other organic compounds suspected of being occupational carcinogens. Coke oven emissions have been associated by NIOSH with skin, lung, kidney, and bladder cancers. NIOSH exposure limits for coke oven emissions have been established at 150 micrograms per cubic meter of benzene solubles averaged over any eight-hour period.

coke plant. A chemical plant for the purification of gases evolved in the carbonization of coal and where commercial products such as ammonia, tar, benzene, and naphtha are recovered. (1) Sludge coking plant. In the petroleum refining industry, a plant for H_2SO_4 recovery from dry acid sludge. The sludge is thermally decomposed under a slight vacuum to form H_2SO_4, water, hydrocarbons, and coke. (2) Byproduct coke plant. A plant used in connection with a distillation process to produce coke in which the volatile matter in coal is expelled, collected, and recovered. It consists of coal and coke handling equipment, a byproduct chemical plant, and other equipment associated with the coking chambers or ovens.

cold front. The line of discontinuity at the earth's surface along which a mass of cold air is displacing a warmer air mass.

coliform index. A rating of the purity of water based on a count of fecal bacteria.

coliform organism. An organism found in the intestinal tract of humans and animals; its presence in water indicates pollution and potentially dangerous bacterial contamination.

collection efficiency. The percentage of a specified substance retained on passage through a gas cleaning or sampling device.

collector. (1) A device for removing and retaining contaminants from air or other gases. Usually this term is applied to cleaning devices in exhaust systems. (2) A sandwich-like structure used in solar systems including a black-surface absorber to absorb solar heat. It absorbs radiant energy from the sun and transfers it to another source (air or liquid). Below are typical air and liquid collectors.

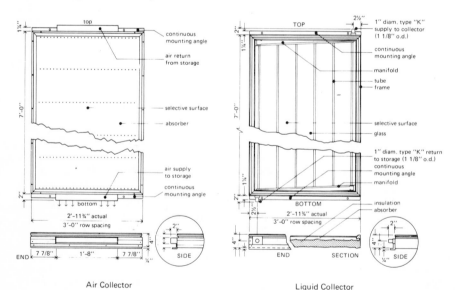

Air Collector   Liquid Collector

Typical collector construction.

collector, air medium. A collector normally composed of a thick slab of insulation on the bottom, an absorber plate painted black resting on that, and double panes of glass or plastic on the top acting as a cover. Air from the interior of the house circulates through the collector between the black-painted surface of the absorber and the enclosing cover. As the air is heated by the sun, it rises and is drawn in through ducts at the top to a storage or distribution zone inside the house. The accompanying diagram (p. 78) shows details of an air medium solar collector by the Solaron Corporation. Air channels are underneath the absorber plate. The glass cover on top is double paned.

collector, centrifugal inertial separator. A variation of the vane axial cyclone wherein both the inner and outer vortices flow in the same horizontal direction.

collector, chip trap. A collector which combines gravity settling with inertial separation resulting from an interposed single or multiple baffle. Such traps are useful for collecting very large particles or protecting a more efficient collector from large particles.

collector, coolant characteristics.

	Air	Water	85% Ethylene glycol/water	Thermia 15 paraffinic oil	UCON (polyglycol) 50-HB-280-X
Freezing point	—	32°F	−33°F	—	—
Pour point	—	—	—	10°F	−35°F
Boiling point (at atm. press.)	—	212°F	265°F	700°F	600°F
Corrosion	Noncorrosive			Noncorrosive	Noncorrosive
Fluid stability	—			Good[a]	Good[b]
Flash point	—	None	None	455°F	500°F
Bulk cost ($/gal)	—	—	2.35	1.00	4.40
Thermal conductivity (Btu/hr °F at 100°F)	0.0154	0.359	0.18	0.76	0.119
Heat capacity (Btu/lb °F at 100°F)	0.24	1.0	0.66	0.46	0.45
Viscosity (lb/ft hr at 100°F)	0.046	1.66	15.7	28.5	143.1

[a]Requires an isolated cold expansion tank or nitrogen-containing hot expansion tank to prevent sludge formation.
[b]Contains a sludge formation inhibitor.

collector, dynamic precipitator. A combination centrifugal fan and dust collector.

collector, flat plate. A flat plate solar heater collector system can supply up to 85 percent of the hot water needed in a temperate or warm climate, at a reasonable cost during the summer, and reduce winter fuel costs. Flat plate heaters are generally made of copper or steel tubing placed in a serpentine or parallel arrangement on a sheet metal plate. The plate is set in a shallow box and backed with insulation. The box is covered with a glass

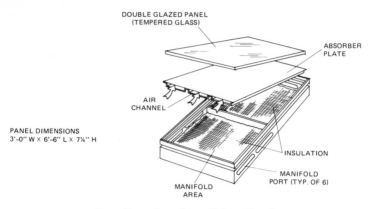

DOUBLE GLAZED PANEL
(TEMPERED GLASS)

ABSORBER
PLATE

AIR
CHANNEL

PANEL DIMENSIONS
3'-0" W × 6'-6" L × 7¼" H

INSULATION

MANIFOLD
PORT (TYP. OF 6)

MANIFOLD
AREA

Air medium solar collector (Solaron Corp.).

or plastic cover. As solar radiation strikes the plate, it is absorbed by the plate and transferred to water circulating through the tubes. Flat plate heaters are usually used in conjunction with a standard water heater and adapt easily to standard plumbing arrangements. They operate on line pressure or forced circulation and lose only a fraction of their efficiency when used with a heat exchanger. Their best feature is the ability to circulate without the aid of a pump when used in a thermosyphon arrangement. Such a system, where cooler water coming from an elevated tank displaces the warm water from the collector tubes, cuts the cost of the system and still delivers ample hot water. Most flat plate collectors consist of the same general component, as illustrated here: (a) *Batten.* Battens serve to hold down the cover plates and provide a weather-tight seal between the enclosure and the cover. (b) *Cover plate.* The cover plate usually consists of one or more layers of glass or plastic film or combinations thereof. The cover plate is separated from the absorber plate to reduce reradiation and to create an air space, which traps heat by reducing losses. This space between the cover and absorber can be evacuated to further reduce convective losses. (c) *Heat transfer fluid passage.* Tubes are attached above, below, or integral with an absorber plate for the purpose of transferring thermal energy from the absorber plate to a heat transfer medium. The largest variation in flat plate collector design occurs with this component and its combination with the absorber plate. Tube on plate, integral tube and sheet, open channel flow, corrugated sheets, deformed sheets, extruded sheets, and finned tubes are some of the techniques used. (d) *Absorber plate.* Since the absorber plate must have a good thermal bond with the fluid passages, an absorber plate integral with the heat transfer media passages is common. The absorber plate is usually metallic, and normally treated with a surface coating which improves absorptivity. Black or dark paints or selective coatings are used for this purpose. The design of this passage and plate combination helps determine a solar system's effectiveness. (e) *Insulation.* Insulation is employed to reduce heat loss through the back of the collector. The insulation must be suitable for the high temperatures that may occur under no-flow or dry-late conditions, or even normal collection operation. Thermal decomposition and outgassing of the insulation must be prevented. (f) *Enclosure.* The enclosure is a container for all the above components. The assembly is usually weatherproof. Preventing dust, wind, and water from coming in contact with the cover plate and insulation is essential to maintaining collector performance.

collector, flat plate (dry). Dry flat plate collectors are less efficient in terms of heat per square foot of collector surface. Instead of water, air is circulated to remove the heat. Since the air circulates between the surface of the collector and the transparent cover,

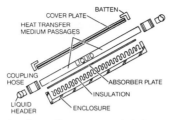

Flat-plate collectors: an exploded view.

the collecting surface can be made from almost anything, regardless of how poorly it might conduct heat. Plywood covered by black-painted aluminum foil is commonly used. Although it lacks in efficiency, the dry collector is far less expensive to build, there is no danger of freezing in the system, no leakage or clogging or pipes to contend with, and the need for a heat exchanger to turn the warm water to warm air is eliminated.

collector, flat plate (wet). In "wet" collectors (those through which water circulates), sheets of aluminum or copper are normally used. The more sophisticated models have copper tubing attached to the back of the collecting surface and water circulated through these tubes carries the heat from the collector to a storage vessel or to the heat circulating system. The less complicated and less expensive models use sheets of corrugated aluminum, eliminating the need for tubing. The water flows over the front surface of the collector and is collected in a trough after being heated. The wet collector can replace the water heater, heat a swimming pool in the summer, and power various types of air conditioning systems.

collector, fluid-medium. The fluid-medium collector is based loosely on the "roof pool collector." From the bottom up, it is composed of a thick slab of insulation to keep the heat inside the collector box. Directly above the insulation backing is the absorber plate. Embedded in the absorber plate is a circuit of pipes, running either vertically up and down the collector box, or in S shapes and other configurations. The circuit carries the liquid through the absorber plate for maximum heat transfer. As the sun shines through the cover, the black absorber plate inside the box warms up. The fluid passing through the conduit in the absorber plate collects the heat and carries it through pipes to the storage zone. In the fluid-medium collector, the absorber plate is attached directly to a separate component, the conduit. The conduit is the heart of the fluid-medium flat plate collector. It is composed of metal tubing and fins of various designs. The conduit carries the fluid medium through or under the absorber plate, where it collects the heat and conveys it to the storage zone. The diagram (p. 80) shows details of a typical fluid-medium solar collector absorber plate. Fluid flows through S-shaped rectangular copper tubes.

collector, grit. A device placed in a grit chamber to convey deposited grit to one end of the chamber for removal.

collector, louvre separator. A collector which is a variation of the recirculating baffle collector and includes the V-pocket and conical-pocket dust louvre. Some such collectors increase collection efficiency by using a cyclone collector in the recirculating cycle. Their collection efficiency falls between those of the baffle collector and the ordinary cyclone collector.

collector, pillow-type. Pillow-type heaters are made of heat-sealed plastic sheets much like a water-bed bag. The pillow can be covered with another sheet of clear plastic to increase efficiency. The life span of the bag is short due to ultraviolet degradation of the plastic, but the bags are fairly cheap.

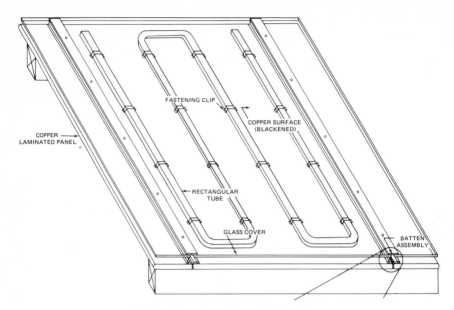

Fluid-medium solar collector absorber plate.

collector shading.   Shading is an issue related to both collector orientation and tilt. Solar collectors should be located on the building or site so that unwanted shading of the collectors by adjacent structures, landscaping, or building elements does not occur. In addition, considerations for avoiding shading of the collector by other collectors should also be made.

As in the diagram, avoid all self-shading for a bank of parallel collectors during useful collection hours (9 AM and 3 PM). This results in designing for the lowest angle of incidence with large spaces between the collectors. It may be desirable therefore to allow some self-shading at the end of solar collection hours, in order to increase collector size or to design a closer spacing of collectors, thus increasing solar collection area.

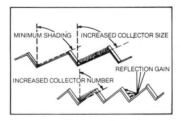

Parallel collector shading.

Building elements such as chimneys, parapets, fire walls, dormers, etc., can cast shadows on adjacent roof-mounted solar collectors, as well as on vertical wall collectors. The drawing shows a house with a 45° south-facing collector at latitude 40° north. By mid-afternoon, portions of the collector are shaded by the chimney, dormer, and the offset between the collector and the garage. Careful attention to the placement of building elements and to floor plan arrangement is required to assure that unwanted collector shading does not occur.

House with a 45° south-facing collector.

collector, sludge. A mechanical device for scraping sludge off the bottom of a settling tank to a sump pump, from which it can be drawn by hydrostatic or mechanical action.

collector subsystem. The assembly used for absorbing solar radiation, converting it into useful thermal energy, and transferring the thermal energy to a heat transfer fluid.

collector, water-trickling. An example of a water-trickling collector is the one developed by Harry Thomason, using a corrugated aluminum roofing panel as an absorber, painted black and covered with one sheet of glass. Water is pumped to a horizontal feeder pipe "manifold" which runs along the top ridge of the collectors. The manifold is pitched slightly to allow the water to flow down its entire length and then enter through holes located at a spacing which coincides with the corrugation or channels in the absorber sheets. To pick up the heat from the metal absorber effectively, the water should flow in a thin film over the entire surface of the corrugated sheets, rather than in the valleys or corrugations alone. The water, heated in its passage down the absorber, is collected in a gutter at the bottom and piped to a thermal storage unit.

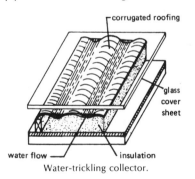

Water-trickling collector.

collector, wet cyclone. A large-diameter type of collector in which the capture and retention of dust is increased by spraying a liquid such as water into the inlet. This spray is captured, as is the dust, and wets the walls of the cyclone. The technique is sometimes used to keep the temperature of the cyclone within structural design limits when cleaning very hot gases.

colligative property. A property that is numerically the same for a group of substances, independent of their chemical nature.

colloid. (1) A finely divided dispersion of one material called the "dispersed phase" (solid) in another material which is called the "dispersion medium" (liquid). It is normally negatively charged. (2) A phase dispersed to such a degree that the surface forces become an important factor in determining its properties. In general, particles of colloidal dimensions are approximately 10 Å to 1 μm in size. Colloidal particles are often best distinguished from ordinary molecules due to the fact that colloidal particles cannot diffuse through membranes which allow ordinary molecules and ions to pass freely.

color body. A complex molecule which imparts color (usually undesirable) to a solution.

combination frequency. A member of the set of frequencies arising from the motion of a nonlinear (distorting) device to which two vibrations of arbitrary frequencies are simultaneously applied. Two simultaneous vibrations of arbitrary frequencies f_1 and f_2 will excite a nonlinear device to a motion containing not only the original frequencies, but also members of a set of "combination" frequencies given by $f_c = mf_1 + nf_2$, where m and n are integers. A resonator sharply tuned to any one of the frequencies which may be produced in the nonlinear device will resound to it with an amplitude depending on the type of nonlinearity. The superheterodyne radio receiver depends on this phenomenon.

combined sewers. A system that carries both sewage and storm water runoff. In dry weather all flow goes to the waste treatment plant. During a storm only part of the flow is intercepted owing to overloading. The remaining mixture of sewage and storm water overflows untreated into the receiving stream.

combining volumes. Under comparable conditions of pressure and temperature, the volume ratios of gases involved in chemical reactions are simple whole numbers.

combining weight. The combining weight of an element or radical is its atomic weight divided by its valence.

combining weights, law of. If the weights of elements which combine with each other be called their "combining weights," then elements always combine either in the ratio of their combining weights or of simple multiples of these weights.

combustion. The act or process of burning. Chemically, it is a process of rapid oxidation, caused by the union of oxygen in the air, which is the supporter of combustion, with any material which is capable of oxidation.

combustion, air. The air introduced to the primary chamber of an incinerator through the fuel bed by natural, induced, or forced draft.

combustion chamber. A chamber where combustible solids, vapors, and gases from the primary chamber are burned and the settling of fly ash takes place.

combustion equipment. Any furnace, incinerator, fuel burning equipment, boiler, apparatus, device, mechanism, fly ash collector, electrostatic precipitator, smoke-arresting or prevention equipment, stack, chimney, breeching, or structure used for the burning of fuel or another combustible material, or for the emission of products of combustion, or used in connection with any process which generates heat and emits products of combustion. Included are process furnaces, byproduct coke plants, coke baking ovens, mixing kettles, cupolas, blast furnaces, open hearth furnaces, heating and reheating furnaces, puddling furnaces, sintering plants, Bessemer converters, electric steel furnaces, ferrous and nonferrous foundries, kilns, stills, driers, roasters, and equipment used in connection therewith, and other equipment used in manufacturing, in chemical, metallurgical, or mechanical processing which may cause the emission of smoke, particulate, liquid, gaseous, or other matter.

combustion gas. A gas or vapor produced in furnaces, combustion chambers, or open burning.

commensalism. A relation existing between members of different species in which one organism definitely benefits from the association, but the other individual does not benefit or is adversely affected under normal conditions.

comminution. The mechanical shredding or pulverizing of waste, used in solid waste management and waste water treatment.

comminutor. A machine that grinds solids to make waste treatment easier.

compaction. The act of compressing metallic or nonmetallic powders, as a briquette in a die.

competition. From an ecological point of view, a struggle between organisms for food, space, mates, or other limited resources. In introspecific competition the struggle is between individuals of the same species. This experience is common to all species whose numbers are increasing in a limited environment. If two species that occupy the same niche (way of life) come together in space and time, as a general rule, three possible results occur: (a) Extinction. One species becomes extinct because its competitor was more successful. (b) Competitive exclusion. When one species is forced out of part of the habitat when it meets with its competitor, but continues to survive in other adjoining portions of the habitat. (c) Character displacement. When two potentially competing species occur sympatrically (together), they have greater differences in their feeding adaptations than in areas where they do not coexist.

complex organic mixture. A term employed by the EPA and other public-health-oriented agencies to identify environmental threats posed by a source containing a number of potentially hazardous chemicals. An example is tobacco smoke, which may contain hundreds of known and unknown substances, some of which have been implicated as carcinogens, mutagens, or other adverse health agents. A 1975 EPA survey of public water supplies in the United States identified more than 700 organic chemicals. The complex organic mixtures in drinking water were subjected to bioassay screening tests, in which they were shown to be mutagenic in bacteria as well as producing transformations in mammalian cell culture lines. When the transformed tissue culture cells were injected into mice, tumors were produced.

component substance, law of. Every material consists of one substance or a mixture of two or more substances, each of which exhibits a specific set of properties, independent of the other substances.

composites. A mixture of particles in a matrix or a combination of two very different materials such as a ceramic with a metal. A composite gives a combination of the properties of the two separate materials.

compost. A relatively stable decomposed organic material.

composting. A method of recycling any organic materials (materials that are or were once alive). The organic material is exposed to air and water so bacteria can begin to break down the wastes.

compressor. A reciprocating, rotary, or centrifugal pump (or mechanism) for raising the pressure of a gas.

Compton effect (Compton recoil effect). The elastic scattering of photons by electrons, resulting in decreases in frequency and increases in wavelength of x rays and gamma rays when scattered by free electrons.

Compton electron. An orbital electron of an atom which has been ejected from its orbit as a result of an impact by a high-energy quantum of radiation (x ray or gamma ray).

concentrating collector. A collector which concentrates solar radiation on a small collection area, resulting in higher temperatures than in flat plate collectors.

concentration. The amount of a given substance in a stated unit of a mixture, solution, or ore. Common methods of stating concentration are percent by weight or by volume and, normally, weight per unit volume.

concentration, hydrogen ion. The weight of hydrogen ions in grams per liter of solution. Commonly expressed as the pH value that represents the logarithm of the reciprocal of the hydrogen ion concentration.

concentrator. An optical system used to concentrate the sun's rays on a focal point or plane so as to increase solar heating in that focal area.

condensate. A liquid produced from a gas, usually by cooling action in a condenser.

condensation. The physical process of converting a material from a gaseous phase to a liquid or solid state by decreasing temperature, increasing pressure, or both. Usually, in air sampling only cooling is used.

condensation level. The level where clouds (visible water droplets or ice crystals) form.

condensation nuclei. A particle, either liquid or solid, upon which condensation of water vapor begins in the atmosphere. Two general types are "neutral particles" (sizes range from 0.1 to 30 μm and larger), commonly known as dust and fumes; and condensation nuclei (sizes range from 0.001 to 0.1 μm) made up of hydroscopic substances.

condensation rate. The number per square centimeter per second at which molecules condense on a surface.

condensation sampling. A gas sampling technique in which the gas is drawn through trapping tubes immersed in refrigerants. By use of different refrigerant solutions, various fractions of the components of the gas can be measured.

condenser. A device for cooling a gas or a vapor in order to convert it to a liquid.

conductance. The reciprocal of resistance, measured by the ratio of the current flowing through a conductor to the difference of potential between its ends. The practical unit of conductance, the mho, is the conductance of a body through which one ampere of current flows when the potential difference is one volt. The conductance of a body in mhos is the reciprocal of the value of its resistance in ohms.

conduction. The transfer of energy within and through a conductor by means of internal particular or molecular activity and without any net external motion. Conduction is to be distinguished from convection (of heat) and radiation (of all electromagnetic energy).

conduction of heat. The transmission of heat by means of a medium without any movement of the medium itself.

conductivity, electrical. A measure of the quantity of electricity transferred across a unit area per unit potential gradient per unit time. The reciprocal of resistivity. For volume conductivity or specific conductance, $k = 1/g$ where g is the volume resistivity. For mass conductivity, $x = k/d$, where d is density. Equivalent conductivity $A = k/c$, where c is the number of equivalents per unit volume of solution. Molecular conductivity $\mu = k/m$, where m is the number of moles per unit volume of solution.

conductivity, thermal. The time rate of transfer of heat by conduction, through unit thickness, across unit area for unit difference of temperature. It is measured as calories per second per square centimeter for a thickness of one centimeter and a difference of temperature of 1°C.

conductor. A substance or entity which transmits electricity, heat, sound, etc; any material through which electrons flow easily when an electrical force is applied. Most metals are good conductors. At normal temperatures, silver is the best solid conductor; copper and aluminum follow in that order. Any conducting object, properly isolated in space, charged by conduction always acquires a residual charge of the same sign as that of the body touching it. When an isolated conductor is given a residual charge by induction, the charge is opposite in sign to that of the object inducing it.

conjugate foci. Upon proper conditions light divergent from a point on or near the axis of a lens or spherical mirror is focused at another point; the point of convergence and the position of the source are interchangeable and are called conjugate foci.

conservation. The protection, improvement, and use of natural resources according to principles that will assure their highest economic or social benefits.

conservation of energy. In a chemical change, a transformation of energy from one form to another, without gain or loss of energy.

conservation of energy, law of. Energy can neither be created nor destroyed and therefore the total amount of energy in the universe remains constant.

conservation of mass. In all ordinary chemical changes, the total mass of the reactants is always equal to the total mass of the products.

conservation of momentum, law of. For any collision, the vector sum of the momenta of the colliding bodies after collision equals the vector sum of their momenta before collision. If two bodies of masses m_1 and m_2 have before-impact velocities v_1 and v_2 and after-impact velocities u_1 and u_2,

$$m_1u_1 + m_2u_2 = m_1v_1 + m_2v_2.$$

Consol synthetic fuel. *Process*: The Consol synthetic fuel process can produce liquid hydrocarbon fuel from coal. Crushed coal (minus 14 mesh), is dried and preheated in a fluid bed to 450°F. It is then slurried with a coal-derived solvent and pumped at 150 psig through a tubular furnace where it is heated to an extraction temperature of 765°F. Extraction mainly occurs in a stirred extraction vessel. Vapors produced in the extraction section are sent to the fractionation section. Unreacted coal and liquid products are separated in hydrocyclones. Liquid passes to the fractionation section. Solids are sent to a low-temperature carbonization unit. The recycle solvent recovery unit is returned to the slurry system. Distillate-oil product is taken overhead from the fractionation section. Bottoms, along with tars from the carbonization unit, are treated with hydrogen in a catalytic reactor operating at 800°F and 3000 psi. This step converts the heavy liquids to donor solvent and naphtha product. Hydrogen for this operation is produced by the partial oxidation of char from the carbonization unit. Gas steams from the carbonization and hydrotreating operations are treated for removal of H_2S, NH_3, CO_2, and light hydrocarbon liquids. This produces a clean fuel gas and a light liquid product (Consolidation Coal Co.).

constants and conversion factors. 1 langley (ly) = 1 gram calorie/square centimeter
1 langley/minute = 3.68 Btu/square foot/minute = 220 Btu/square foot/hour
1 Btu = 2520 gram calories
1 gram calorie (mean) = 3.9685×10^{-3} Btu (mean)
1 gram calorie/second = 14,285 Btu/hour
1 Btu/hour = 0.07 gram calories/second
1 therm = 100,000 Btu (10^5 Btu)
1 kilowatt (kW) = 1.341 horsepower (hp) = 56.92 Btu/minute
1 kilowatt-hour (kWhr) = 3413 Btu

1 inch = 0.0254 meter (length)
1 foot = 0.3048 meter (length)
1 pound force = 3.448 newton (force)
1 ton = 907.2 kilograms (mass)
1 pound = 0.4536 kilograms (mass)

contact pesticides. A chemical that kills pests when it touches them, rather than by being eaten (stomach poison).

contamination. A general term signifying the introduction into water of microorganisms, chemicals, waste, or sewage which renders the water unfit for its intended use.

continuous air monitoring program (CAMP). A program undertaken by the U.S. Public Health Service to increase knowledge about the nature of air pollution by providing information in the form of comparable, continuous, concurrent data on significant gaseous pollutants from a number of urban areas. Monitoring stations have been established in or near the principal business districts of such cities as Chicago, Cincinnati, Philadelphia, San Francisco, St. Louis, and Washington to monitor gaseous pollutants— SO_2, NO, NO_2, CO, hydrocarbons, O_3, and oxidants. Measurements comparable with the CAMP operations are also made in cities at stations operated by local groups (such as Detroit, New York, and Los Angeles).

contour plowing. A farming method that breaks ground following the shape of the land in a way that discourages erosion.

contrail. A long narrow cloud caused when high-flying jets disturb the atmosphere.

control rod. A rod, plate, or tube containing a material that readily absorbs neutrons; used to control the rate of fission in a nuclear reactor. Control rods are interspersed between the fuel rods in order to regulate the rate of fission. The production of too much heat energy would burn up the reactor. Cadmium and boron are commonly used as control rods, since they capture some of the free neutrons and slow down the fission reactions. Control rods are inserted or withdrawn from the reactor by electric motors. When the control rods are completely lowered into the reactor, the number of neutrons they absorb is greatest and the least amount of fission reaction takes place. As the control rods are withdrawn, the number of fission reactions is correspondingly increased.

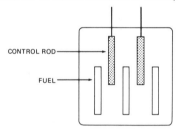

CONTROL ROD

FUEL

Control rods, lowered into a reactor core, catch neutrons to slow the chain reaction.

control subsystem. The assembly of devices used to regulate the processes of collecting, transporting, storing, and utilizing solar energy.

controlled atmosphere. An atmosphere which may be used as a primary standard for the calibration of analytical techniques or as a simulated environment for the study of biological responses or the resistance of materials. The first requirement of a controlled atmosphere is a source of "pure air"; other components are the test substance, or substances; a container, duct, or other boundary for the system; and devices to limit, define, or measure the flux of the amount of air and test substances. Atmospheres may be classified as static or dynamic; two intermediate types are quasistatic and quasidynamic.

convection. The process of heat transfer from one place to another by the actual mass motion of heated liquid or gas from one place to another. At ordinary temperatures the principal method of heat transfer in liquids and gases is convection. Convection is possible only in fluids or gases, because they alone have internal mass motions. In convection, the moving masses carry heat acquired by conduction in their previous positions. Since the atmosphere is a medium in which mass motions are easily started, convection is found to be one of the chief ways in which heat is transferred. This transfer may be accomplished either by vertical or by horizontal motions. Horizontal convection transport is called advection. The mechanism by which heat is transferred from a hot water or steam "radiator" to the air, walls, and objects in a room is almost entirely convection. An open fireplace is very inefficient because most of the heat is lost by convection up the chimney to the outdoors.

convection current. A closed circulation of material that sometimes develops during convection. Convection currents normally develop in pairs; each pair is called a convection cell.

convector. A finned, heated tubular unit (e.g., steam or water radiator) used in space-heating systems.

conversion. The process in which a fissionable material is made by neutron capture while another is being consumed inside the reactor, as plutonium in a uranium-fueled reactor; the term is usually applied when the amount of fissionable material thus made is less than the amount consumed.

converter. (1) A machine or device that changes alternating current to direct current or the reverse. (2) A device that changes the frequency of periodic phenomena. (3) A device that changes one type of signal to another.

coolant. A material used in a nuclear reactor to remove the heat energy resulting from fission. In a power reactor this heat energy is used to generate electrical power.

cooling. The act of processing highly radioactive materials to attain less radioactivity for subsequent uses or handling.

cooling rate. The rate of cooling of a body, substance, or gas environment, obtained by plotting time against temperature under constant conditions.

cooling tower. A device that aids in heat removal from water used as a coolant in electric power generating plants.

COP (coefficent of performance). A number that describes the performance of a heat pump. When a pump can provide 3 Btu of heat energy for every 1 Btu of electrical energy required to drive its compressor, the pump's COP is said to be 3.

copper (Cu). A metallic element. Control of copper wastes and iron, manganese, and zinc wastes is most commonly effected by precipitation as metal hydroxides. Precipitation is usually effected at a pH between 7 and 9. Public water supplies may tolerate as a permissible criterion 1.0 mg/liter. In general, copper is highly toxic to algae, sea plants, and invertebrates, but only moderately toxic to mammals. The relationship between concentrations of different heavy metals in a water supply are extremely important in the determination of ultimate copper toxicity.

core. The uranium-containing heart of a nuclear reactor, where energy is released.

corpuscular radiation. A stream of charged particles. The sun emits at times of increased activity a neutral stream of charged particles with velocities of about 1000 km/s. If and when the stream meets the earth, it produces a magnetic storm about 30 hours after the effects associated with the simultaneously emitted ultraviolet radiation.

correlation coefficient. A measure of the persistence of eddy velocity as a function of time and space.

cosmic ray. A beam of particles from outer space, entering the atmosphere at great speeds. Studies have shown that there are always some ions in the atmosphere, most of which are caused by high-speed particles coming into the atmosphere from outer space. The source of these particles is not definitely known, but it is believed they come from beyond the solar system, and are therefore called cosmic rays. Cosmic rays have great penetrating power, the intensity varying with latitude, indicating that they are charged and are affected by the earth's magnetic field. It is believed that cosmic rays consist largely of the nuclei of elements of low atomic weight, protons (hydrogen nuclei) being the most abundant type. Scientists believe that gigantic magnetic fields in space, in galaxies, and near certain stars may be influential in producing cosmic rays.

cosmotron. A particle accelerator capable of giving them energies of billions of electron volts.

coulomb. A unit of quantity of electricity; it is the quantity of electricity which must pass through a circuit to deposit 0.001 118 0 g of silver from a solution of silver nitrate. An ampere is one coulomb per second. A coulomb is also the quantity of electricity on the positive plate of a condenser of one-farad capacity when the electromotive force is one volt.

count median size. The median measurement of particle diameters of particulate matter of a gas sample.

countercurrent efficiency. The unique advantage of a carbon column, permitting partially spent activated carbon to adsorb impurities before the semiprocessed stream comes in contact with fresh carbon. This allows the maximum capacity of the activated carbon to be utilized.

couple. Two equal and oppositely directed, parallel but not collinear forces acting upon a body. The moment of the couple, or torque, is given by the product of one of the forces by the perpendicular distance between them.

covalent bond. A bond formed by electron sharing. Not all bonds are formed by the complete transfer of electrons from one to another, often they are formed by electron sharing. Covalent bonds are usually formed between two nonmetallic elements. Electrons are held tightly in covalent bonds and these compounds are not good conductors of electricity; instead, they are examples of covalent substances.

cover. The vegetation or another material providing protection.

cover material. The soil used to cover compacted solid waste in a sanitary landfill.

cracking. A thermal decomposition process, such as the cracking of NH_3 to N_2 and H_2, and natural gas hydrocarbons, such as CH_4 to C and H_2, or into other carbons.

creep. The slow, but continuous, downward migration of soil and mantle rock under the force of gravity. The rate of creep, which is usually imperceptible except over a long period of time, depends on factors such as temperature changes, amount of rainfall, angle of slope, type of soil, and nature of parent material. In aeolian soil, such as loess, which has a tendency to stand nearly vertical, creep is exceedingly slow; whereas in loosely consolidated , sandy soils with a high proportion of rounded grains, the response to the force of gravity and creep is more pronounced. In areas where there is alternate freezing and thawing, the rate of creep is increased by the wedging and heavy action of ice. Other agents that assist creep are active in warmer climates. Clayey slopes may contract and crack during dry seasons, and when such desiccation fissures are closed or filled, there is a greater movement down-slope than up.

criteria.　The standards EPA has established for certain pollutants, which not only limit the concentration, but also set a limit to the number of violations per year.

critical bed depth.　In a carbon column, the bed depth is the depth of granular carbon which is partially spent. It lies between the fresh carbon and the spent carbon and is the zone where adsorption takes place. In a single-column system, the critical bed depth is the amount of carbon that is not completely utilized.

critical inversion height.　The altitude above ground level to which a plume of smoke or similar emission will rise until it reaches an atmospheric inversion layer. Generally, a low inversion height will tend to trap flue gases and increase their concentration at levels close to the ground. In some cases, the effluent may have enough buoyancy and momentum as it leaves the stack to penetrate a very low inversion. The height of the stack and wind speed are among factors that influence the dispersion of a plume of effluent in the presence of an inversion. Neighboring terrain that is equivalent in height to the effective stack height can result in an increased concentration of pollutants during inversion entrapment. However, high terrain in the path of a plume can also counter the inversion trap condition through a so-called chimney effect caused by sunlight warming the air and producing an updraft that carries the effluent over the hill or mountain. The critical inversion height and ground-level concentrations of pollutants can sometimes be changed by increasing the height of the stack, or by increasing the speed and buoyancy of the effluent plume, or both.

critical mass.　The minimum mass the fissile material must have in order to maintain a spontaneous fission chain reaction. For pure ^{235}U it is computed to be about 20 lb. If the material surrounding a fission reaction is pure enough, as well as massive enough, the reaction is self-sustaining. The amount of a particular fissionable material required to make a fission reaction self-sustaining is called the critical mass. If the critical mass is exceeded, and if the emitted neutrons are not absorbed by nonfissionable material, the reaction runs out of control and a nuclear explosion results.

critical temperature.　The temperature above which a gas cannot be liquefied by pressure alone. The pressure under which a substance may exist as a gas in equilibrium with the liquid at the critical temperature is the critical pressure.

cross section.　A measure of the probability of a particular process. The nuclear cross section is expressed by a/bc, where a is the number of processes occurring, b is the number of incident particles, and c is the number of target nuclei per square centimeter. There are nuclear cross sections for fission, slow neutron capture, Compton collision, and ionization by electron impact.

cross-sectional bed area.　The area of activated carbon through which the stream flow is perpendicular.

crown fire.　One of the most destructive types of fire. A crown fire incinerates the forest canopy, as well as surface debris, and causes wholesale destruction of vegetation. Temperatures that prevail while the fire is in progress may reach 1300°F in the litter environment, but three quarters of an inch below this level, in the mineral soil, the temperature may be between 70 and 217°F, making the surface a short distance below incompatible with life. A crown fire can practically incinerate most of the surface biota that cannot escape. The destruction of vegetation results in massive erosion.

crude petroleum. A petroleum that is almost entirely a mixture of hydrocarbons. There are many thousands of separate and distinct hydrocarbons contained in crude oil, and except for a few of the lightest ones, the differences between boiling points of one compound and its neighbor is only a fraction of a degree. They fall into two chemical classes: the paraffins or carbon-chain compounds and various carbon ring compounds, which are collectively called naphthenes.

crude shale oil. A term applied to several kinds of organic and bituminous shales, most of which consist of varying mixtures of organic matter with shale and clay. The organic matter is chiefly in the form of a mineraloid called kerogen, which is of indefinite composition, insoluble in petroleum solvents, and of uncertain origin. For this reason these shales are better called kerogen shales. The oil distilled from most of them does not occur as oil in the shale, but is formed by heating, the distillation of the vapors beginning at temperatures around 1350°C (2462°F); hence kerogen shales are classed as pyrobitumens.

cryogenic insulation materials. (1) Foamed plastics with densities between 0.015 and 0.030 g/cm³ and an apparent thermal conductivity of 100–200 at cryogenic temperatures. (2) Evacuated powders with densities between 0.100 and 0.200 g/cm³ and a thermal conductivity of 10–20. (3) Evacuated foil laminates with densities between 0.140 and 0.350 g/cm³ and a thermal conductivity of 0.5–2.0, made with many sheets of aluminum foil, or aluminized plastic material interspersed with layers of glass fiber mesh.

cryogenic temperature. In general, a temperature range below the boiling point of nitrogen (−195°C); more particularly, temperatures within a few degrees of absolute zero.

cryogenics. The study and use of devices utilizing properties of materials with temperatures near absolute zero. The opposite effect of heat resistance is to cool a metal down to a very low temperature to reduce its resistance. Near absolute zero, at 0°K or −273°C, some metals abruptly lose practically all their resistance.

cryohydrate. The solid which separates when a saturated solution freezes. It contains the solvent and the solute in the same proportions as they were in the saturated solution.

cryopump. (1) An exposed surface refrigerated to cryogenic temperature for the purpose of pumping gases in a vacuum chamber by condensing the gas and maintaining the condensate at a temperature such that the equilibrium vapor pressure is equal to or less than the desired ultimate pressure in the chamber. (2) The act of removing gases from an enclosure by condensing the gases on the surfaces at cryogenic temperatures.

cryoton. An apparatus in which a low temperature can be maintained.

crystal. A solidified form of a substance consisting of atoms arranged in a regular, three-dimensional pattern. Most solid substances are composed of crystals. The three-dimensional pattern is called a lattice structure. The lattice structure for a crystal of table salt is in the form of a cube. Crystals can be held together by ionic or covalent bonds. Some crystals are partly ionic and partly covalent, as in silicon dioxide (quartz). In covalent cyrstals, like the diamond, all the electrons in the outer shells take part in the bond. The atoms of some solids, like charcoal, have no crystalline structure or definite pattern and are called amorphous, or without form.

cullet. Recycled glass, usually broken into small, uniform pieces about the size of a pea.

culm pile. Anthracite that has disintegrated into coarse powder (≈ 1/8-in. screen) underground because of a disturbance in the strata.

cultural eutrophication. Increasing the rate at which water bodies "die" by pollution from human activities.

cumulative frequency function. A statistical analysis with reference to the cumulative frequency distribution of pollutants.

cumulative sample. A sample obtained over a period of time with the collected atmosphere being retained in a single vessel, or a separated component accumulating into a single whole. Examples are dust sampling, in which all the dust separated from the air is accumulated in one mass of fluid; the absorption of acid gas in an alkaline solution; and collection of air in a plastic bag or gasometer. Such a sample does not reflect variations in concentration during the period of sampling.

cupola. (1) A vertical shaft furnace used for melting metals, as distinct from a blast furnace in which ore is melted. Metal, coke, and flux are charged from the top of the furnace onto a bed of hot coke, through which hot air is blown. In hot blast cupolas the hot blast is derived from a recuperator fed by hot gases drawn from the cupola stack, or from a separately fired blast heater. This allows higher melting rates, and grit and dust emissions from the recuperative types of cupolas tend to be lower than from the conventional plant. The amount of metallurgical fume emitted is higher than that produced in a cold blast cupola. (2) A small type of shaft furnace, in which iron and steel are melted, usually for recovery from scrap metal, and cast into ingots or castings of special shape. The fuel is metallurgical coke, and the flux is usually limestone. The hotter waste gases from the hot blast cupola contain metallurgical fumes. Both hot blast and cold blast cupolas are sources of grit and dust. Heavy smoke emissions may be inevitable if the scrap iron to be milled is contaminated with oil, grease, tar, or other volatile matter.

curie. A unit used to indicate the strength of radioactive sources in terms of the number of disintegrations per second in the source. One curie is equal to 3.7×10^{10} disintegrations per second. This disintegration rate is very close to the radioactivity of one gram of radium.

Curie point. The transition temperature of ferromagnetic substances at which the phenomena of ferromagnetism disappear and the substances become merely paramagnetic. This temperature is usually lower than the melting point.

Curie–Weiss law. The Curie law modified by Weiss, stating that the susceptibility of a paramagnetic substance above the Curie point varies inversely as the excess of the temperature above that point. This law is not valid at or below the Curie point.

Curie's law. The intensity of magnetization $I = AH/T$, where H is the magnetic field strength, T is the absolute temperature, and A is Curie's constant. It is used for paramagnetic substances.

current. The rate of transfer of electricity. The transfer at the rate of one electrostatic unit of electricity in one second is the electrostatic unit of current. The electromagnetic unit of current is a current of such strength that one centimeter of the wire in which it flows is pushed sideways with a force of one dyne when the wire is at right angles to a magnetic field of unit density. The practical unit of current is the ampere, a transfer of one coulomb per second, which is one tenth the electromagnetic unit. The international ampere is the unvarying electric current which, when passed through a solution of silver nitrate in accordance with certain specifications, deposits silver at the rate of 0.001 118 00 g/s. The international ampere is equivalent to 0.999 835 absolute ampere. The ampere-turn is the magnetic potential produced between the two faces of a coil of one turn carrying one ampere.

curtain wall. A partition wall between chambers in an incinerator under which combustion gases pass. A drop arch is a refractory construction or baffle which serves to deflect gases in a downward direction.

cutie pie. An instrument used to measure radiation levels.

cyanide. A substance composed of a cyanogen group in combination with some element or radical. The cyanides, represented by hydrocyanic acid and its salts, are important and ubiquitous industrial chemicals. They are extremely toxic, especially at low pH. It is thought that cyanide acts by inhibiting the phosphorylative oxidation reactions which permit cellular respiration. Many lower animals and fishes seem to be able to convert cyanide to thiocyanate ion, which does not inhibit respiratory enzyme activity. Cyanide compounds formed by the reaction of CN^- with heavy metals may be even more toxic. For these reasons, control of cyanide in industrial effluents is extremely important. Cyanide removal from industrial effluents is commonly effected by the stagewise application of lime and chlorine, which progressively oxidize cyanides to cyanates and then to carbon dioxide and nitrogen. A permissible criterion of 0.20 mg/liter and a desirable criterion of complete absence from public waters are the recommendations of the Federal Water Pollution Control Agency Water Quality Committee.

cyclone. An area of low barometric pressure around which winds circulate counterclockwise in the Northern Hemisphere and clockwise in the Southern Hemisphere. Cyclones have centers of low barometric pressure, as well as rotating wind patterns. They do not originate in the tropics; they are usually confined to the middle latitudes. In general, in weather reports they are simply referred to as centers of low pressure, but they can provide heavy downpours of rain or severe snowfalls that continue for long periods. Cyclones are the largest storm centers created by unsettled weather conditions, often embracing an area more than 1000 square miles in extent. This is twice the expanse covered by an average hurricane, and many times that of a tornado. The winds in temperate zone cyclones seldom, if ever, reach 50 mph. Cyclones are most frequent in the winter months.

cyclone collector. A structure without moving parts in which the velocity of an inlet gas stream is transformed into a confined vortex, where centrifugal forces tend to drive the suspended particles to the wall of the cyclone body. Types in common use may be classified as follows: tangential inlet with axial dust discharge (the common cyclone); tangential inlet with peripheral dust discharge; axial inlet (through swirl vanes) with axial dust discharge; axial inlet (through swirl vanes) with peripheral dust discharge.

cyclone collector, industrial. Basically there are two types of mechanical collectors in commercial applications: tangential inlet cyclones and axial inlet collectors. Axial inlet cyclones are frequently referred to as multicyclones. Generally of small diameter, they are grouped together in a common housing with a common inlet and a series of gas outlet tubes that discharge at a common plenum. Because of their small diameter, they are usually of higher efficiency than the larger-diameter tangential cyclones. In principle, however, both types of collectors are alike. Whether the dust-laden gas enters tangentially through a duct or is directed by "turning" vanes, it is caused to spin in a downward spiral. The spinning action causes the suspended particles to migrate toward the inner wall of the cyclone. The vertical force component of the downward spiral causes the particulates to be carried down through the cone section and into the hoppers; as the gases move downward, the cyclonic phenomenon takes place. An ascending vortex is developed by gas migrating along the entire length of the cyclone. This vortex has its terminal vertex and apex at the apex of the cone. Because of centrifugal and inertial forces, the migrating gas is generally particulate free. The ascending vortex rises through the center of the cyclone and is finally exhausted through the gas outlet tip and/or vertex finder.

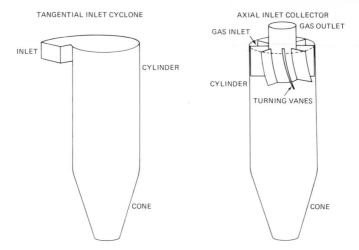

Tangential inlet cyclone (left) and axial inlet collector (right).

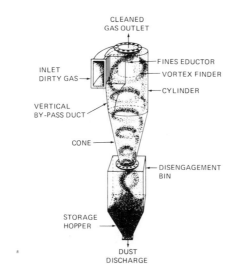

Flow pattern in a mechanical cyclone with vertical fines eductor.

cyclone collector, with involute inlet.   The accompanying diagram represents a single involute cyclone and illustrates the manner in which the dirty gas is injected into the cylinder. The rectangular involute inlet nozzle has its right-hand wall tangent to the cylinder through a fully open, 180° involute. In this design the descending vortex will be formed with a minimum of turbulence, thereby facilitating the precipitation of the dust under the influence of centrifugal force to the walls of the cylinder and cone. In this cyclone the dirty gas induces the cyclonic spin of the cylinder. This energizes the descending outer vortex arrester, where the rapidly rotating dust is flung rapidly outward, settling into the storage hopper. The ascending inner vortex of cleaned gas starts within the vortex arrester and spirals upward, being fed and enlarged by the cleaned inner surfaces of the descending outer vortex. The cleaned gas is kept away from the inlet and conducted out of the cylinder by the vortex finder.

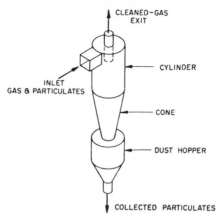

Single high-efficiency involute cyclone.

cyclone collector, vane axial. The accompanying diagram shows a typical collector, consisting of 25 cyclones arranged in a common shell, having a common inlet, common gas outlet, and common dust storage hopper. In this typical arrangement the dirty gas

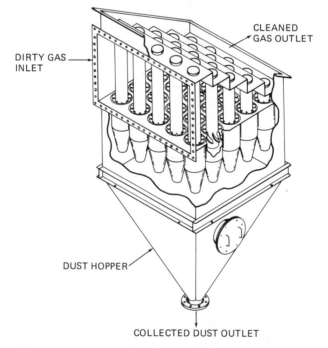

Vane axial cyclone collector.

enters through the common dirty gas inlet into the dirty gas plenum, from which it turns downward to feed the individual cyclones. Cyclonic spin is imparted to the descending gas by the interposed vanes, which cause the gas and dust to spiral downward in the descending outer vortex. The centrifugal forces generated by the high rate of spin move the dust outward, then hold it against the walls of the cylinder and cone until it is flung radially outward through the dust outlet into the storage hopper. The ascending inner vortex of cleaned gas starts slightly below the dust outlet and is fed along its entire length by the cleaned inner surface of the descending vortex. The cleaned gas from each cyclone is picked up and piped through the dirty gas plenum into the cleaned gas plenum by vortex finder tubes of varying lengths.

cyclone separator.   A separator which utilizes centrifugal force to arrest all or part of the particles suspended in a current of gas.

cyclonic spray tower.   A liquid scrubbing apparatus where sprays are introduced countercurrent to gases for removal of contaminants. Axial gas velocities can be as high as 3–6 fps, because drops which might be entrained collect on the cyclone wall. The transfer unit's height is lower than that in a simple spray tower, but the pressure drop is correspondingly higher.

cyclotron.   A device in which a powerful permanent magnet and an alternating electric current are used to accelerate charged subatomic particles along a spiral path until the particles have enough energy to be useful for bombarding atomic nuclei or for initiating special nuclear reactions; an accelerator in which charged particles are whirled in a spiral path. Starting near the center, each time the particles pass a certain point they receive a boost in energy which causes them to move in ever-widening circles. They whirl out toward the rim in a thousandth of a second and are then shot off at a target through an opening. The energies attained by particles in a cyclotron are limited to roughly 25 MeV by the increase in their mass as they approach the speed of light. This additional mass makes the particles harder to move and the applied voltage becomes less and less effective.

D

dalton. A unit of weight equal to that of a single hydrogen atom.

Dalton's law of partial pressures. The pressure exerted by a mixture of gases is equal to the sum of the separate pressures which each gas would exert if it alone occupied the whole volume. This fact is expressed in the following formula: $PV = V(p_1 + p_2 + p_3 + \ldots)$.

damage risk criterion. A criterion that specifies the maximum allowable exposure for people if risk of hearing impairment is to be avoided. A damage risk criterion may include in its statement a specification of such factors as the length of exposure, the noise intensity and frequency, the amount of hearing loss that is considered significant, the percentage of the population to be protected, and the method of measuring the noise.

daughter. The atom that is made by radioactive disintegration of another atom, which is called the "parent."

DDT. The first chlorinated hydrocarbon insecticide [chemical name: 1,1,1-trichloro-2, 2-bis(p-chlorophenyl)ethane]. It has a half-life of 15 years and can collect in fatty tissues of animals. EPA banned registration and interstate sale of DDT for virtually all but emergency uses in the United States in 1972 because of its persistence in the environment and accumulation in the food chain.

debris slide. A variety of erosion by mass wasting; rapid downslope movement of a mass of unconsolidated material. Debris slides do not move as single blocks as in the case of a slump.

Debye length. A theoretical length which describes the maximum separation at which a given electron will be influenced by the electric field of a given positive ion.

decay. (1) The disintegration of a radioactive substance due to nuclear emission of alpha or beta particles, positrons, or gamma rays. (2) Decay rate. In air pollution sampling, the gross beta activity of a weighed sample is measured after it is two days old to allow decay of the natural alpha and associated beta and gamma activities. The samples may be recounted after four to seven days to establish decay in terms of apparent half-life. Decay values are given as gross beta radioactivity in picocuries per cubic meter of air on the date of sample collection. The decay rate is expressed as the half-life beginning on the date of the first count.

decay element. The rate of spontaneous disintegration of a radioactive element, unaffected by heat or pressure.

decay product. A nuclide resulting from the radioactive disintegration of a radionuclide, being formed either directly or as a result of successive transformations in a radioactive series.

decay, radioactive. The spontaneous transmutation of one atomic nucleus into another, usually by the injection of a subatomic particle.

decibel (dB). A dimensionless measure of the ratio of two powers, equal to 10 times the base 10 logarithm of the ratio of two powers p_1/p_2.

decimal notation. A numerical notation based on the number 10. The speed of light is 3×10^8 m/s (three times ten to the eight meters per second), to an accuracy of 0.1 percent. This notation is easier to write and read than 300,000,000 m/s, and facilitates calculations; the 8 is called the exponent, and to multiply two such numbers you merely add their exponents. Very small numbers are similarly written, with the help of negative exponents; e.g., the elementary charge (the charge of a proton) is 1.60×10^{-19} C, a more compact expression than 0.000 000 000 000 000 000 160 C. To some extent, the use of very large and very small numbers can be avoided by the use of units with prefixes. Thus the prefixes k (kilo), M (mega), and G (giga) indicate that the unit is to be multiplied by 10^3 (a thousand), 10^6 (a million), and 10^9 (a billion), respectively; division of the unit by those same factors is indicated by the prefixes m (milli), µ (micro), and n (nano). Thus 1 nm is one nanometer or one millionth of a millimeter; 1 GeV is one giga-electron-volt, or a billion electron volts. Other prefixes exist, but are not much used, except for c for one hundredth, as in cm (centimeter).

declination. The angle between the vertical plane containing the direction of the earth's field at any point and a plane containing the geographic north and south meridians.

decomposer. An organism such as a carrion beetle or a fungus that feeds upon and breaks down dead organic matter.

decomposition. The chemical separation of a substance into two or more substances which may differ from each other and from the original substance.

dedusting. An operation or process arresting all or part of the dust present in a gas. A distinction is usually implied between dedusting and filtration.

definite proportions, law of. In every sample of each compound substance, the proportions by weight of the constituent elements are always the same.

defoliation. The removal of leaves from growing plants due to the effect of liquid or solid chemical agents (herbicides, as well as some pesticides).

deformation thermometer. A thermometer formed by two metal strips having different coefficients of thermal expansion that are welded and rolled together; temperature changes will cause this bimetallic strip to become distorted by the shearing stress between the two faces of the strip. With most metals at ordinary temperatures, this bending can be reproduced time and time again in exactly the same amounts and it can thus be calibrated in terms of temperatures. Invar, a metal having an extremely low coefficient of expansion, is commonly used for the inner portion of the arc-shaped strip, with brass or steel in the outer face. Deformation thermometers are not used for direct measurements of temperature, but rather for writing a record of it. An instrument designed to make a temperature record in this way is called a thermograph.

degasification. The removal of the last trace of a gas from wires used in vacuum tubes, metals to be plated, and substances to be used in other specialized applications.

degradation. The mass movement of water over any type of substrate, causing an erosion or wearing away of the parent material. The finer particles are carried as suspended materials, but the larger or coarser objects may be moved along gradually, as a traction or a bed load. If the current along a stream loses some of its initial velocity, the larger particles of suspended material settle out first, along with a larger portion of the bed load.

degreasing. The process of removing greases and oils from sewage, waste, and sludge.

degree day concept. The degree day concept is used as an aid in calculating the heat required to maintain a living space at a desired temperature, in an environment of changing outside conditions. The heat required will be equal to the heat lost through the walls to the atmosphere. Heat lost = K − surface area − temperature gradient − time, where K is the heat flow factor per unit wall thickness, as defined below, and the temperature gradient is the difference between the inside (room) air temperature and the average outside air temperature (OAT) for the period considered. By selecting a datum level of average daily OAT's of +12°C, below which space heating begins and above which heating is no longer required, a "heating day" (HT) is defined as any day when the average OAT is below +12°C. If time in the above equation is reckoned in heating days (HT), then to find the heat lost, and hence the space heating required for a given period, it is necessary to include a summation of the daily average temperature gradient over the period. Thus the heat requirement = K − surface area − HT $\times \sum(t_i - t_a)$. The term $(t_i - t_a)$ is called the "degree day" (GT) and it represents the sum, reckoned over all the heating days of the period, of the measured daily difference between room temperature t_i and the average daily OAT t_a.

degree day cooling. A unit used in determining the cooling demand based upon the difference, in one day, between the average temperature, with some consideration for relative humidity, and some specified temperature representing ideal comfort conditions. It may be used to estimate the cooling capacity of the air conditioning plant for a building, in conjunction with other heat input factors.

degree day heating. A unit used in determining heating demand based upon the difference, in one day, between the average temperature and 65°F. This unit is used in estimating fuel consumption and specifying the nominal heating load for a building during the heating season.

degree of freedom. The number of variables determining the state of a system (usually pressure, temperature, and concentrations of the components) to which arbitrary values can be assigned.

de-inking. A process in paper recycling that uses detergents and chemicals, such as caustic soda, to remove old inks from the millions of tiny wood fibers in used paper.

deliquesce. The property of certain substances to take up water from the air to form a solution. Such substances are very soluble in water. Their concentrated solutions have aqueous vapor pressures that are low compared to the normal range of partial pressures of water vapor in air. Consequently, such solutions absorb water vapor from the air more rapidly than they give it off. The absorption of water and the resulting dilution of the solution occurs until the aqueous vapor pressure of the solution and the partial pressure of water vapor in the surrounding air are equal.

demand, biochemical oxygen (BOD). The quantity of oxygen utilized in the biochemical oxidation of organic matter in a specified time period and at a specified temperature. It is not related to the oxygen requirements in chemical combustion, being determined entirely by the availability of the material as a biological food and by the amount of oxygen utilized by the microorganisms during the oxidation.

demister. An apparatus made of wire mesh or glass fiber used to eliminate acid, as in the manufacture of H_2SO_4.

density. (1) The concentration of matter, measured by the mass per unit volume. (2) In ecology, the number of individuals per unit area. Crude density is the number (or biomass) per unit of total space, including the whole environment within the described boundaries. In specific or ecological densities, the ecologist may measure the number (or biomass) of organisms of a species per unit of habitat space. It states the number of individuals in the area or volume actually available to be colonized by the population.

deodorizer. An apparatus used for the removal of noxious gases and odors. It may consist of a combustion, an absorption, or an adsorption unit.

depletion curve (hydraulics). A graphical representation of water depletion from storage-stream channels, surface soil, and groundwater. A depletion curve can be drawn for base flow, direct runoff, or total flow.

deposit gage. A general name for an air pollution instrument used to measure the amount of material deposited on a given area during a given time.

deposited matter. The particulate matter from the atmosphere which rapidly settles out of the air.

dermal toxicity. The ability of a pesticide or toxic chemical to poison people or animals by touching the skin.

DES (diethylstilbestrol). A synthetic estrogen used as a growth stimulant in food animals. Residues in meat are thought to be carcinogenic.

desalination. The removal of salt, especially from seawater. Also called desalinization.

desalination (distillation method). All distillation methods for saline water utilize the fact that only water and the gases dissolved in it are volatile, whereas the salts are not. Distillation is the best-developed seawater purification method. The principle of distillation involves boiling seawater in an evaporator by passing hot steam through a steam chest, where the steam condenses on the inside of the tubes of the chest and is returned to the boiler. The vapors rising from the seawater are cooled in the condenser and thus converted into pure liquid water which is collected in a storage vessel. The system is vented through a pump or ejector and thus the amount and pressure of air in it can be regulated. The brine concentrate is continuously or intermittently withdrawn from the evaporator. Instead of passing the steam through tubes surrounded by seawater, it is also possible to design the evaporator so that the seawater is on the outside of a bank of tubes, the outer surface of which is heated by steam. The temperature of the boiling seawater is a very important process variable, for it sets limits on the inside of a bank of tubes, the outer surface of which is heated by steam.

desalination (electrodialysis). Electrodialysis is a unit operation in which the partial separation of the components of an ionic solution is induced by an electric current. This separation is accomplished by placing across the path of current flow one or more sheets of a material in which the transport numbers of the ions differ from the values that prevail in the bulk solution on either side of the sheet. In electrodialysis, the ions forming the salt are pulled out of the saltwater by electrical forces and concentrated in separate compartments. The higher the salinity of the raw water, the more the electrical power needed for the process. Electrodialysis units consist of a number of narrow compartments through which saline water is pumped. These compartments are separated by alternating kinds of special membranes which are permeable to positive ions (cations) or negative ions (anions). The terminal compartments are bounded by electrodes for passing currents through the whole stack. In saline water the salts exist as separate positive and negative ions. Electrodialysis may be classified as a "selective transport" method, in which salt or solvent is transported away from the feed solution through some physical barrier, with no change in state of any component in the system.

desalination (flash distillation). In the principle of flash distillation, saltwater enters a bundle of tubes located in the vapor space of the flash chamber where it is preheated. It then passes into a heater consisting of a bundle of tubes externally heated by steam, where the temperature of the salt is raised to 100°C, but since the pressure is kept higher than 1 atm, no boiling occurs. The hot seawater then enters a flash chamber kept under reduced pressure. Part of the water evaporates, the vapors condensing on the tube carrying the incoming cold seawater to the heater. Distillate and brine are restored to atmospheric pressure by pumps. The tubes across which heat exchanges between condensing steam and incoming seawater occur play an important role in the economy of the process. In a four-stage unit, the flashing process is carried out in four stages, reducing by 10°C the temperature of the evaporating brine and increasing by the same amount that of the incoming saltwater. The brine is discharged at 30°C and the amount of heating steam required is only one-fourth the amount in the single-stage unit, because the heat raises the temperature of the saltwater by only 10°C.

desalination (freezing process). (1) All practical desalination methods fall into one of two categories: those that remove the freshwater and leave behind a concentrated brine or those that remove the salt and leave behind the freshwater as residue. When saltwater is reduced to a certain critical temperature, which is a function of its salinity, ice crystals can be mechanically separated from the mother liquor and remelted to obtain pure water. If seawater is reduced in temperature to a value very close to that of the freezing point, ice crystals will begin to form within the solution. All freezing desalination processes are based upon this phenomenon. (2) Separation of freshwater from salt solutions by freezing is based on the fact that the ice crystals which form when saltwater is cooled are essentially salt free. In this respect, ice formation is analogous to distillation, which leads to salt-free vapor, although the liquid may hold a higher concentration of salts. However, the two processes are different. Distillation is carried out well above the ambient temperatures and hence the equipment must be designed for minimal heat losses. Only liquid and vapor have to be moved and purified. Both direct and indirect freezing processes have been developed. In the former process, water acts as its own refrigerant. In the latter process, a more volatile liquid, e.g., butane, is used as a refrigerant.

desalination (ion exchange). The complete removal of salts from water by ion exchange is a routine water treatment operation for high-pressure boilers which require completely salt-free water. In the principle of ion exchange desalting, raw water is passed through a column containing the active form of a solid cation exchanger, which is an organic resin containing hydrogen ions, H^+, and capable of exchanging them against the positive ions contained in the raw water. If the resin and the water were merely mixed in a tank, instead of the water being passed through the column, the exchange would only be partial. In this column, however, the water first exchanges ions with the top layer of the resin, whose spaces, separated by solid heat-transfer walls, it partially exhausts. In these units, the temperatures of both fluids vary with position, but remain constant in any particular location of the unit.

desalination (multiple-effect distillation). In multiple-effect stills, the vapors from the first evaporator condense in the second evaporator, and their heat of condensation serves to boil the seawater in the latter; the task of these vapors is like that of the heating steam in the first evaporator. Similarly, the third evaporator acts as a condenser for the second, and so on. Each evaporator in such a series is called an "effect."

desalination (reverse osmosis hyperfiltration). In hyperfiltration, desalination is achieved by forcing salt solution under pressure through a membrane which generally passes water more readily than salts. It differs from electrodialysis in that water is removed from salt, rather than salt from water, and the driving force is pressure, rather than electrical potential. By using extremely dense filtering media, an almost complete removal of salt, even from concentrated solutions, can be achieved with special synthetic membranes. The process is called hyperfiltration or reverse osmosis. Membranes which are permeable to water but not to salt are called semipermeable. When a semipermeable membrane is placed between seawater and pure water, which are both at the same pressure, diffusion of freshwater into seawater will occur because of nature's tendency to equalize concentrations. This process, called osmosis, is exactly the opposite of the desired action, namely, the transfer of water from the salt solution into the freshwater reservoir. To effect the process in the desired direction, pressure must be exerted on the saltwater. The pressure to be exerted on the seawater to filter it through the membrane and thus convert it to freshwater must be at least slightly larger than 24.8 atm, which is defined as the osmotic pressure of seawater, hence the name.

desalination (solar evaporation). In a simple explanation of a solar still, saltwater contained in a black pan covered with a sloping glass roof is heated by the sun. Water vapor rises to the glass, where it condenses, forming a film which runs off into a collecting trough and from there, to storage. In this type of still, water does not boil, but vaporizes slowly, and the vapors reach the cooler glass surface by convection. The rate of evaporation is controlled primarily by the intensity of the incoming solar radiation. Although the energy for solar evaporation is free of charge, the installation costs are considerable, since only a few liters per day can be produced per square meter of pan area.

desalination (vapor-compression distillation). Vapor compression distillation literally uses its own steam, after it has been compressed, as a heat source. The principle is that seawater, preheated in a tubular heat exchanger by the outgoing streams of brine and freshwater, boils in the tubes of the still. The vapors are compressed and led back to the still to condense outside the tubes, thus providing the heat necessary for the boiling process. The noncondensing gases are withdrawn from the steam condensation space by a suitable vent pump or ejector. The heat of the installation is the compressor.

desalination (vapor-reheating process). In the vapor-reheating, flash-evaporation process the vapors in the flash chambers are condensed on a spray of film of cold distilled water. The contact between the vapors and the distilled water film is direct and no metallic surfaces separate them, but the saltwater and the distilled water in the flash chambers are not allowed to mix. The temperature of the distilled water rises as a result of the condensation of vapor. The distilled water steam is pumped from each stage to the adjacent stage operating at a higher temperature and pressure. Its volume increases continuously as condensation of the vapor takes place and reaches a maximum in the first flash chamber. This hot distillate is now passed through a countercurrent, liquid heat exchanger in which it first transfers the heat gained in the passage through the flash chambers to an immiscible oil, and then the latter, in turn, transfers the heat to the incoming cold seawater in another exchanger. These heat-transfer processes take place by direct contact and without metallic surfaces. In the heat exchangers the light oil rises through the center of the water, which flows downward. An amount of distilled water equal to the flash steam condensed in all of the stages is withdrawn as a product. The balance is recycled to act again as a condensing medium for the vapor in the flash chamber.

desalinization. The removal of salt from sea or brackish water. Also called desalination.

desert soil. A soil that develops in temperate to cool areas under very dry conditions; the surface soil is light or brownish gray, with a calcium carbonate layer very close to the surface. Desert soils have scanty vegetation.

desiccant. A (hygroscopic) substance capable of absorbing moisture and therefore used as a drying agent. Salt and calcium chloride are two examples. Dessicant-type spacers may be used to keep condensation from forming in a solar collector.

desorption. The opposite of adsorption. A phenomenon where an adsorbed molecule leaves the surface of the adsorbent.

destructive distillation. A process of distillation in the absence of air which continues until all liquids or fractions have been vaporized and only a carbonaceous residue remains in the still, such as in the cracking process when cracking is continued until only coke remains.

desulfurization. The removal of sulfur. (1) In coal processing, a process involving elutriation, froth flotation, laundering, and magnetic separation. (2) In petroleum refining, removing sulfur or sulfur compounds from a charge stock (oil that is to be treated in a particular unit). (3) Removing sulfur from iron, metal, or an ore.

detector. A sensitive device that produces a signal when it is struck by a photon or particle from a radioactive atom that disintegrates.

detergent. A cleaning agent that removes foreign matter by dispersion or a washing-away action, rather than by a dissolving action. Detergents may be either oil-soluble or water-soluble. The unusual cleaning powers of detergents are not the result of a single molecule, but of many components, each tailored for a specific cleansing function, yet working together so that the result is greater than the potential of any single part. Typically, a detergent contains a surfactant, a builder, a silicate, and carboxymethylcellulose. The surfactant, or sudsing agent, is an organic agent that is able to penetrate between soil particles and cloth or fabric. Once in the wash, dirt must be kept in suspension; this is one function of the builder; another of its functions is to soften water, eliminating the scummy precipitate that soap produces in hard water. The remaining detergent ingredients, though present in small amounts, complement and complete the cleaning action of the surfactant and builder. Sodium silicate is added as a corrosion inhibitor and carboxymethylcellulose is an antiredeposition agent that helps keep removed dirt in suspension until the wash is flushed away. The builder, which comprises up to half of package detergent, contains a large amount of phosphorus in the form of tripolyphosphate. Since this, as well as the other detergent ingredients, must be water-soluble to perform its cleaning function, it is also available to microorganisms when it enters the environment in sewage effluent.

detinning. A process in can recycling that separates tin and steel. The recycled cans move through a shredding machine that cuts the cans into tiny strips, like a kitchen grater cuts up cabbage. The shredded cans are then added to a chemical solution and prepared for electrolysis. In electrolysis, an electric current passes through the shredded steel and chemical solution, forcing the tin to separate.

deuterium. A rare form (isotope) of hydrogen whose atom contains one proton and one neutron; also called heavy hydrogen, or ^2H (the atomic nucleus contains two particles, in contrast to that of ordinary hydrogen which contains one). Deuterium slows down neutrons passing through it and is used as a moderator to adjust the speed of neutrons in atomic furnaces.

deuteron. The nucleus of the deuterium atom or the ion of deuterium. Its structure consists of one neutron and one proton.

dew. The water vapor that condenses on solid surfaces cooled below the condensation point of the air in contact with them. Dew and frost are the results of the earth's cooling by radiation. Dew is formed as warm air comes into contact with cool surfaces.

dew point. The temperature to which the air must be cooled, at constant pressure and with constant water vapor content, for saturation to occur. The temperature of the dew point changes in the same way as does the vapor pressure. It is conservative for nonadiabatic processes involving no pressure change, provided that there is no evaporation or condensation.

diabatic process. A process in a thermodynamic system in which there is a transfer of heat across the boundaries of the system.

diamagnetic material. A material within which an externally applied magnetic field is slightly reduced due to its alteration of the material's atomic electron orbits. Diamagnetism is an atomic-scale consequence of the Lenz law of induction. The permeability of diamagnetic materials is slightly less than that of empty space.

diaminodiphenylmethane (DDM). $C_{12}H_{13}N_3$, an important chemical intermediate in the manufacture of polyurethane foams used in automobile safety cushioning and in thermal insulation. DDM reacts with phosgene to form isocyanates and polyisocyanates as a step toward polyurethane production. More than 200 million pounds of diaminodiphenylmethane, also known as *p,p'*-methylenedianiline, are manufactured in the United States each year. More than 2500 workers are exposed to DDM through occupational contact on construction sites and in the manufacture of isocyanates and polyisocyanates. A possible safety hazard in the handling of DDM was first observed in Great Britain in 1965 when an outbreak of hepatotoxic effects occurred, involving 84 persons who had eaten bread accidentally contaminated with the chemical. The disorder was given the name of "Epping jaundice" at the time. Later experiments showed that rats fed DDM developed liver lesions and liver degeneration (in separate studies), as well as spleen lesions. In 1972, six men working with an epoxy resin manufactured from DDM developed liver disease while employed in the construction of a nuclear power plant in Alabama. A NIOSH survey found that 13 cases of hepatitis developed between 1966 and 1972 among a group of workers using DDM to produce a hard-plastic insulating material. There have been no reported cases of human cancers associated with DDM, but miscellaneous types of tumors developed in rats fed DDM in amounts ranging from 5 to 20 mg; feeding was by intubation. The tumors were observed in the liver, kidney, and uterus. Some of the tumors were benign. The Center for Disease Control has projected, on the basis of past experience, that between 25 and 50 cases of toxic hepatitis per year can be expected among workers using DDM in their occupations.

diamond. Pure carbon crystallized in the isometric system with a hardness of 10 on the Mohs' scale. It is an important industrial mineral, much of it off-color, gray, or black, used in drills, abrasives, and dies.

diatomaceous earth (diatomite). A chalk-like material used to filter out solid wastes in wastewater treatment plants; also found in powdered pesticides.

diatomic gas. A gas whose molecule is made of two atoms, for instance, hydrogen (H_2), oxygen, nitrogen, chlorine, and many others. Gases whose molecules consist of a single atom, like some vapors and the noble gases helium, neon, argon, krypton, and xenon, are called monatomic. The terms "triatomic," etc., are seldom used.

diesel engine. A type of internal combustion engine that burns heavy oil and needs no spark to ignite the fuel. The diesel engine differs from the spark-ignited gasoline engine in that (a) the air supply is unthrottled; that is, its flow into the engine is unrestricted. Thus a diesel normally operates at a higher air-to-fuel ratio than does the gasoline engine. (b) The fuel is injected directly into the combustion chamber, and a carburetor is not required. The power output is changed by the rate of fuel injection. (c) There is no spark ignition system. The air is heated by compression: The air in the engine cylinder is squeezed until it exerts a pressure high enough to raise the air's temperature to about 1000°F, which is enough to ignite the fuel as it is injected into the cylinder. Because of its construction, the diesel's exhaust pipe accounts for about 100 percent of its emissions. Since diesel engines operate with an excess of air, theoretically little unburned fuel is normally exhausted. However, improper operation and maintenance, as well as an overloaded engine, actually make diesels worse polluters than gasoline engines. In addition, the highly heated excess air results in the formation of more nitrogen oxides than the typical automobile engine, even in a well-maintained diesel engine.

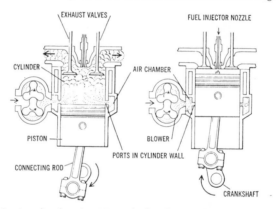

Combustion in a diesel engine. (One cylinder of a typical diesel engine shown.)

diethylcarbamoyl chloride (DECC). A compound used in the synthesis of diethylcarb-amazine citrate, an anthelmintic (worming) agent. Annual production of DECC has averaged less than 15,000 pounds. In 1976, The New York University Institute of Environmental Medicine advised NIOSH that, on the basis of laboratory studies, diethylcarbamoyl chloride should be regarded as a potential mutagen. The studies found that DECC caused chromosome changes in two strains of *Escherichia coli* bacteria commonly used to screen substances that may produce adverse health effects in humans and other higher forms of life. However, the DECC was found to be less of a mutagenic hazard than its close analog, dimethylcarbamoyl chloride (DMCC). In addition to being a potential mutagen, DMCC has been found to cause cancer in laboratory animals.

differential thermostat. An automatic starting and shut-off device that is triggered by the difference between the temperatures recorded at two separate points. In a typical solar heating system, liquid or air type, the compared temperatures are measured at the outlet point of the solar collector and at the outlet of the thermal storage system.

diffraction. The phenomenon produced by the spreading of waves around and past obstacles which are comparable in size to their wavelength; the splitting of a beam into two or more beams on passing through or being reflected by a grating or a crystal. Its occurrence demonstrates the wave property of the beam. It can be used to measure the wavelength of the beam.

diffraction analysis. The application of diffraction techniques (of x rays, electrons, neutrons) to study the structure of matter, gases and solids. This may involve the diffraction of electromagnetic radiation or particle beams.

diffraction grating. A plate of glass or polished metal ruled with a series of very close, equidistant, parallel lines, used to produce a spectrum by the diffraction of reflected or transmitted light. If s is the distance between the rulings and d is the angle of diffraction, then the wavelength where the angle of incidence is 90° is (for the nth-order spectrum) $\lambda = (s \sin d)/n$. If i is the angle of incidence, d is the angle of diffraction, s is the distance between the rulings, and n is the order of the spectrum, the wavelength is $\lambda = (s/n)(\sin i + \sin d)$.

diffuse radiation. The scattered radiation (by clouds, airborne moisture, etc.) that reaches the earth's atmosphere.

diffuse sound field. A field in which the time average of the mean-square sound pressure is the same everywhere and the flow of energy in all directions is equally probable.

diffused air. A type of aeration that forces oxygen into sewage by pumping air through perforated pipes inside a holding tank.

diffuser. A porous plate or tube through which air is forced and divided into minute bubbles for diffusion in liquids. Commonly made of Carborundum, Alundum, or silica sand.

diffusion. The spreading or scattering of a gaseous or liquid material. (1) *Eddy diffusion.* The most important mixing process in the atmosphere, it causes a parcel of polluted air to occupy larger and larger volumes. (2) *Molecular diffusion.* A process of spontaneous intermixing of different substances, attributable to molecular motion, tending to produce uniformity of concentration.

If the concentration (mass of solid per unit volume of solution) at one surface of a layer of liquid is d_1, and d_2 at the other surface, and the thickness of the layer is h and the area under consideration is A, then the mass of the substance which diffuses through the cross section A in time t is $m + \Delta A [(d_2 - d_1)t/h]$, where Δ is the coefficient of diffusion.

diffusion, ridge and furrow air. A method of air diffusion in an aeration tank of an activated sludge process, where porous-tile diffusors are placed in rows in depressions at right angles to the direction of flow created by the sawtooth construction of the tank bottom.

diffusion, spiral flow air. A method of air diffusion in an aeration tank of an activated sludge process, where a spiral helical movement is given to both the air and the liquor in the tank by means of properly designed baffles and properly located diffusers.

diffusivity. The coefficient of diffusion given by Δ in the equation

$$\frac{dQ}{dt} = -\Delta \left(\frac{dc}{dx} \right) dy \, dz,$$

where dQ is the amount passing through an area $dy \, dz$ in the direction of x in a time dt, dc/dx being the rate of increase of the volume concentration in the direction of x.

digester. In wastewater treatment a closed tank, sometimes heated to 95°F, where sludge is subjected to intensified bacterial action.

digestion. The biochemical decomposition of organic matter which results in the formation of minerals and simpler organic compounds.

dilution ratio. The relationship between the volume of water in a stream and the volume of incoming waste. It can affect the ability of the stream to assimilate waste.

dimensional formula. An expression involving only the dimensions of a physical quantity. For example, if mass, length, and time are considered fundamental quantities, the relation of other physical quantities and their units to these three may be expressed by a formula involving the symbols l, m, and t, respectively, with appropriate exponents. The dimensional formula for volume would be expressed as $[l^3]$; velocity, $[lt^{-1}]$; force, $[mlt^{-2}]$. Other fundamental quantities used in dimensional formulas may be indicated as follows: the dielectric constant of a vacuum, ϵ; temperature, θ; and the magnetic permeability of a vacuum, μ.

dimethylcarbamoyl chloride (DMCC). A compound prepared by the reaction of phosgene with trimethylamine. It is used mainly in the synthesis of pharmaceuticals for the treatment of myasthenia gravis and as a reagent for the synthesis of carbamates in chemical research laboratories. It may also be employed in the synthesis of pesticides, dyes, and a rocket fuel, unsymmetrical dimethylhydrazine. NIOSH advised in 1976 that DMCC may also be formed in side reactions during the manufacture of other products, as in the example of DMCC formation during production of phthaloyl chlorides. Production is limited to about 3000 pounds per year in the United States. It is also manufactured in Germany. Only about 200 persons in the United States are at risk of occupational exposure to DMCC, according to NIOSH. Toxic effects observed in laboratory animals include irritation to the eye membranes, respiratory organs, and, after repeated contact, inflammation of the skin. A carcinogenic potential was reported by the New York University Institute of Environmental Medicine in 1972 and again in 1974 after skin tumors and subcutaneous sarcomas, along with papillary lung tumors, were observed in Swiss mice following subcutaneous and intraperitoneal injection of DMCC. In 1976, NIOSH reported that a study of rats exposed to inhalation levels of 1 ppm of DMCC developed squamous cell carcinomas of the nose within seven months. Because of the very high incidence of tumors found in 89 of 93 rats exposed, NIOSH issued a warning that DMCC posed a potentially serious hazard for workers exposed to the chemical.

Dimilin. An insecticide designed to act in a specific toxic mode by inhibiting the formation of the chitin, the hard covering shell or exoskeleton, of an insect. The chemical is used primarily in the protection of cotton and soybean crops. Because it is a chitin inhibitor, the EPA has conducted laboratory experiments that indicate Dimilin to be a potential threat to marine crustaceans, including crawfish, crabs, lobster, and shrimp. EPA studies conducted with a small mysid shrimp indicated that at concentrations of one microgram per liter of water an average of 2.4 offspring per female survived, compared to an average of 21.4 in a control group, and the percentage of survival of adults was 62 for a concentration of one microgram per liter, as compared to a 92.5 percent survival rate among controls. In addition to demonstrating a toxic level at 2.0 parts of Dimilin per billion parts of water, the EPA study found that Dimilin is apparently accumulated in the eggs of shellfish so that a higher incidence of mortality and reduced fertility was observed in a second generation of the shrimp, which had not been exposed to the insecticide, although the third generation was not severely affected.

diminution of pressure p at the side of a moving stream. For a fluid of density d moving with a velocity v, and neglecting viscosity, $p = \frac{1}{2}dv^2$.

dioxide. A molecule containing two atoms of oxygen to one atom of another element.

dioxin. $C_{12}H_4Cl_4O_2$, a group of compounds formed from other halogenated hydrocarbons in the presence of oxygen and at high temperatures. A common source of the chemical is the burning or heating of polychlorinated biphenyls (PCB's) through trash incineration or during the welding of transformers or capacitors. A common form of dioxin, tetrachlorodibenzodioxin (TCDD), is released in the process of converting trichlorophenol into hexachlorophene, the disinfectant. When the process is not kept under control, the reaction has a tendency to form increasing percentages of dioxin as the temperature of the mixture increases. The ultimate hazard is an explosion that releases large quantities of dioxin into the environment, as occurred on July 10, 1976, in a chemical factory near Seveso, Italy. Evacuation of the town was ordered after domestic and wild animals in the area began to die and children developed a distinctive skin rash. Other residents developed signs of damage to the eyes, nerves, blood, and liver. About 100 women underwent abortions because of a fear of teratogenic effects. The dermal exposure effects were of an acne form that did not occur until two years after the accident. Animal carcasses, vegetation, and even the topsoil from the contaminated area were removed in plastic bags and stored, awaiting a decision that was to come several years later concerning the safest method of disposal. More than 200 residents were unable to return to their homes because of the danger of contamination and the Italian government filed a $135 million damage suit against the chemical company (Givaudan, a subsidiary of Hoffmann-La Roche) to reimburse the government for its clean-up and health monitoring expenses. Dioxin is a powerful suppressor of the immune system of animals and has caused atrophy of the thymus gland in every species tested. Very small doses increase the susceptibility of animals to bacterial infections. It is also one of the most powerful poisons ever discovered, as the LD_{50} dose for male rats is only 22×10^{-6} g, making dioxin approximately one million times as toxic as PCB's.

dipole. (1) A combination of two electrically or magnetically charged particles of opposite sign which are separated by a very small distance. (2) Any system of charges, such as a circular current, which has the properties such that (a) no forces act on it in a uniform field; (b) a torque proportional to $\sin\theta$, where θ is the angle between the dipole axis and a uniform field, acts on it; (c) it produces a potential proportional to the inverse square of the distance from the dipole.

dipole moment. A mathematical entity; the product of one of the charges of a dipole unit by the distance separating the two dipolar charges. In terms of the definition of a dipole, the dipole moment **p** is related to the torque **T** and the field strength **E** (or **B**) through the equation $\mathbf{T} = \mathbf{p} \times \mathbf{E}$. Certain molecules have permanent dipole moments, found from measurements of the dielectric constant (i.e., by a molecule's temperature dependence, as in the Debye equation for total polarization). These moments are associated with the transfer of charge within the molecule and provide valuable information as to the molecular structure.

direct radiation. The radiation that strikes the surface of the earth without being scattered by the atmosphere.

direct solar radiation. That portion of the radiant energy received by an actinometer directly from the sun, as distinguished from diffuse sky radiation, effective terrestrial radiation, or radiation from any other source.

discharge measurement. Several methods are used to measure the quality and quantity of industrial discharges into neighboring streams. One approach is to collect random "grab samples" of the discharge into the water. The samples may represent the content of effluent at a particular time and location. Several such samples may be combined as a composite representation indicating an average of the content of the discharge. Composite samples are often preferred for studying the effluent of a plant from which there are variations in the discharge due to batch operations or a rapid and varied flow of discharge. Grab samples may be considered satisfactory when the discharge is slow and comparatively uniform in content. The velocity and volume of the discharge are measured by simple floats that travel along the surface at an easily observed velocity. A more sophisticated method involves installation of devices that meter the velocity of the discharge by propellers or rotating cups. The deployment of several automatic discharge meters at various positions in the discharge flow can provide continuous data.

disinfection. (1) The killing of the larger portion of the harmful and objectionable microorganisms in or on a medium by means of chemicals, heat, ultraviolet light, etc. (2) The use of a chemical additive or other treatment to reduce the number of bacteria, particularly the pathogenic organisms.

disintegration. The process in which a radioactive nucleus gives off radiation and changes into another form.

dispersant. A chemical agent used to break up concentrations of organic material such as spilled oil.

dispersion. (1) The internal distribution pattern of individuals in a population. (2) The dilution of a pollutant by diffusion or turbulent action, etc. Technically a two-phase system of two substances, one of which, the dispersed phase, is uniformly distributed in a finely divided state through the second substance, the dispersion medium. The dispersion medium may be a gas, liquid, or solid, and the dispersed phase may be any one of these. (3) The resolution of a complex electromagnetic radiation into components in accordance with some characteristic, for example, wavelength. The difference between the index of refraction of any substance for any two wavelengths is a measure of the dispersion for these wavelengths, called the coefficient of dispersion.

dispersion force. The force of attraction between molecules possessing no permanent dipole. The interaction energy is given by $U_0 = -\frac{3}{4}h\,[V_0 a^2/r^6]$, where h is Planck's constant, V is a characteristic frequency of the molecule, r is the distance between the molecules, and a is the polarizability.

dispersion rate. A diffusion parameter of gas plumes or stack effluents.

dispersive power ω. For n_1 and n_2 the indices of refraction for two wavelengths and n the mean index for the wavelength of sodium light, $\omega = (n_2 - n_1)/(n - 1)$.

displacement. (1) A reaction in which an elementary substance displaces and sets free a constituent element from a compound. (2) A vector quantity that specifies the change in position of a body or particle and is usually measured from the mean position or position of rest. In general, it can be represented by a rotation vector or translation vector, or both.

dissolved oxygen (DO). The oxygen dissolved in sewage water, or other liquid, usually expressed in milligrams per liter or percent of saturation. It is the test used in BOD determination.

dissolved solids. The total of disintegrated organic and inorganic material contained in water. Excesses can make water unfit to drink or use in industrial processes.

distillation. A process in which a portion of a liquid is first evaporated and subsequently condensed. The end product of distillation is in the same phase as it was initially, i.e., in the liquid phase. The process is of use only when the original liquid is a mixture of substances, one of which may be evaporated, at given conditions of temperature and pressure, in greater proportions than the others. Hence, by distillation, an end product is achieved in which the more volatile substances are preponderate, as compared with the original liquid.

distributed collector. A modular solar-electric concept; solar energy collected and processed at several small fields for transmission to a central facility.

distribution law. A substance distributes itself between two immiscible solvents so that the ratio of its concentrations in the two solvents is approximately a constant (and equal to the ratio of the solubilities of the substance in each solvent). The above statement requires modification if more than one molecular species is formed.

diurnal. Recurring daily. Diurnal variations, for example, in the concentration of air contaminants, implies a distinctive pattern which recurs from day to day.

doldrums. An equatorial area of low pressure (trough) characterized by light surface winds.

dominant species. A species of plant or animal that is particularly abundant or controls a major portion of the energy flow in a community.

Doppler effect. An effect on the apparent frequency of a wave train produced by motion of the source toward or away from a stationary observer and by motion of the observer toward or from the stationary source, the motion in each case being with reference to the (supposedly stationary) medium. For sound waves, the observed frequency f_0 in Hz is given by

$$f_0 = \frac{v + w - v_0}{v + w - v_s} f_s$$

where v is the velocity of sound in the medium, v_0 is the velocity of the observer, v_s is the velocity of the source, w is the velocity of the wind in the direction of sound propagation, and f_s is the frequency of the source.

dosage. The concentration of pollutants in an exposure chamber.

dose. In radiology, the quantity of energy or radiation absorbed.

dosimeter. An instrument for measuring the ultraviolet radiation in solar and sky radiation.

dosing ratio. The maximum rate of application of sewage to a filter of any unit of area, divided by the average rate of application on that area.

down point. The temperature at which useful heat can no longer be taken from storage. In an air-type solar heating system, useful heat can be taken from a rock storage bin or eutectic bin at temperatures as low as 75°F. In liquid-type solar heating systems, except those including a heat pump, the down point is at least 90°F.

downdraft. A current of air with a marked downward or reversed motion in a chimney or flue.

downwash. A phenomenon occurring when chimney gases are drawn downward by a system of vortices or eddies which form in the lee of a chimney when wind is blowing. It affects the appearance of a plume and causes blackening of stacks. It may also bring flue gases prematurely to the ground. Downwash may be prevented by discharging gases with a velocity 1½ times that of the wind passing over the top of the chimney.

draft. A gas flow resulting from the pressure difference between an incinerator, or any component part, and the atmosphere, which moves the products of combustion from the incinerator to the atmosphere. (1) *Natural draft.* The negative pressure created by the difference in density between hot flue gases and the atmosphere. (2) *Induced draft.* The negative pressure created by the action of a fan, blower, or ejector which is located between the incinerator and the stack. (3) *Forced draft.* The positive pressure created by the action of a fan or blower, which supplies the primary or secondary air.

drag coefficient. A dimensionless ratio of the component of force parallel to the direction of flow (drag) exerted on a body by a fluid to the kinetic energy of the fluid multiplied by a characteristic surface area of the body.

dredging. To remove earth from the bottom of water bodies using a scooping machine. This disturbs the ecosystem and causes silting that can kill aquatic life.

drizzle. A fairly uniform precipitation composed exclusively of fine drops of water (diameter less than 0.5 mm) that are very close to one another. Freezing drizzle is recorded when the drops freeze on impact with the ground or with objects on the earth's surface.

droplet. A small spherical liquid particle that may fall free under still conditions, but which may remain suspended in the atmosphere under turbulent conditions; a liquid cloud particle. The size of a droplet is usually less than 0.2 mm in diameter. A drop of liquid precipitation is usually larger than 0.2 mm.

dry impingement. The process that pushes particulate matter carried by a gas stream against a retaining surface. The retaining or collecting surface may be coated with an adhesive film.

dry limestone process. An air pollution control method that uses limestone to absorb the sulfur oxides in furnaces and stack gases.

dry tower. In a dry tower, the heated effluent is contained in a system of pipes much like the radiator of an automobile. Air is passed over the pipes by a large fan, facilitating heat exchange by radiation and convection. Water loss and, to a lesser extent, fog are controlled by this method, but installation costs are much higher, and so are maintenance costs.

drying agent. A substance that removes water—such as heat, certain chemicals, or certain organic liquids (alcohols); these include metallic sodium, calcium oxide, metallic calcium, barium oxide, and aluminum amalgam.

dual laser. A gas laser having Brewster windows and concave mirrors at opposite ends, the mirrors having different reflectivities so as to produce two different visible or infrared wavelengths from a helium–neon laser beam.

duct. (1) A tube or passage that confines and conducts a fluid, as a passage for the flow of air to the compressor of a gas turbine engine or a pipe leading air to a supercharger. (2) A plastic or metal pipe, carefully insulated, used for carrying air in an air-type solar heating system.

dump. An open land site where waste materials are burned, left to decompose, rust, or simply remain. The burning of waste materials generates smoke fumes and ash particles. Dumps can cause water pollution and attract flies, rats, and mosquitoes.

dust. Solid particles projected into the air by natural forces, such as wind, a volcanic eruption, or an earthquake, and by mechanical or man-made processes such as crushing, grinding, milling, drilling, demolition, shoveling, conveying, screening, bagging, and sweeping. Generally, dust particles are about 1–100 μm in size. When smaller than 1 μm, particles are classed as fumes or smoke. Dusts do not tend to flocculate, except under electrostatic forces; they do not diffuse, but settle under the influence of gravity.

dust collector. A device for monitoring dust emissions; equipment to remove and collect dust from exhaust gases. These may employ the following approaches: sedimentation (dustfall jars, coated slides, papers, settled dust samples), inertial separation (cyclones, impactors, impingers, sticky tapes), precipitation (thermal and electrostatic), and filtration.

dust counter. An instrument for counting the dust particles in a known volume of air. In Aitken's dust counter, condensation is made to occur on the nuclei present by adiabatic expansion of the air, and the number of drops is counted. In Owen's dust counter, a jet of damp air is forced through a narrow slit in front of a microscope coverglass. The fall of pressure due to the expansion of the air passing through the slit causes the formation of a film of moisture on the glass, to which dust adheres, forming a record which can be studied under a microscope.

dust deposit. The quantity of solid matter deposited on the ground from the external air in a given time and over a given area.

dust flow. The rate at which dust is carried in a gaseous medium.

dust horizon. The top of a dust layer which is confined by a low-level temperature inversion in such a way as to give the appearance of the horizon when viewed from above against the sky. The true horizon is usually obscured by the dust in such instances.

dust loading. The amount of dust in a gas, usually expressed in grains per cubic foot or pounds per thousand pounds of gas; the engineering term for dust concentration.

dustfall. Particulate matter in the air which falls to the ground under the influence of gravity.

dustfall jar. An open container used to collect large particles from the air for measurement and analysis.

dynamic head. The pressure differential between two points in a fluid system based on the vertical distance between these points, measured while the system is dynamic and the fluid is flowing.

dyne. The force necessary to give an acceleration of one centimeter per second to one gram of mass.

dystrophic lake. A shallow body of water that contains much humus and organic matter. It contains many plants, but few fish, and is almost eutrophic.

E

ear. The organ of hearing. The human ear responds to sound waves in frequency ranges of about 16–16,000 Hz, the upper extreme of which approaches 20,000 Hz, in the very young. The ear converts the energy of sound waves in the air into neural impulses, which are interpreted by the brain. The process involves the conversion of (a) wave motions in the air into mechanical vibrations, (b) mechanical vibrations into wave motions in fluid, and (c) wave motions in fluid into neural impulses. The external ear consists of the external auditory canal, which penetrates the temporal bone and ends at the tympanic membrane. The inner two-thirds of the canal forms a tunnel through the temporal bone (osseous portion), while the outer third is composed of cartilage. The entire canal is lined with skin. The external structure, the pinna, commonly called the ear, does not function to any degree in the process of hearing. A vibrating source generates sound waves in the air, which pass through the external canal and impinge on the tympanic membrane. Pressure from the sound waves causes the membrane to vibrate, converting the wave motions into mechanical vibrations, the first step in the process of hearing.

earthquake. A group of elastic waves in the solid earth, generated when internal earth stress exceeds crustal strength, causing rocks to break (fracture) or slip along fault planes or zones.

ecologic niche. The status of an organism within a community or ecosystem, depending on the organism's structural adaptations, physiologic responses, and behavior.

ecological efficiency. The percentage of available energy utilized by a trophic level from the next lowest level.

ecological impact. The total effect of an environmental change, natural or man-made, on the community of living things.

ecology. A word derived from the Greek root *oikos* meaning "house," ecology is literally the study of "homes" or "environment"; the study and function of nature and the interrelationships between organisms and their environment. Ecology is one of several basic divisions of biology concerned with the principles fundamental to all life. Physiology, genetics, embryology, and evolution are examples of some basic divisions that deal with the structure, physiology, ecology, etc., of specific kinds of organisms. In ecology, the term "population" was originally developed to denote a group of people, then broadened to include groups of individuals of any kind. A community in the ecological sense includes the entire population of a given area. The community and the nonliving environment function together as an ecological system, or ecosystem.

economic poison. A chemical used to control pests and to defoliate cash crops such as cotton.

ecosphere. *See* **biosphere.**

ecosystem. The biotic community and physical environment in an area. The community and nonliving environment function together as an ecological system, or ecosystem, each influencing the properties of the other and both necessary for the maintenance of life on earth. From a structural standpoint it is convenient to recognize four constituents as comprising the ecosystem: (a) abiotic substances, basic elements and compounds of the environment; (b) producers, the autotrophic organisms, largely the green plants; (c) the large consumers, or macroconsumers, heterotrophic organisms, chiefly animals that ingest other organisms or particular organic matter; (d) the decomposers, or microconsumers, heterotrophic organisms, chiefly the bacteria and fungi that break down the complex compounds of dead protoplasm, absorb some of the decomposition products, and release simple substances usable by the producers. The amount of abiotic materials such as phosphorus, nitrogen, etc., that is present at any time can be considered as the standing state or standing quantity. The amount and distribution of both inorganic chemicals and organic materials present either in the biomass or the environment are important factors in any ecosystem. The ultimate source of energy powering the ecosystem is sunlight. But this energy is dissipated as it flows through the system, from the photosynthesizing primary producers, algae or grass, through a series of consumers, hunters, and harvesters. Without continued recharging from sunlight, ecosystems would soon run down and their component organisms die.

ecotone. In nature, where there is never a discernible sharp line or point indicating the beginning of one community and the end of another, a zone of transition or tension, in which the conditions for each of the adjacent communities becomes more adverse and there is often an intermingling of species from both communities; also called a tension zone. A general characteristic of the ecotone is that there are very often a greater number of species. In addition, the densities of many of these "edge species" is higher than it is for the neighboring communities.

ectoparasite. A parasite that lives on the body surfaces of the host and passes modifications for secure attachment to the host species. Examples include ticks, lice, fleas, and bedbugs.

eddy current. A current induced in a mass of conducting material by a varying magnetic field; also called a Foucault current.

edge retaining system. A channel holding in place the various layers of a solar collector.

eductor. (1) A device for mixing air with water; a liquid pump operating under a jet principle, using liquid under pressure as the operating medium.(2) A device with no moving parts used to force an activated carbon water slurry through pipes to the desired location.

efficiency, thermal. In a boiler, the ratio of the heat of generated steam to the heat supplied to a boiler in the form of fuel. The ratio of work done to the heat energy received by an engine is the thermal efficiency of the engine.

effluent. A general term denoting any fluid emitted by a source. (1) The various spent liquors that are allowed to flow away as waste from plating shops, pickling tanks, etc. Effluent treatment usually involves (a) neutralization of acid liquids, usually with lime; (b) precipitation of the salts of heavy metals as hydroxides by treatment with lime; (c) treatment of cyanides either by removal as hydrocyanic acid or by conversion into cyanates or prussian blue; (d) reduction of chromates with ferrous sulfate and precipitation with lime. (2) Sewage, water, or other liquids, partially or completely treated, or in its natural state, flowing out of a reservoir, basin, or treatment plant.

Einstein's theory for mass – energy equivalence. The equivalence of a quantity of mass m and a quantity of energy E by the formula $E = mc^2$. The conversion factor c^2 is the square of the velocity of light.

elastic limit. The maximum unit stress which can be obtained in a structural material without causing a permanent deformation.

elasticity. The property where a body, when deformed, automatically recovers its normal shape when the deforming forces are removed.

electric charge. The force with which a small electrified body attracts or repels another one carrying a standard charge, at a standard distance. In the cgs system, a unit charge repels an equal one at a distance of 1 cm with a force of 1 dyn (= 1 cm g/s²); in the mks system, an additional constant enters, and the unit (1 C) is 3×10^9 times larger. The passage of 1 C/s through a conductor represents 1 A. The terms "positive" and "negative" charge are purely conventional: Nuclei have a positive charge, electrons, a negative charge, always whole multiples of the "fundamental charge" $e = 1.60 \times 10^{-19}$C.

electric field intensity. The intensity of an electric field measured by the force exerted on a unit charge. A field of unit intensity is a field which exerts the force of one dyne on a unit positive charge.

electrochemical equivalent of an ion. The mass liberated by the passage of a unit quantity of electricity.

electrodialysis. A process that uses electrical current applied to permeable membranes to remove minerals from water; often used to desalinize salt or brackish water.

electrolysis. The decomposition of an electrolyte by the action of an electric current passing through it. If a current i flows for a time t and deposits a metal whose electrochemical equivalent is e, the mass deposited is $m = eit$. The value of e is usually given for m in grams, i in amperes, and t in seconds.

electromagnetic spectrum. A whole family of electromagnetic radiations that have been arranged in order of frequency and wavelength. Beginning with the lowest frequency, these radiations go from radio waves to infrared, visible light, ultraviolet, x rays, and, finally, to gamma rays. Although each type of radiation falls into a group which is assigned a specific place on the spectrum, they all have essentially the same nature and adjacent frequencies may overlap.

electromagnetic wave. A wave propagated through space or matter by the oscillating electric and magnetic field generated by an oscillating electric charge. An electric charge at rest is surrounded by an electric field. When the charge is in motion, a magnetic field is also produced. If the moving electron increases its speed or changes its direction, it accelerates and its energy is increased. The accelerating electron oscillates, or vibrates, back and forth like a plucked violin string, owing to its increasing energy. These oscillations disturb the magnetic field which has been generated by and follows the electron's motion. The disturbed magnetic field changes the electric field which accompanies it. These two interrelated oscillating fields radiate outward from the accelerating electron, traveling through space as an oscillating electromagnetic wave.

electromotive force. That force which causes a flow of current. The electromotive force of a cell is measured by the maximum difference of potential between its plates. The electromagnetic unit of potential difference is that against which one erg of work is done in the transfer of an electromagnetic unit quantity. The volt is that potential difference against which one joule of work is done in the transfer of one coulomb. One volt is equivalent to 10^8 electromagnetic units of potential. The international volt is the electrical potential which, when steadily applied to a conductor whose resistance is one international ohm, will cause a current of one international ampere to flow. The international volt = 1.000 33 absolute volts. The electromotive force of a Weston standard cell is 1.0183 international volts at $20°C$.

electron. A small particle having a unit negative electrical charge, a small mass, and a small diameter. Its charge is $(4.802\ 94 \pm 0.0008) \times 10^{-10}$ absolute electrostatic units, its mass is 1/1837 that of the hydrogen nucleus, and its diameter is about 10^{-12} cm. Every atom consists of one nucleus and one or more electrons. Cathode rays and beta rays are electrons. The electron's theoretical rest mass is equal to 9.1091×10^{-28} g and its rest energy is equal to $0.511\ 006 \times 10^6$ eV. The term "electron" is usually reserved for the orbital or extranuclear particle, whereas the term "beta particle" refers to a nuclear electron.

electron affinity. The measure of the energy released when an electron is added to a neutral atom. Whether an atom will lose electrons or take electrons from another atom depends on electron affinity. An atom with a high electron affinity holds its electrons tightly and has a high ionization energy. Atoms with low electron affinities have low ionization energies.

electron beam. Specifically, a focused stream of electrons used for neutralization of the positively charged ion beam in an ion engine. It is also used to melt or weld materials with externally high melting points.

electron charge. The smallest possible negative electric charge ($4.802\ 98 \times 10^{-10}$ electrostatic units).

electron shell. A group of electrons in an atom, all having approximately the same average distance from the nucleus and approximately the same energy.

electron temperature. The temperature T in degrees Kelvin given by $2E = kT$, where E is the average kinetic energy per degree of freedom of the constituent molecules of a gas and k is Boltzmann's constant.

electron theory. The theory stating that all matter consists of atoms which, in turn, comprise a positive nucleus and a number of negative electrons, which may be detached from the atom under certain conditions, leaving it positively charged.

electron valence. The number of electrons an atom must gain or lose to acquire stability.

electron volt (eV). The energy acquired by any charged particle carrying a unit electronic charge when it falls through a potential difference of one volt; 1 eV = $(1.602\ 07 \pm 0.0007) \times 10^{-12}$ erg. Multiples of this unit are also in common use: the kilo-, million, and billion electron volt, respectively 1 keV = 10^3 eV, 1 MeV = 10^6 eV, and 1 BeV = 10^9 eV. An electron volt is associated through the Planck constant with a photon of wavelength 1.2395 μm.

electronegative gas. When an electron collides with a neutral gas molecule, it may attach itself and form a negative ion. The likelihood of doing so depends on the so-called electronegative nature of the gas and on the energy of the electron, slower-moving electrons remaining for a longer time within the range of the atomic field of the gas molecule. In addition, the probability of electron attachment is markedly affected by the presence of certain gases and vapors as impurities. The inert gases and hydrogen, if very pure, do not form negative ions at all by electron attachment. Certain gases which have no electron affinity by themselves are still capable of forming negative ions by an indirect process. The molecule is first dissociated by impact with an energetic electron and then attachment occurs to one of the fragments and subsequent dissociation occurs with the halide-acid gases HCl, HBr, and HI. SO_2, a constituent of importance in many precipitator applications, attaches electrons directly.

electroplating. A process of putting a metallic coating or plating on a base material, usually metal or plastic by electrodeposition.

electrostatic desalting. To remove salts, solids, and formation water from unrefined crude before the crude oil is given subsequent processing. The charge is crude oil. The product is crude oil from which have been removed most water-soluble and solid contaminants such as chlorides, sulfates, bicarbonates, sand, silt, and tar.

electrostatic filter. A filter where an electrostatic charge is applied to the filter element. Fibrous filter material is pleated in the conventional manner between V-shaped supports consisting of metal rods which are insulated from the supporting frame. The rods are electrostatically charged with respect to the frame and the filter material fibers, which have dielectric properties, become charged. The small-particle collection efficiency is improved.

electrostatic precipitation. A process used to separate particulate matter from air or other gases using an electrostatic field. The electrostatic precipitator is a highly efficient instrument for collecting airborne dust. An ionizing wire charged at a corona voltage is situated coaxially in a cylinder open at one end. Since a space charge exists in the vicinity of the wire, fine particles entering the cylinder and passing through this region acquire a net electric charge of the same polarity as that of the wire. A charged particle is promptly attracted to the oppositely charged cylinder walls and collected there. Electrostatic precipitators sampling up to eight cubic feet per minute are about 99 percent efficient for the removal of particles down to submicron sizes.

electrostatic precipitation (particle charging). When gases laden with suspended particulate matter are passed through an electrostatic precipitator, the great bulk of the particles acquire an electric charge of the same polarity as that of the discharged electrodes. This preferential charging occurs because the region of corona, that is the region of intensive ion-pair generation, is limited to the immediate vicinity of a discharge wire, and so occupies only a small fraction of the total cross section of the precipitator. The remaining cross-sectional area contains a concentration of unipolar ions of the same sign as the wires. Two distinct particle-charging mechanisms are generally considered to be active in electrostatic precipitation: bombardment of the particles by ions moving under the influence of the applied electric field, and attachment of ionic charges to the particles by ion diffusion, in accordance with laws of kinetic theory.

electrostatic precipitator. An instrument for collecting airborne dust, based on the simple principle that a charged body attracts one oppositely charged. Electrostatic attraction and repulsion are accomplished in the electrostatic precipitator via a set of electrodes. Between these is a voltage potential of sufficient magnitude to create a negative corona effect on the negatively charged electrode so that ionization occurs. Both positive and negative ions are generated. Of these, the positive ions remain on the negatively charged electrode, while the negative ions pass over the grounded electrode along the force lines of the electrostatic field between the electrodes. As dust-laden gas passes between the electrodes, the dust in the suspended particles intercepts the negative ions and is electrostatically charged, thereupon being attracted to the grounded electrode. The dust remains there until removed, and the gas which has thus been cleaned moves on to recovery or exhaust. Single-stage precipitators are those in which gas ionization and particulate collection are combined into a single step. Basic designs for these are plate-type and pipe-type units. In the pipe-type units the collecting electrodes are formed by a nest of parallel tubes. In the plate-type units the collecting electrodes consist of parallel plates. The diagrams show electrostatic precipitators: (a) tubular (1 m diameter) and (b) duct type. In both cases the corona discharge and precipitating field extend over the full length of the apparatus.In a two-stage unit, ionization is achieved by one element of the unit and the collection by the other. Electrostatic precipitators are highly efficient collectors for minute particles. A diagram of a two-stage electrostatic precipitator is shown. The charging zone is confined to the region about the corona wire. The

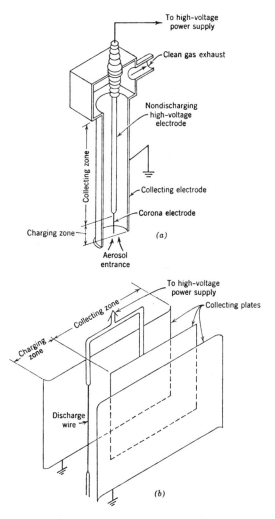

Two-stage electrostatic precipitators: (a) tubular, (b) duct type. The charging zone in each model is confined to the region about the corona wire. The downstream collecting zone provides a pure electrostatic field in which the previously charged particles are precipitated on the collecting surface.

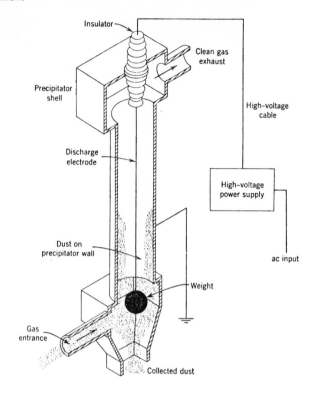

Single-stage electrostatic precipitator, tubular type.

downstream collecting zone provides a pure electrostatic field in which the previously charged particles are precipitated on the collecting surface.

element. A substance which cannot be broken down into simpler substances. Calcium, magnesium, sodium, hydrogen, oxygen, nitrogen, iron, manganese, sulfur, fluorine, chlorine, and iodine are some elements. Each element is given a symbol, usually one or two letters from the English or Latin name of the element. Thus the symbol for calcium is Ca and for magnesium, Mg.

element, rare-earth. The fourteen elements following lanthanum, with atomic numbers from 58 to 71, also called rare earths or lanthanides. All the rare earths are transition elements with two electrons in the outer shell. They are built by the addition of electrons primarily to the fifth, or O, shell. Such changes deep within the atom do not strongly affect chemical properties and all the lanthanides are nearly identical in behavior. In addition, most rare earths are found in combined form and, because of the similarity of their chemical properties, are difficult to isolate from each other.

elementary particle. One of the 34 particles from which all matter is made up. The list includes the photon, two types of neutrinos, the electron, the mu meson, two pi mesons, two K particles, the proton, the neutron, the lambda particles, three sigma particles, and two xi particles, each with its antiparticle.

elution. The process of moving a substance through a packed porous bed or chromatograph by means of a slow moving stream of liquid or gas.

elutriation. (1) The separation of the lighter particles of a powder from the heavier ones by means of an upward stream of fluid. This is especially useful for very fine particles below the usual screen size and is used for pigments, clay dressing, and ore flotation. (2) A process of sludge conditioning in which certain constituents are removed by successive flushings with freshwater or plant effluent, thereby reducing the need for using conditioning chemicals.

emergency episode. *See* **air pollution episode.**

emission. The total amount of a solid, liquid, or gaseous pollutant emitted into the atmosphere from a given source in a given time, and indicated in grams per cubic meter of gas or by a relative measure (e.g., smoke density) upon discharge from the source. Primary emissions may be characterized as follows: fine solids (less than 100 µm in diameter), coarse particles (greater than 100 µm in diameter), sulfur compounds, organic compounds, nitrogen compounds, oxygen compounds, halogen compounds, radioactive compounds. Secondary emissions may be characterized as the products of polluted air reactants, such as those which occur in atmospheric photochemical reactions. Secondary pollutants include O_3, formaldehyde, organic hydroperoxides, free radicals, NO, O, etc. Emissions are commonly reported in terms of weight of pollutant per unit of time.

emission control equipment. Equipment to control air pollution by collecting the pollutants. These include gravity settling chambers, inertial separators, cyclonic separators, filters, electrical precipitators, scrubbers (spray towers, jet scrubbers, Venturi scrubbers, inertial scrubbers, mechanical scrubbers, and packed scrubbers). Control equipment used for gases, vapors, and odors includes combustion, absorption, and adsorption units. Incineration equipment with a single combustion chamber is generally used for the combustion processes. Absorption equipment includes spray chambers, mechanical contactors, bubble cap or sieve plate contactors, and packed towers. Adsorption equipment includes packed beds, and sometimes fluidized beds.

emission factor. The relationship between the amount of pollution produced and the amount of raw material processed. For example, an emission factor for a blast furnace making iron would be the number of pounds of particulates per ton of raw materials.

emission inventory. The systematic collection of detailed information concerning the air pollution emissions in a given area. Inventories list the types of sources, as well as their contributions, in terms of the composition and rates of discharge of the individual pollutants. Supplemental information may include the number and geographical distribution of sources, description of processes, raw materials, and control measures.

emission monitoring. The collection of data on the air quality of the atmosphere. This may be done at a specific emission source, as well as for a general area.

emission point. The point where a pollutant is discharged temporarily, intermittently, or continuously into the atmosphere.

emission source. A process, building, furnace, incinerator, or plant releasing foreign matter into the air.

emission standard. The maximum amount of discharge legally allowed from a single source, mobile or stationary.

emissive power. The power of a surface to emit heat or radiation, measured by the energy radiated from a unit area of surface in a unit of time for a unit difference of temperature between the surface in question and surrounding bodies. Also called emissivity. For the cgs system the emissive power is given in ergs per second per square centimeter, with the radiating surface at 1° K and the surroundings at 0° K. In meteorology the minute is often used as the unit of time and the gram-calorie per square centimeter, or langley, is used as the energy unit. One may speak of the total emissive power, referring to all wavelengths, or the monochromatic or characteristic emissive power, referring to a particular wavelength or spectral band.

emissivity. (1) A property of a material, measured as the emittance of a specimen of the material that is thick enough to be completely opaque and has an optically smooth surface. (2) The relative power of a surface to emit heat by radiation, or the ratio of radiant heat energy emitted by a surface to that emitted by a black body (considered to have almost perfect heat absorption and therefore a very low emissivity) when the given surface and black body are at the same temperature. (3) The ratio of the observed emissive power to that of a black body or surface under identical conditions. It varies with wavelength and temperature. This quantity may be designated as a coefficient of emission, since it is a dimensionless number between zero and one.

emitter. Any technical equipment discharging foreign matter into the open air.

encapsulant. A sealant protecting the solar cell from chemical and physical attack by the environment.

endoparasite. A parasite that lives inside the host, frequently in the digestive tract. Endoparasites exhibit adaptations for resisting the internal movements and defensive measures of the host's body, such as tapeworms and flukes which have outer skins protecting them against digestive secretions and cellular enzymes.

energy. (1) The capacity of doing work. Potential energy is energy due to the position of one body with respect to another, or to the relative parts of the same body. Kinetic energy is energy due to motion. The cgs unit of energy is the erg, the energy expended when a force of one dyne acts through a distance of one centimeter; a joule is 10^7 ergs. The potential energy of a mass m, raised through a distance h, where g is the acceleration due to gravity, is $E = mgh$. The kinetic energy of a mass m, moving with a velocity v, is $E = \frac{1}{2}mv^2$. Energy will be given in ergs if m is in grams, g is in centimeters per second squared, h is in centimeters and v is in centimeters per second. (2) That which holds matter together. Energy can become mass, or can be derived from mass. It takes on such forms as kinetic, potential, calorific, chemical, electrical, and atomic energy; one form of energy can be changed to another.

energy absorption. A process whereby some or all of the kinetic energy of electromagnetic radiation is lost to another substance.

energy density. The sound energy per unit volume in a sound wave. The unit is the erg per cubic centimeter.

energy distribution. The dispersion of energy. Any disturbance that can be analyzed into two or more harmonic wave trains will undergo dispersion, each component being propagated at its own group velocity, provided this velocity is not zero. The meteorological importance of this dispersion arises from the fact that energy is, in a certain sense, propagated at the group velocity. The synoptic phenomenon of "downstream intensification" following development farther upstream has been attributed to horizontal energy dispersion in the atmosphere.

energy level. (1) The distribution of the electrons in an atom among the various orbits they can occupy. When an electron jumps from one orbit to a smaller one, it loses energy and emits radiation; an electron absorbs energy when jumping into a greater orbit. The lowest energy level of an atom is called its ground state. (2) Any one of the different values of energy which a particle, atom, or molecule may adopt under quantum-mechanical conditions.

energy plantations. The growing of plant materials for its fuel value; a renewable source of energy-rich, fixed carbon produced by photosynthesis.

enforcement. Concerns the prohibition of discharges into the atmosphere from any source constructed or operated without a permit (if one is required), or in violation of the terms of a permit, rules, regulations, or orders of an agency. This may involve civil and criminal penalties, abatement authority, and judicial review of agency orders.

enrichment. The process of sewage effluent or agricultural runoff adding nutrients (nitrogen, phosphorus, carbon compounds) to a water body, greatly increasing the growth potential for algae and aquatic plants.

enthalpy. A mathematically defined thermodynamic function of state, $h = u + pv$, where h is specific enthalpy, u is specific internal energy, p is pressure, and v is specific volume; the heat content of a substance per unit mass; the internal energy plus the product of the volume and pressure of a working substance.

entrainment. The mist or fog droplets of liquid carried off by the vapor of a boiling liquid, or more frequently from a liquid through which bubbles of gas or vapor are passing rapidly.

entropy. The capacity factor for isothermally unavailable energy. The increase in the entropy of a body during an infinitesimal stage of a reversible process is equal to the infinitesimal amount of heat absorbed, divided by the absolute temperature of the body. Thus for a reversible process $dS = dQ/T$. Every spontaneous process in nature is characterized by an increase in the total entropy of the bodies concerned.

environment. An external condition or the sum of such conditions, in which people, living organisms, equipment, or a system operates, as in a temperature environment, vibration environment, or space environment. Environments are usually specified by a range of values, and may be either natural or artificial.

environment, human ideal. The ideal human environment can be specified as a gravitational field of 1 g, an atmosphere consisting of oxygen and nitrogen in a 4:1 ratio and containing between 7 and 13 mm Hg water vapor, a total pressure of about 760 mm Hg, a temperature such that the mean skin temperature remains at 33°C and with regional skin temperature not differing from this figure by more than ± 3°C, a place not exposed to electric or magnetic fields, nor to bombardment by nuclear particles.

environmental impact assessment (EIA). The analysis of detailed data to determine whether there is a need for a full-scale environmental impact statement.

environmental impact statement (EIS). A statement preparing for actions which constitute a major federal effort with the potential for significant environmental impact.

Environmental Protection Agency (EPA). An independent agency established in the executive branch of the federal government in 1970 to permit coordinated action by several separate departments that were previously under the jurisdiction of the Department of Agriculture, the Bureau of Reclamation, and other federal offices. Pesticide regulation, for example, had been under the control of the Department of Agriculture, but its activities were transferred to the EPA. Among other responsibilities of the EPA are support of research and antipollution activities by state and local governments, establishing standards for a healthy environment, monitoring air and water pollution, radiation levels, noise, and solid waste disposal, and enforcing regulations pertaining to pollution from all sources.

eolian. Pertaining to the wind.

epiclastic detrital sediment. Physically weathered (disintegrated) and insoluble products of chemically weathered (decomposed) material from older rock (igneous, metamorphic, or sedimentary) which have been transported and deposited by running water (streams, oceans, currents), wind, or ice. In the main it consists of minerals not readily soluble (e.g., quartz, feldspar, and clay minerals), which are transported as particles.

epidemiology. The study of diseases as they affect populations.

episode, pollution. An air pollution incident in a given area caused by a concentration of atmospheric pollution reacting with meteorological conditions that may result in a significant increase in illnesses or deaths.

epm (equivalents per million). A unit chemical weight of solute per million unit weights of solution. The epm of a solute in solution is equal to the ppm (parts per million) divided by the equivalent weight.

Eppley pyrheliometer. A pyrheliometer of the thermoelectric type. Radiation is allowed to fall on two concentric silver rings, the outer covered with magnesium oxide and the inner covered with lamp black. A system of thermocouples is used to measure the temperature difference between the rings. Attachments are provided so that measurements of direct and diffuse solar radiation may be obtained.

equilibrium, chemical. A state of affairs in which a chemical reaction and its reverse reaction are taking place at equal velocities, so that the concentration of reacting substances remains constant.

equilibrium constant. The product of the concentrations (or activities) of the substances produced at equilibrium in a chemical reaction, divided by the product of concentrations of the reacting substances, each concentration raised to that power which is the coefficient of the substance in the chemical equation.

equivalent weight (combining weight) of an element or ion. The atomic weight or formula weight of an element or ion divided by its valence. Elements entering into combination always do so in quantities proportional to their equivalent weights. In oxidation–reduction reactions, the equivalent weight of a reacting substance is dependent upon the change in oxidation number of the particular substance.

erg. The unit of energy of work in the cgs system; the work performed by a force of one dyne acting through a distance of one centimeter.

erosion. Principally, the mechanical transportation of sediment by wind, water, ice, or mass wasting. It includes the cutting and carving away of land by mechanical wear effected by the pounding and grinding of sediment or bedrock by particles in transport. When water is involved, mechanical transport and wear may be supplemented by chemical decomposition and solution transport.

escarpment. A steep slope or cliff separating gently sloping areas; an extended line of cliffs or bluffs; the high steep face of a mountain or ridge.

estuary. The portion of a stream valley influenced by the tide of the body of water into which it flows. Estuaries are characterized by water whose salt content is between that of fresh and marine environments, and by a distinct population of animals and plants.

ethane. C_2H_6, the second member of the family having two carbons and six hydrogen atoms, and a boiling point of $-120°F$; one of the chief constituents found in commercial grades of natural gas.

ether soluble. A substance which is soluble in a specific ether, i.e., when a substance is chemically analyzed, various liquids such as ether, chloroform, benzol, etc., are used as solvents to determine the percentages of certain materials. Those soluble in ether are termed ether solubles.

2-ethoxyethanol. $C_4H_{10}O_2$, a colorless, nearly odorless liquid often identified commercially as Cellosolve. It was identified by NIOSH in 1979 as one of a group of 19 "multiple-target" substances selected from more than 100,000 chemicals to which industrial workers may be exposed. A multiple-target chemical is one that is known to affect more than one animal organ system. 2-Ethoxyethanol has been found to cause toxic effects to the kidney, liver, central nervous system, reproductive system, skin, and respiratory system. The chemical is absorbed through the skin, but may also enter the body via the oral or inhalation routes. No human poisoning cases have been reported as a result of industrial exposure, but animal experiments show that exposure for periods of 18 to 24 hours to air saturated with a 0.6 percent 2-ethoxyethanol vapor can result in lung edema, congestion, and renal congestion. Human exposure to the same concentrations produced eye irritation. The oral LD_{50} of 2-ethoxyethanol is 3 g/kg.

ethyl alcohol (ethanol). C_2H_5OH, a clear, mobile, water-white liquid of molecular weight 46.04; specific gravity, 0.789; freezing point, $-117°C$; and boiling point 78.5°C.

ethylene. C_2H_4, a colorless flammable gas. *Sources:* gasoline fumes from motor vehicles, combustion of natural gases, various chemical manufacturing processes. In relation to plants, it interferes with the normal action of plant hormones or growth regulators producing epinasty and leaf abscission, and abscission and abnormal flower development of some plants, i.e., epinasty in tomato plants after exposure to 0.1 ppm for 48 hours, poor flower development in carnations after six hours exposure, dry sepal injury to orchids.

ethylene dibromide (EDB). CH_2BrCH_2Br, a sweet-smelling, heavy, colorless liquid that has been used for a variety of purposes, ranging from an antiknock compound for gasoline to a fumigant for fruits and vegetables. In 1974, the National Cancer Institute issued an alert warning that bioassays found a strong carcinogenic activity of ethylene dibromide in both rats and mice, producing squamous cell carcinomas of the stomach. The LD_{50} for oral doses of ethylene dibromide in rats was established at 140 mg/kg. It had been associated with loss of fertility among industrial workers employed in jobs that required exposure to the chemical. But there had not been any reports of any association between the substance and cancers in humans. Direct contact with EDB causes irritation and injury to the eyes and skin. Exposure to the vapor has caused the development of respiratory inflammation along with anorexia and headache, although recovery followed discontinuance of exposure. Weakness and rapid pulse have also been associated with exposure to ethylene dibromide, as well as cardiac failure leading to death. The stomach cancers in test animals appeared as early as 10 weeks after exposure to ethylene dibromide, but the tumors were not observed in controls. OSHA established the standard for exposure to EDB of 20 ppm as an eight-hour time-weighted average and 50 ppm as a maximum peak exposure with five minutes duration. No causal evidence of adverse health effects from consumption of food products treated with ethylene dibromide are reported, although the chemical is strongly absorbed by wheat and wheat products, decomposing to ethylene glycol and inorganic bromide when the grain or grain products are heated.

ethylene glycol. $C_2H_6O_2$, a colorless fluid with a characteristic sweet odor and taste, derived from the cracking of petroleum; a thick, clear liquid obtained from petroleum which lowers the freezing point of water, used as a refrigerant and as a permanent-type antifreeze fluid.

Ettinghausen's effect (Von Ettinghausen's effect). When an electric current flows across the lines of force of a magnetic field, an electromotive force is observed at right angles to both the primary current and the magnetic field; a temperature gradient is observed in the direction opposite to the Hall electromotive force.

eutectic. Refers to a mixture of solids, in a specific ratio by weight percent (eutectic ratio), exhibiting a minimum melting temperature (eutectic temperature). The melt has the same composition as the bulk composition of the mixed solids.

eutectic material. A chemical having the property of changing from a solid to a liquid at a relatively low temperature, while maintaining a constant temperature. A eutectic liquid then stores the heat energy which caused the transformation until the liquid returns to solid form and gives up heat. Eutectic salts, stored in plastic tubes or traps, are used as thermal reservoirs to store and then release solar heat.

eutrophic lake. A shallow, murky water body that has lots of algae and little oxygen.

eutrophication. (1) The promotion of plant growth in an aquatic ecosystem by rapidly adding substantial amounts of nutrients. Available oxygen is used up by the plants and substantial fish kills may result. (2) For natural waters such as lakes and reservoirs, a process of ecological aging stimulated by increased levels of mineral nutrients, especially orthophosphates and nitrates. These factors stimulate the growth of algae and other plants at very low levels (phosphorus at 0.05 ppm stimulates profuse growth). As a result of the decay of these plants and associated organisms, the oxygen utilization of the water, the biochemical oxygen demand (BOD), is greatly increased and fish and other aquatic animals may find it impossible to exist. The most serious aspect of eutrophication is the great length of time required to reverse the process in large bodies of water.

evaporate. To change a liquid into a vapor by means of heat. Boiling represents a high rate of evaporation.

evaporation. The physical process by which a liquid or solid is transformed into the gaseous state; the opposite of condensation. In meteorology, evaporation is usually restricted in use to the change of water from liquid to gas, while sublimation is used for the change from solid to gas, as well as from gas to solid. According to the kinetic theory of gases, evaporation occurs when liquid molecules escape into the vapor phase as a result of the chance acquisition of above-energy, outward directed, translational velocities at a time when they happen to be within about one mean free path below the effective liquid surface. It is conventionally stated that evaporation into a gas ceases when the gas reaches saturation. In reality, net evaporation does cease, but only because the numbers of molecules escaping from and returning to the liquid are equal, that is, evaporation is counteracted by condensation. Energy is lost by an evaporating liquid, and when no heat is added externally, the liquid always cools. The heat thus removed is called the latent heat of vaporization.

evaporation pond. An area where sewage sludge is dumped and allowed to dry out.

evaporation rate. (1) The mass of material evaporated per unit time from a unit surface of a liquid or solid. (2) The number of molecules of a given substance evaporated per second per square centimeter from the free surface of the condensed phase.

evaporator. Usually a vessel which receives the hot discharge from a heating coil and by a reduction in pressure flashes off overhead the light products and allows the heavy residue to collect in the bottom.

evolution. A change in gene frequency in a population, usually involving a visible change in the species characteristics.

exfoliation. For smooth rocks exposed to weathering, the process of chipping off in thin slabs, sheets, or scales concentric with the surface. Exfoliation is caused by changes in the volume of a rock, resulting from ice wedging, alternate heating and cooling, expansion due to chemical changes in the rock, or all of these combined. The outer part of the rock pulls away from the inner part until it finally falls off and exposes a fresh surface to attack.

exhaust emissions, crankcase. Crankcase emissions arising from gases escaping past the pistons (blowby gas) represent a considerable proportion of potential hydrocarbon emissions. Analysis has shown that crankcase emissions are a principally unburned fuel–air mixture and thus hydrocarbon concentrations are in the region of 10,000 –20,000 ppm (as hexane). The olefin content of the blowby emissions is of particular importance, since higher olefins are known to be major contributors in the development of petrochemical smog. A smaller proportion (20–30%) of the crankcase emissions consists of combustion products and thus carbon monoxide concentrations are less than 10 percent of typical exhaust emissions.

exhaust emissions, diesel engine. In diesel engines, the medically harmful pollutants such as carbon monoxide, benzopyrene, and aldehydes are emitted only in low concentrations. The oxides of nitrogen are present in much lower proportions than in gasoline engines. However, black diesel exhaust smoke is a potential safety hazard. Unburned carbon, appearing as visible black smoke, is an indication of inefficient operation and should be eliminated. The composition of exhaust smoke has been reported as 75–95 percent carbon, showing significant variations with engine loading. Particle size varies in the 0.1–0.3-μm range, with smaller particles predominating. A fine mist of partly vaporized fuel and water droplets is often produced in "cold start" conditions or on misfire; this is the "white smoke" and is a powerful irritant, partly because of accompanying aldehydes in the exhaust gases. "Blue smoke" does not become visible until several feet from the exhaust and is probably the result of a cooling and condensation process. Precipitation of the droplets in blue smoke yields a dark amber liquid with the viscosity of light lube oil. Mass spectrometric analysis has shown this to be a mixture of hydrocarbons. Blue smoke represents a particular fraction of the unburned fuel in the exhaust, that fraction which will condense in the colder conditions some feet away from the exhaust pipe. The production of blue smoke is thus a function of both engine conditions and fuel specifications. It is heaviest at medium load, the maximum emission occurring at 40 percent rated load with straight-run fuels, and at 60 percent with cracked fuel.

exhaust emissions, gasoline engine. Gasoline engine exhaust is likely to consist of the following gases in varying concentrations, according to driving conditions: carbon monoxide, hydrocarbons, oxides of nitrogen, sulfur dioxide, aldehydes, carbon dioxide, hydrogen, oxygen, water vapor, and nitrogen. Carbon monoxide, hydrogen carbons, and oxides of nitrogen are recognized as the most serious pollutants. Engine conditions (speed, fuel ratio, manifold, etc.) exert a significant influence on the exhaust composition, and these conditions are largely determined by the various driving modes. While modern carburetors are designed to cope with these varying conditions, the main pollutant concentrations are significantly affected. Hydrocarbons are most prominent during deceleration. Hydrocarbon emissions are seen to increase in a nearly linear fashion when expressed on a weight (lb/hr) basis, but decrease significantly when volumes (ppm) are related to air flow. Carbon monoxide, because of its toxicity, is still the subject of many exhaust control studies. Approximately 99 percent of the oxides of nitrogen present in exhaust emissions have been shown to be in the form of nitric oxide, showing considerable variations with the fuel–air ratio. The polynuclear aromatic compound 3,4-benzopyrene has received less attention than other exhaust pollutants. However, the danger to health from this recognized carcinogen is now well established and with the rapid rise in the incidence of lung cancer, all possible sources of this material are receiving greater scrutiny.

exosphere. The outer earth's atmosphere, from about 1600 to 3000 km, where there is no longer any absorption of electromagnetic radiation and ions follow the magnetic force lines of the Van Allen radiation belt. The region is composed essentially of protons, but is distinct from the interplanetary hydrogen and its temperature is some 1500°K.

exposure. The subjection of a person, animal, plant, or material to an environment containing a harmful concentration of air pollutant.

exposure chamber. A chamber for exposing a person, plant, animal, or material to a particular environment. Various parameters, such as pollutants, humidity, temperature, pressure, noise, movement, radiation, and fluid contents, may be controlled.

exposure dose. A measure of radiation exposure based upon its ability to produce ionization. The unit of exposure dose is the roentgen (R). Exposure dose rates are measured in roentgens per unit of time.

exposure rate. The concentration of pollutant or radiation per unit of time to which an animal, plant, or material is exposed.

eye irritation. A soreness or itchiness of the eye caused by air pollutants, such as gases contained in smog and automotive exhausts (SO_2, SO_3, NO_2, O_3, olefins, formic acid, oil, gasoline), secondary pollutants caused by the photochemical reactions of NO or NO_2 and hydrocarbons such as PAN (peroxyacetyl nitrate), formaldehyde, or acrolein, alone, or in combination with aerosols (such as NaCl and silica dust).

F

fabric filter. A collector designed to remove particles from a carrier gas by filtration of the gas through a porous medium. Two basic types of filter are (a) a fibrous medium utilized as the collecting element, and (b) where the medium is utilized as a support for a layer of collected particles, relying on the coat of collected particles to serve as the principal collecting medium. The most common type of fabric collector is the tubular type, consisting of a structure in which cylindrical fabric bags are suspended vertically over a tube sheet, with the open end of the bag attached to the sheet. Some of the fabric materials include cotton, wool, nylon, asbestos, and Orlon. Particles equal to or greater than 1 µm can be collected.

fabric filters, automatic baghouse with reverse airflow cleaning. In an automatic baghouse with reverse airflow cleaning, the casing is divided into two parts by a tube sheet. The upper portion contains the woven-fabric tubular bags and the lower portion contains the hopper. The hopper may be a continuous trough-type hopper running under many compartments, or a series of individual pyramidal hoppers, each serving one compartment. The dirty gas can be fed through a single inlet into the trough-type hopper or through a common inlet manifold into the individual hoppers, upward into the tubes through the fabric from the inside to the outside, then out the common outlet manifold. The hopper serves as a settling chamber for the coarse dust. The diagram shows an automatic conventional baghouse with mechanical shaking.

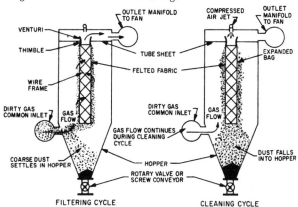

Automatic baghouse with mechanical shaking.

fabric filters, intermittent baghouse. A fabric filter consisting of a casing divided into an upper and lower part by a tube sheet. The upper portion contains the woven-fabric filter which may be in the shape of one or more tubes, or in a great variety of other shapes, all designed to squeeze sufficient filtering area into the available space. The lower portion may be in the form of one or more pyramidal hoppers, or if the casing is long, there may be a single trough-type hopper running the full length of the casing. Under normal operations, the dirty gas is introduced below the tube sheet and into the upper portion of the hopper. Since the gas velocity is reduced upon entry, the coarse particles of dust will settle directly into the hopper, while the finer particles and all of the gas will pass upward into the fabric tubes. The gas will pass outward through the fabric; the dust will be trapped within the cloth tube. The cleaned gas is collected by the casing and discharged into a duct leading to the fan.

facultative. Having the power to live either with or without oxygen.

fahrenheit scale. A temperature scale in which the freezing point of water is 32° and the boiling point is 212° under standard conditions of pressure.

fallout. A measurement of air contamination, consisting of the mass rate at which solid particles deposit from the atmosphere. Fallout is used in the same sense as the older terms "dust fall" and "soot fall," but without any implications as to the nature and source of the particles. Another example is radioactive fallout from an atomic or thermonuclear explosion.

fan–coil system. A space heating unit containing a coil, usually of copper, filled with water at a typical temperature of 140°F, and a fan which blows air over the hot-water coil on command of a thermostat.

fanning. In blast furnace operations, the idling period between the blowing periods when the blast pressure is reduced to a minimum. During this period no charging or tapping of slag or metal is carried out.

Faraday effect. (1) The rotation of the plane of polarization produced when plane-polarized light is passed through a substance in a magnetic field, the light traveling in a direction parallel to the line of force; the rotation is proportional to the thickness of the material and to the strength of the magnetic field. (2) The rotation of the plane of polarization of a wave due to the effect of the earth's magnetic field on the refractive index.

Faraday's laws. In the process of electrolytic changes, equal quantities of electricity charge or discharge equivalent quantities of ions at each electrode. One gram equivalent weight of matter is chemically altered at each electrode for 96.501 int. C, or one faraday, of electricity passed through the electrolyte.

fecal coliform bacteria. A group of organisms found in the intestinal tracts of people and animals. Their presence in water indicates pollution and possible dangerous bacterial contamination.

feeder, dry chemical. A mechanical device for applying dry chemicals to water or sewage at a rate controlled manually or automatically by the rate of flow.

feedlot. A relatively small, confined area for raising cattle that results in lower costs but which may concentrate large amounts of animal wastes. The soil cannot absorb such large amounts of excrement, and runoff from feedlots pollutes nearby waterways with nutrients.

fen. Low-lying land partly covered with water.

Fermat's principle. The path followed by light (or other waves) passing through any collection of media from one specified point to another is that path for which the time of travel is least.

fermentation. The decomposition of organic material to alcohol, methane, etc., by organisms, especially bacteria and yeasts, usually in the absence of oxygen.

Fiberglas. A trade name for a light-weight material made of fine glass filaments bonded together with a polyester resin product. It is nonflammable, tough, erosion resistant, and can be transparent.

fiberglass. The generic name for a manufactured nonflammable fiber fabricated from glass; a continuous filament or staple fiber having unusual resistance to heat and chemicals. Fiberglass is the strongest fiber known and is perfectly elastic up to its ultimate strength. It is attacked by hydrofluoric acid and alkalis, but is resistant to most other chemicals and solvents.

fields of force. Electric charge, as represented by the proton and electron, is a basic property of matter. The proton is the fundamental unit of positive charge, while the electron is the basic unit of negative charge. The electric charge of one proton exactly balances the electric charge of one electron, even though the proton is a larger particle. Most of the matter in the universe is therefore electrically neutral, since matter contains equal amounts of positive protons and negative electrons. Charged particles create an electric field that distorts the space around them. As the electron revolves around the nucleus, it creates not only an electric field, but a magnetic field as well. The combined fields are called an electromagnetic field. It is this electromagnetic field which fills the space between the nucleus and its orbiting electrons. Through this field is exerted the electromagnetic force which holds the electrons to the nucleus, binds atoms together, and is basically responsible for the properties of all living and nonliving things.

filling. Depositing dirt and mud, often raised by dredging, into marshy areas to create more land for real estate development. It can destroy the marsh ecology.

film badge. A pack of photographic film used for approximate measurement of radiation exposure for personnel monitoring purposes. The badge may contain two or three films of differing sensitivity, and it may contain a filter which shields part of the film from certain types of radiation.

filter. (1) A porous material on which solid particles are largely caught and retained when a mixture of liquids and solids is passed through it. (2) A device used for separating components of a signal on the basis of their frequency. It allows components in one or more frequency bands to pass relatively unattenuated, and it greatly attenuates components in other frequency bands.

filter collector. A mechanical filtration system for removing particulate matter from a gas stream for measurement analysis and control; also called bag collector.

filter, diatomite. A filter in which the filter medium is diatomaceous earth, or diatomite, composed of the siliceous (silica) skeleton of diatoms, minute plants growing in water. In practice, a thin layer of diatomite is formed on a supporting screen of some sort (frequently in the shape of a cylinder). This is done by filtering a water suspension of diatomite through the screen support. The filter is now ready to operate. It is not necessary to use a coagulant with a diatomite filter, because the layer of diatomite is an extremely fine filter in itself and can filter out the ameba cysts which cause amebic dysentary. Diatomite filters are in common use in swimming pools, recirculating systems, and, in special situations, in community water plants, especially smaller ones, where the turbidity of the raw water is not excessive. When the layer of diatomite eventually becomes plugged with the suspended matter it has removed, it is washed off the supporting screen by reversing the flow of water through the filter. Normally the diatomite is discarded after one use.

filter fabric characteristics.

Fiber	Max. temp. °F	Combustion[a]	Abrasion	Resistance		
				Mineral acids	Organic acids	Alkali
Cotton	180	P	G	P	G	G
Wool	200	G	G	F	F	P
Nylon	200	P	E	P	F	G
Orlon	240	P	G	G	G	F
Dacron	275	P	E	G	G	G
Polypropylene	200	P	E	E	E	E
Nomex	425	G	E	F	E	G
Fiberglass	550	G	F	E	E	P
Teflon	450	G	F	E	E	E

[a]P = poor; F = fair; G = good; E = excellent.

filter, fibrous. A mass of randomly oriented fibers which act as targets for the collection of fine particles. When airborne particles enter a filter and flow around the fibers, they are subjected to aerodynamic forces which result in their collection on the fibers. The principal mechanisms are gravitation, inertia, interception, and diffusion. In practice, two types of fibrous filters are used for collecting particles for weight or size determination or chemical analysis. These consist of the usual Whatman filters composed of cellulose fibers, and filters consisting of microfine fibers made of glass or synthetic plastics.

filter, high-rate. A trickling filter operated at a high daily average dosing rate, usually between 10 and 30 mgd/acre, sometimes including recirculation of effluent.

filter, intermittent. A natural or artificial bed of sand or other fine-grained material through the surface of which sewage is intermittently applied in flooding doses, opportunity being given for filtration and maintenance of aerobic conditions.

filter, low-rate. A trickling filter designed to receive a small load of BOD per unit volume of filtering material and to have a low dosage rate per unit of surface area (usually 1–4 mgd/acre); also called standard rate filter.

filter membrane. A filter consisting of a thin plastic film having an extremely large number of very fine pores. The filter behaves as a sieve but, because of its inherent electrostatic characteristics, it can collect particles much finer than the port size. Since collection occurs on the filter surface, membrane filters lend themselves well to such special analyses as radioactivity counts, morphological examinations of dust, and particle-size frequencies.

filter plant process. Raw turbid or colored water enters a mixing tank or basin where it is mixed rapidly, with a coagulating chemical or chemicals. The rapid mixing ensures prompt and uniform dispersion of the chemicals in the water. The water then passes through a coagulating or flocculating basin, spending possibly 20 minutes to an hour in its passage. The purpose is to give the coagulated material an opportunity, aided by slow, mechanical stirring or by baffles, to build up the desired large and heavy flocs. The water then flows through one or more settling basins. There it moves so slowly that practically all the flocculated material falls out, leaving relatively few solid particles to be removed by filtration, the next and last step in the clarification process. The filter medium of community water purification plants is almost always a layer of sand. The sand layer is supported on several layers of gravel. The gravel in the top layer is not much coarser than the sand above it. Each lower layer of gravel is successively coarser; in a large filter the bottom gravel may be as much as 1 ½ or 2 in. in diameter.

filter, pressure. Most industrial water filters and household ones are of the "pressure type." The general principle of the pressure filter and the sand filter is the same as far as filtration and backwash are concerned, but the pressure filter has a closed tank. The pressure filter can therefore be installed almost anywhere in a water line, without having to repump the filtered water, as is usually done with a gravity filter.

filter (radiology). Primarily a sheet of material, usually of metal, placed in a beam of radiation to remove as much as possible of the less-penetrating components. Secondarily a sheet of material of low atomic number, relative to that of the primary filter, placed in the filtered beam of radiation to remove characteristic radiation produced by the primary filter.

filter tape sampler (AISI smoke and haze sampler). An automatic sampler consisting of a continuous Whatman filter tape through which air is drawn at approximately 0.25 ft³/min for a preset time period. Suspended particles collected on the filter form a dark spot, the shade varying with the concentration of particulates. At the end of the sampling period, usually two to four hours in duration, the tape is automatically advanced to a new sampling position and the cycle is repeated. The sampler can run unattended for months. To analyze the spots, the tape is removed and placed in a spot evaluator which determines the light transmission for each spot. These readings are converted to optical densities and eventually are expressed as COH (coefficient of haze) units. The filter tape sampler is useful in measuring the soiling index of the air.

filter, trickling. A filter consisting of an artificial bed of coarse material, such as broken stone, clinkers, slate, slats, or brush, over which sewage is distributed and applied in drops, films, or spray from troughs, drippers, moving distributors, or fixed nozzles, and through which it trickles to the underdrains, giving opportunity for the formation of zoogleal slimes which clarify and oxidize the sewage.

filter, vacuum. A filter consisting of a cylindrical drum mounted on a horizontal axis, covered with a filter cloth, revolving partially submerged in liquid. A vacuum is maintained under the cloth for the larger part of a revolution to extract moisture, and the cake is scraped off continuously.

filtration. (1) The process of separating suspended solids from a liquid or gas, usually by forcing a carrier gas or liquid through a porous medium. (2) A process in waste treatment used for the removal of particulate matter from plant effluent and the polishing of effluent from sedimentation basins and/or clarifiers. Filtration involves the passage of water through a packed bed for removal of suspended solids. The suspended solids fill interstices in the bed, and it gradually requires increasing pressures to pass the rated quantity of water through the same bed area. When the pressure drop across the bed reaches a partial limiting value, the bed is taken out of service and backwashed, and then placed back on line.

filtration, biological. The process of passing a liquid through a biological filter containing media on the surfaces of which zoogleal films develop, which absorb and adsorb fine suspended, colloidal, and dissolved solids, and release various biochemical end products.

filtration, inherent. The filtration introduced by the wall of an x-ray tube and any primary and secondary filters.

fines. Fine particulates.

Fischer–Tropsch process. The process that basically converts carbon monoxide and hydrogen to liquid hydrocarbon. The two chemical equations which generalize the formation of hydrocarbons are: $nCO + 2nH_2 = (CH_2)_n + nH_2O$ and $2nCO + nH_2 = (CH_2)_n + nCO_2$. Products from the fluid-bed synthesis are mainly low boiling hydrocarbons $(C_1–C_4)$ and gasoline, with little medium- and high-boiling material. Substantial amounts of oxygenated products and aromatics are made. A portion of fixed-bed and fluid-bed tail gas is removed and used for utility gas.

fission. The splitting of a heavy nucleus, as of uranium or plutonium, into two approximately equal parts, accompanied by the conversion of mass into energy, the release of this energy, and the production of free neutrons, gamma rays, and other radiation. In nuclear fission, the nucleus of a heavy element is split into two lighter elements. Fission occurs when the nucleus is bombarded with neutron bullets. After fission occurs, the resulting free neutrons may strike and be absorbed into other ^{235}U nuclei to create additional unstable ^{236}U nuclei, which will also undergo fission. The process whereby free neutrons strike other nuclei, causing them to split and release more free neutrons, is called a self-sustaining chain reaction.

fission fragments. Two parts, approximately equal in weight, into which the uranium or plutonium atom splits in the process of fission. These parts are usually highly radioactive, disintegrating several times before reaching stability. In slow neutron fission the fragments are seldom equal in mass, but generally fall into a heavier group, with masses around 140, and a lighter group, with masses around 95.

fission products. Any of the new atoms made by fission. The nuclides produced by the fission of a heavy element nuclide such as ^{235}U or ^{239}Pu. Thirty-five fission product elements from zinc through gadolinium have been identified from slow neutron fission.

flame photometry. Emission spectroscopy in the ultraviolet and visible regions which makes use of flame sources. Approximately seventy elements can be qualitatively or quantitatively determined based on the emission spectra of these elements when they are excited in an arc or high-voltage spark.

flare. (1) An arrangement of piping and a burner to dispose of surplus combustible vapors; situated around gasoline plants, refineries or producing wells. (2) A bright eruption from the sun's chromosphere.

flash drying system. A recycling-type flash drier used to incinerate wet sludges. This system is similar to the standard flash drier used for drying many chemicals. The solids in the system are dried, separated in a cyclone, and sent to a secondary incinerator for final destruction. The gases from the drier often require afterburning. Sewage sludge may be dried in this manner, and the dry sludge can be used as fertilizer.

flash point. The temperature at which a liquid heated in a specified type of apparatus gives off sufficient vapor to flame momentarily on the application of a small flame; the lowest temperature at which a substance, such as fuel oil, will give off a vapor that will flash or burn momentarily when ignited.

Fleming's rule. A simple rule for relating the direction of the flux, motion, and emf in an electric machine: The forefinger, second finger, and thumb, placed at right angles to each other, represent, respectively, the directions of flux, emf, and motion or torque. If the right hand is used, the conditions are those obtained in a generator and if the left hand is used, the conditions are those obtained in a motor.

float air vent. A valve placed at the outlet of an array of liquid-type solar collectors and containing a float so that the pressure of the sun-heated liquid automatically purges air from the system. Air escapes from this valve usually when the collectors are first filled with liquid.

floc. A very fine, fluffy mass formed by the aggregation of fine suspended particles in a liquid.

flocculation. (1) A process of contact and adhesion whereby the particles of a dispersion form larger-size clusters; synonymous with agglomeration and coagulation. (2) The process of converting a finely divided or colloidally dispersed suspension of a solid into particles of such size that reasonably rapid settling occurs. This is usually done by adding the salt of a bi- and trivalent metal, e.g., when alum, aluminum sulfate, and ferric sulfate are added in water being clarified of suspended impurities.

flocculator. An apparatus designed for the formation of floc in water or sewage.

floccule. A small, loosely aggregated mass of material suspended in or precipitated from a liquid; a cluster of particles.

flotation. A method of raising suspended matter to the surface of a liquid in a tank as scum by aeration, vacuum, evolution of gas, chemicals, electrolysis, heat, or bacterial decomposition, and the subsequent removal of the scum by skimming. The flotation process takes advantage of the natural tendency of oil globules to rise to the surface. It is also used for the removal of some finely divided suspended solids. Air is inducted into a pressurized water stream to achieve saturation. The air-laden water is then passed through a pressure-reducing valve, and the air is released from the water as small bubbles. The bubbles attach themselves to oil globules or suspended particles and float them to the surface, from which they are removed for further handling. Chemicals such as coagulants, polymer coagulant aids, acids, and/or alkalis are often added ahead of the system to promote a more complete removal.

flow meter. An instrument for measuring the rate of flow of a fluid moving through a pipe or duct system. The instrument is calibrated to furnish the volume or mass rate of flow.

flow rate. A measure of the volume of fluid per unit of time.

flow register. A device used to maintain constant airflow for a given period of time in particulate sampling equipment.

flue. A passage for conducting combustion gases in an incinerator installation; also used synonymously with chimney.

flue gas. Waste gas from combustion processes.

flue gas scrubber. Equipment for removing fly ash and other objectionable materials from the products of combustion by means of sprays or wet baffles. It also reduces excessive temperatures of effluent.

fluid flow. A stream or movement of air or fluid, or the rate of fluid movement in the open or in a duct, pipe, or passage. Various types of fluid flow are: (a) *Uniform flow*. A flow that is steady in time, or the same at all points in space. (b) *Steady flow*. A flow whose velocity at a fixed point with respect to a fixed system of coordinates is independent of time. (c) *Rotational flow*. A flow with an appreciable vorticity and which cannot be described mathematically by a velocity potential function. (d) *Turbulent flow*. A flow in which the fluid velocity at a fixed point fluctuates with time in a nearly random way. The motion is essentially rotational and is characterized by rates of momentum and mass transfer considerably larger than in the corresponding laminar flow. (e) *Laminar flow*. A flow in which the mass of fluid may be considered separate laminae (sheets) with simple shear existing at the surface of contact of laminae, should there be any difference in the mean speed of the separate laminae. If turbulence exists, its effect is confined to individual laminae and there is no exchange of momentum between laminae. (f) *Streamline flow*. A flow in which fluid particles move along the streamlines. This motion is characteristic of viscous flow at low Reynolds numbers or of inviscid, irrotational flow. (g) *Secondary flow*. A less rigorously defined term than many of the foregoing types of flow. Flow in pipes and channels is frequently found to possess components at right angles to the axis. These components, which take the form of diffuse vortices with axes parallel to the main flow, form the secondary flow.

fluidity. The reciprocal of viscosity. The cgs unit is the rhe, the reciprocal of the poise.

fluidized bed. First used in industrial waste situations, especially in the paper industry, the waste sludge is fed onto a bed of sand that is fluidized with air at 3–5 psig. The air has been preheated to 1000°F or more and evaporation and combustion of the waste takes place on the surface of the bed. Often, auxiliary burners are located above the bed to provide supplemental heat and destroy noxious gases formed in the chamber. Excess air is held to about 25 percent and combustion temperatures reach 1500–2000°F. The preheated air is generated by exchange with the stack gases. Ash passes out with the effluent gas and is removed in a cyclone.

flume. A natural or man-made channel that diverts water.

fluorescence. The property of emitting radiation as the result of absorption of radiation from some other source. The emitted radiation persists only as long as the exposure is subjected to radiation, which may be either electrified particles or waves. The fluorescent radiation generally has a longer wavelength than that of the absorbed radiation. If the fluorescent radiation includes waves of the same length as that of the absorbed radiation, it is termed resonance radiation.

fluoride. A compound of fluorine and one or more elements or radicals. Fluorines occur as either gaseous or solid emissions of such industrial processes as the manufacture of fertilizer and aluminum, iron ore smelting, and ceramics production. Airborne fluorides can damage vegetation, with some plants concentrating and accumulating these fluorides, and the livestock eating the plants as forage can become ill. Fluorides ingested at low levels are good for bones and teeth of both animals and man. In normal amounts, fluoride does not appear to cause any adverse effects on health. Close to an industrial source of fluoride pollution, however, many may be subjected to eye and skin irritation, inflammation of the respiratory tract, and breathing difficulties.

Fluorides are a special menace to livestock, even when they are not to people, because certain of the plants used for fodder have the ability to store the fluorides they take in. Thus the plants build up far greater concentrations than would remain deposited on their surfaces, and they accomplish this without harm to themselves and with no external indications of their contents. When these plants become a meal for forage-consuming animals, the enormous overdose can be devastating. The animal's teeth become mottled. Then, as they feed further on this food, they lose weight, give less milk, and grow more slowly. Eventually spurs grow on their bones, and they become so crippled they have to be killed.

Fluoride is a deadly enemy of vegetable life. Aluminum, brick, ceramic, chemical, and fertilizer industries, glassworks, smelters, and steel mills may all release fluorides. Fluorides enter the leaf through the stomata. From there they move to the edges and the tip of the leaf. The body of the leaf, although it may continue to absorb low concentrations of fluoride, remains relatively unharmed, while lethal amounts of the substance pile up on the edges. Continued exposure spreads the killing inward from the edge and tip. Leaves exposed to fluoride generally have burnt, dried out edges, with a narrow, reddish-brown line of dead tissue distinctly marking off the healthy part of the plant. Although all farming soils contain appreciable amounts of fluoride, plants take up little from the soil. When plants contain more than a few parts per million of fluoride, it can be assumed that polluted air supplied it. Gladiolus, prune, apricot, and peach plants are so sensitive to fluorides that they are injured by extremely low concentrations. Other plants less sensitive but still susceptible to fluoride damage include the sweet potato, corn, and conifers.

fluorine (F).   A greenish-yellow gas, with atomic weight 10.00; density, 1.69 g/liter; melting point, −220°C; and boiling point, −188°C. It combines with nearly all elements.

fluorite.   Calcium fluoride, CaF_2; it forms cubic crystals, has a hardness of 4 on the Mohs' scale, shows a vitreous luster, may be colorless, green, white, purple, blue, or some other tint. It is used as a flux in steel furnaces and smelters, as a source of hydrofluoric acid and other chemicals, and for making glass and enamel.

fluorocarbon.   A gas used as a propellant in aerosols, thought to be modifying the ozone layer in the stratosphere, thereby allowing more harmful solar radiation to reach the earth's surface.

fluorocarbon spray products ban.   A federal ban against the use of fluorocarbon gas in almost all aerosol sprays, announced by the Food and Drug Administration on March 15, 1978. The action was taken to prevent damage to the earth's atmosphere. Fluorocarbons floating into the atmosphere caused a depletion of the ozone layer—which protects the planet from harmful effects of the sun's ultraviolet rays—according to scientific reports. Without this protection, scientists warned, a higher incidence of skin cancer was possible, as well as changes in climate and harm to plant and other animal life. The ban issued by the Food and Drug Administration, the Environmental Protection Agency, and the Consumer Product Safety Commission, cut off the manufacture of aerosol products using fluorocarbons after December 15, 1978. Entry of the products into the market was prohibited after April 15, 1979. The ban applied to 97–98 percent of all aerosols using fluorocarbon gas as a propellant, and covered such products as deodorants, hair sprays, household cleaners, and some pesticides. The 2–3 percent of products exempted from the ban were those "for which no acceptable substitutes" existed. These included certain insecticides, respiratory drugs, contraceptive foams, and cleaning sprays for electrical and aircraft equipment. The ban covered what the agencies considered "all nonessential uses" of fluorocarbons in sprays. Most of the products involved were expected to be able to convert to the use of mechanical sprayers or other propellants such as carbon dioxide or hydrocarbons. Since the scientific findings of ozone depletion and under the preliminary curbs ordered by the federal agencies, use of fluorocarbons in aerosols has dropped by about 40 percent.

fluorometer.   An instrument used to measure the intensity and color of fluorescent radiations.

flux.   (1) For electromagnetic radiation, the quantity of radiant energy flowing per unit time. (2) For particles and photons, the number of particles or photons flowing per unit time.

fly ash.   Finely divided particles of ash entrained in flue gases, resulting from the combustion of fuel. The particles of ash may contain incompletely burned fuel. Fly ash is predominantly a gas-borne ash from boilers with spreader stokers, underfeed stokers, and pulverized fuel (coal) firing. Fly ash also constitutes all solids including ash, charred papers, cinders, dust, soot, or other partially incinerated matter carried in the products of combustion. Three main constituents are unburned coal, shale, and magnetite. The dust consists largely of a mixture of well-defined hollow punctured spheres, or cenospheres, of a light shiny appearance, and irregularly shaped particles of ash and carbon of a darker appearance (about 10–75 μm). The color of the dust varies from light to dark gray, and the higher the sulfur content, the darker the ash. (Analysis of typical fly ash: silica, 43.2 percent; alumina, 31.6 percent; ferric oxide, 9.8 percent; lime, 6.1 percent; magnesia, 3.4 percent; sulfate, 1.2 percent; alkalis, etc., 0.9 percent; loss in ignition, 3.8 percent.)

fly ash collector. Equipment used to remove fly ash from the products of combustion.

foamed plastic. A plastic material, used primarily for insulation, in which a foaming agent is used to provide minute voids to improve insulating qualities, often foamed in place within the structure.

foehn. A wind originating on the windward side of mountains. Winds are forced up the mountain slopes, become cooler, and lose their moisture through condensation. The moisture is transformed into clouds that float above the mountain tops and frequently assume a distinctive pattern. Because of this, they are referred to as a "foehn wall." The wind continues over the crest of the mountains, flows downward on the protected side, and, as it does, it becomes compressed and heated. When it arrives at the foot of the mountain, it is a warm, dry wind. It is called a chinook in the Rocky Mountains.

fog. A large mass of water vapor condensed to fine particles; a thick, obscuring mist. When the surface air is cooled by contact with the ground, its temperature may be reduced below the dew point, and dew or frost settles on the ground. If, however, a thick layer of air is cooled below the dew point, water condenses within the air to form extremely small droplets, and a fog is produced. Fogs occur most frequently in the early morning during calms or light winds accompanied by an inversion. These are the very conditions which produce high concentrations of pollution. This is why in cities fogs are nearly always accompanied by abnormal densities of smoke and sulfur dioxide. A well-established fog may cover a wide area and extend to several hundreds of feet above the ground. Sunshine falling on the top of the fog is reflected away, so that little heat reaches below to assist in the evaporation of the water droplets. City fogs are usually dissipated by winds, rather than by sunshine. The mixing of fog-laden air has the effect that gases and smoke from even tall chimneys can diffuse down to the ground level, although they are prevented from upward escape by the inversion at the top of the fog.

fog chamber. A confined space in which air or another gas is supersaturated by the reduction of pressure or cooling, used to study the movement and interaction of electrified particles by the condensation (fog tracks) they produce.

fog horizon. The top of a fog layer which is confined by a low-level temperature inversion in such a way as to give the appearance of the horizon when viewed from above against the sky. The true horizon is usually obscured by the fog.

fogging. Applying a pesticide by rapidly heating the liquid chemical so that it forms very fine droplets that resemble smoke. It is used to destroy mosquitoes and blackflies.

food waste. Discarded animal and vegetable matter, also called garbage.

foot-candle. One lumen per square foot, 1.076 milliphots, 10.76 lux.

foot-lambert. A unit of photometric brightness (luminance) equal to $1/\pi$ candle per square foot.

force. That which changes the state of rest or motion in matter, measured by the rate of change of momentum. The absolute unit is the dyne, the force which will produce an acceleration of one centimeter per second per second in a gram mass. The gram weight, or weight of a gram mass, is the cgs gravitational unit. The poundal is that force which will give an acceleration of one foot per second per second to a pound mass.

formaldehyde. HCHO, also called formalin, a clear, slightly acidic liquid or gas with a pungent odor, used in embalming, as a precursor compound in the manufacture of plastics, and in home insulation. Annual production of formaldehyde in the United States exceeded 5500 million pounds in the 1970s. As a hazardous transported chemical, formaldehyde was ranked slightly above propane, vinyl chloride, and hydrochloric acid. Because of its tendency to polymerize in pure concentrations, the compound is generally manufactured in aqueous solutions of 37–50 percent formaldehyde. Formaldehyde is also sold in methanol solutions. In 1980, the Center for Disease Control advised that persons exposed to formaldehyde in their work wear protective clothing and have adequate ventilation of the work area. Workers exposed to formaldehyde tend to develop a form of asthmatic bronchitis, coughs, dryness of the mouth and throat, and other upper respiratory complaints, plus headaches and eye irritation. If swallowed, formaldehyde can cause violent vomitting and diarrhea, followed by collapse. The Consumer Product Safety Commission has warned that improper installation of insulation might cause formaldehyde gas to be released, producing many of these same symptoms. Some industry tests have indicated that formaldehyde might have the potential to cause cancer. Formaldehyde has also caused mutations and chromosome aberrations in laboratory tests. The oral LD_{50} for rats is 800 mg/kg, but the lowest published lethal dose for women was 36 mg/kg. The time-weighted OSHA standard for exposure to formaldehyde is 3 ppm for eight hours and a 10 ppm maximum for 30 minutes. The NIOSH recommended ceiling is 1.2 mg/m³ for 30 minutes.

formula, chemical. A combination of symbols with their subscripts representing the constituents of a substance and their proportions by weight.

fossil fuel. A fuel that is the result of the decomposition of deposited vegetation over a considerable period of time, under the extreme pressure of the overburden of earth which has accumulated since the vegetation was deposited. The hardest and densest of these deposits is coal. Liquid deposits are oil, and those in gaseous forms are found as natural gas.

fractionation. The separation of the constituents of a mixture in successive stages, each stage removing from the mixture some proportion of one of the constituents. This process may be precipitation, crystallization, or distillation. In such an operation, a part of the vapor is condensed and the resulting liquid contacted with more vapor, usually in a fractionating column with plates or packing. The term "fractional distillation" is applied to any distillation in which the product is collected in a series of separate fractions.

Fraunhofer lines. The dark lines visible in the spectrum of sunlight. Kirchhoff conceived the idea that the sun is surrounded by layers of vapors which act as filters of the white light arising from incandescent solids. These vapor filters abstract those rays which correspond in their periods of vibration to those of the components of the vapors. Thus reversed or dark lines are obtained owing to absorption by the vapor envelope, in place of the bright lines found in the emission spectrum.

free field. A sound field in which the sound pressure decreases inversely with distance from the source. For free-field conditions, the power level (PL) of a simple point source may be calculated from a single measurement of the sound pressure level (SPL): PL = SPL + 20 log r + 10.5, where r is the distance in feet from the noise source to the point of measurement, and SPL is the overall sound pressure level re 0.0002 μbar. The power level determined from this equation will be in decibels re 10^{-13} W. In order to predict sound pressure levels at various points in a specified direction from the source, it is often convenient to add a directivity factor Q to the above equation. Q is defined as the ratio of the power of an imaginary point source producing the same observed sound pressure level at the specified place of measurement to the total sound power of the actual source.

free progressive wave. A wave propagated under conditions equivalent to those in an infinite, homogeneous (but not necessarily isotropic) medium.

free sound field. A field in a homogeneous, isotropic medium free from boundaries. In practice, it is a field in which the effects of the boundaries are negligible over the region of interest.

freezing out/combustion. A procedure used to analyze incinerator flue gas involving a series of traps for the collection of pollutants. Coolants used may become progressively lower-temperature substances, e.g., ice, dry ice, acetone, liquid nitrogen are applied to respective traps in series to bring about separation of the gaseous components. This is a method used for the analysis of liquid, vapor, and gaseous organic compounds which cannot be trapped by the filtration procedures used for suspended particulate matter.

freezing point. The temperature at which a liquid solidifies under any given set of conditions. Pure water under atmospheric pressure freezes at 32°F (0°C). However, the freezing point of water is depressed with increasing salinity; both freezing and melting points can be defined as the temperature at which both the solid and liquid forms of a substance can coexist.

Freon. A trade name for various chlorine- and fluorine-containing carbon compounds used as a working fluid in refrigerators and in certain air conditioners.

frequency. In uniform circular motion or in any periodic motion, the number of revolutions or cycles completed in a unit time.

Fresnel lens. A large lens, now usually made of plastic, scored so as to produce many small lenses which concentrate sunlight at a focal point or along a focal point.

friction, coefficient of. The ratio of the force required to move one surface over another to the total force pressing the two surfaces together. If F is the force required to move one surface over the other and W is the force pressing the surfaces together, the coefficient of friction is $K = F/W$.

front. A line or zone of discontinuity between two adjacent air masses of different temperature and moisture content. The air masses involved have different characteristics of temperature, humidity, and wind velocity. The properties of fronts are determined by whichever air mass wins the struggle in the boundary region or frontal zone. Fronts can be classified as four main types. The first is the cold front; since the cold air mass behind this front is denser and heavier than warm air, it displaces the warm air at the ground surface by sliding beneath it and pushing it upward. The warm front, the second type of front, pushes the cold air mass away from an area at the surface when its movement is stronger than that of the cold front. A third type of front is the stationary front; as its name indicates, it has little or no movement. The fourth type is the occluded front, a complex front created when a cold front swiftly overtakes a warm front. There are three kinds of occluded fronts: cold, warm, and neutral. With the cold occluded front, precipitation falls close to and behind the surface front. The warm occluded front produces precipitation ahead of the surface front. Precipitation falls along the front when a neutral occluded front develops.

frontal slope. The slope that develops when a cold air mass displaces a warmer one by sliding under it and pushing it upward, the advancing edge of the cold front being given a downward slant or slope. The slope is in the opposite direction, upward, when a warmer front pushes in on a cooler one.

fuel. The nuclear material in which fissions take place to sustain a nuclear chain reaction. The term is frequently applied to a mixture, such as natural uranium, in which only part of the atoms are fissionable, if it can maintain a self-sustaining chain reaction under the proper conditions.

fuel cell. (1) A galvanic cell in which chemical action is used to produce electricity. Both open-cycle and closed-cycle fuel cells of various types, using hydrogen and oxygen, are being developed. Fuel cells have a relatively high fixed mass, varying from 20 to 100 lb/kW, are dependent on size, and have efficiencies of 50–65 percent. (2) A device (given great impetus by the space program) consisting of a positive and a negative electrode immersed in a conducting solution or electrolyte, which is able to convert chemical energy into electrical energy. The greatest potential use of such devices would be to supply the power needs of individual buildings, eliminating the need for wires, transmission lines, and other equipment associated with centralized power production.

fuel oil characteristics. (1) *Calorific value.* The amount of heat given off by a unit quantity when it is completely burned, usually expressed in British thermal units per pound of oil (Btu/lb). (2) *Viscosity.* The viscosity of any fluid is a measure of its resistance to flow and is defined as the force needed to shear a cube of unit dimensions of the fluid at a unit speed. (3) *Specific gravity.* The ratio of the weight of a given volume of the oil to the weight of an equal volume of water. (4) *Flash point.* The closed flash point is the temperature at which the oil produces sufficient vapor in the closed space of the flash point apparatus to ignite when a flame is introduced. (5) *Pour point.* The lowest temperature at which the oil will just flow under the conditions of a laboratory test. It has an important bearing on the storage and handling of the heavier grades of fuel oil. (6) *Sulfur content.* Different fuels contain from 0 to 4 percent sulfur. Those derived from the residual oils of the distillation process generally contain the highest percentages. (7) *Ash content.* The lighter grades of oil contain virtually no ash; the heavier grades may contain by specification up to 0.2 percent, but contain generally below 0.1 percent. (8) *Water and sediment.* The water and sediment content of light distillate grades of oil is extremely small, but the heavier, more viscous oils tend to keep the sediment and droplets in fine suspension.

fume. Fine solid particles predominantly less than 1 μm in diameter, resulting from the condensation of vapor from chemical reactions.

fumigant. A pesticide that is vaporized to kill pests; often used in buildings or greenhouses.

fumigation. An atmospheric phenomenon in which pollution, which has been retained by an inversion layer near its level of emission, is brought rapidly to ground level when the inversion breaks up. High concentrations of pollutant can thus be produced at ground level.

fuming nitric acid. A highly concentrated solution of nitric pentoxide (H_2O_5) in water, red or brown in color, and more active than clear nitric acid. It is sometimes used as an oxidizer.

fundamental frequency. The lowest component frequency of a periodic quantity.

fungus. A tiny plant that lacks chlorophyll. Some fungi cause disease, others stabilize sewage and break down solid wastes for compost.

furfural. A solvent obtained from oat hulls and used in connection with selective solvent extraction by the furfural solvent refining process of lubricating oils and diesel fuels; also used to recover butadiene during the hydrogenation of butenes. Furfural alcohol ($C_4H_4OCH_3OH$) is a colored liquid of molecular weight 98.05; specific gravity, 1.128; and boiling point, 170.2°C.

furnace. Any enclosed structure in which heat is produced; a combustion chamber. When air and fuel enter a furnace and heat is applied, normal, incomplete combustion is likely to take place because there is too much or too little air, because the temperature is too low, because the time allowed for burning is insufficient, or any combination of these factors. As a consequence the byproducts emitted through the smokestack pollute the air. They may include unburned bits of carbon, carbon monoxide gas, and fly ash and gas from the impurities in the fuel. When the furnace is used for smelting or for other industrial processes requiring heat, still more impurities may be given off. When sulfur is present in the fuel, it is converted to gas—about 98 percent sulfur dioxide and 2 percent sulfur trioxide. In addition to being undesirable themselves, these gases combine with water vapor in the flue or the outside atmosphere to produce sulfuric acid. The diagram shows an oil-burning furnace used to supply steam heat.

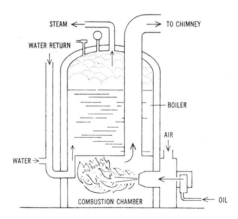

Oil-burning furnace used to supply steam heat.

G

gal. A unit of acceleration equal to one centimeter per second per second. 1 gal = 1 cm/s^2. Therefore where the value of gravity is 980 cm/s^2, this is the same as a value of 980 gals. The milligal is now quite commonly used, since it is approximately one part in a million of the normal gravity of the earth.

gallium arsenide. GaAs, a compound used in making photovoltaic cells.

game fish. Species like trout, salmon, bass, etc., caught for sport. They show more sensitivity to environmental changes than "rough" fish.

gamma photon. A quantum of electromagnetic radiation emitted by a nucleus as the result of a transition between two energy levels.

gamma ray (γ ray). (1) A very-short-wavelength highly penetrating ray with a frequency higher than that of x rays, emitted during the spontaneous disintegration of radioactive substances and during nuclear reactions; they are dangerous to life. (2) Quanta of electromagnetic wave energy similar to but of much higher energy than an ordinary x ray. The energy of a quantum is equal to hv ergs, where h is Planck's constant (6.6254 × 10^{-27} erg s) and v is the frequency of radiation. Gamma rays are highly penetrating, an appreciable fraction being able to transverse several centimeters of lead.

garbage. *See* **food waste.**

garbage grinding. Using a household disposal to crush food waste and wash it into the sewer system.

Garrett's coal pyrolysis. *Process:* Crushed, dried coal is conveyed by recycled product gas to the entrained bed carbonizer (pyrolysis reactor). It is heated by recycled char from the char heater. The temperature of the reactor is maintained at about 1100°F. Effluent from the reactor passes through cyclones to separate char from gas. A portion of this is cooled to be product char. Remaining char is fed to the char heater, an entrained reactor. A portion is burned with air to reheat the remainder to a temperature of about 1400°F. Heated char is recycled to the pyrolysis reactor. The gaseous stream is cooled and scrubbed to remove tar. A portion of the gas stream is used for conveying coal and recycled char to the pyrolysis reactor. After acid gas removal, a portion of the gas is taken as product gas and may be further upgraded to pipeline quality. The remaining gas is used in the production of hydrogen, which is then used to hydrotreat tar obtained from the coal pyrolysis to obtain synthetic crude oil (Garrett Research and Development Co., Inc.).

gas. A state of matter in which the molecules are practically unrestricted by cohesive forces. A gas has neither definite shape nor volume.

gas amplification. As applied to gas ionization radiation detecting instruments, the ratio of the charge collected to that charge produced by the initial ionizing event.

gas black. Finely divided carbon produced by incomplete combustion or thermal decomposition of natural gas. Other names used are carbon black, channel black, and furnace black.

gas plume. A stack effluent composed of gases alone or gases and particulates. The geometrical form and concentration distribution of gas plumes are dependent on turbulence. Because of distinctly different atmospheric conditions, gas plumes may be classified into five types: (a) *Looping plume.* Looping plumes occur with a high degree of turbulence, especially convective turbulence, typical of a daytime condition with intense solar heating of the earth's surface, which causes unstable thermal conditions. Similar plume characteristics may occur with large-scale mechanical turbulences caused by large and sharp hills upwind of the stack. (b) *Coning plume.* Coning plumes occur under nearly neutral thermal conditions when mechanical turbulences of a smaller scale are predominate. (c) *Fanning plume.* Fanning plumes occur under stable conditions when mechanical turbulence is suppressed. The vertical component is suppressed more than the horizontal, with the result that the plume width is greater than its thickness; this is likely to occur at night when the earth is cooled by outgoing radiation. (d) *Fumigating plume.* Fumigating plumes are caused by unstable air below and stable air above, causing a greater downward dispersion of gases and bringing more of the plume to the ground than would occur with unstable conditions above and below. This usually occurs in the morning following a night of marked stability; the rising sun heats the ground, causing an unstable layer to develop from the ground upward. When the unstable layer reaches plume level, the compact, highly concentrated plume quickly diffuses downward and a high ground-level concentration may occur for half an hour. (e) *Lofting plume.* Lofting plumes are caused by conditions inverse to those causing fumigation, the greater diffusion is upward and late afternoons and early evenings are favorable for this condition. Two or more stack effluents from closely grouped stacks tend to merge into one.

gaseous atmospheric contaminants—properties.

Substance	Mol. wt.	B.p. (°C)	F.p. (°C)	Ordinary state	Vapor pressure (mm)[a]					
					20°C	Ice	NH_3	CO_2	CS_2	O_2
Ammonia	17.0	−33.4	−77.7	vapor	6430.0	3192.0	760.0	44.0	0.6 (s)	
Arsine	77.9	−55.0	−113.5	vapor	11,400.0	6384.0	2128.0	338.0	36.0	0.2
Benzene	78.1	80.1	5.5	liquid	75.0	25.0 (s)	1.4 (s)	<0.1 (s)		
Carbon bisulfide	76.1	46.3	−111.8	liquid	295.0	127.0	21.0	0.6		
Chlorine	70.9	−33.8	−102.0	vapor	5016.0	2812.0	760.0	62.0	2.5 (s)	0.1 (s)
Ethylene	28.0	−103.7	−169.4	gas	—	30,856.0			462.0	0.1 (s)
Formaldehyde	30.0	−21.0	−92.0	vapor	b			22.0	1.0 (s)	
Hydrogen chloride	36.5	−84.0	−112.0	vapor	31,616.0	19,380.0	6840.0	988.0	120.0	0.8
Hydrogen sulfide	34.1	−61.8	−82.9	vapor	13,452.0	7752.0	2736.0	259.0	16.0	<0.1 (s)
Methanol	32.0	64.7	−97.8	liquid	96.0	30.0	2.7	0.1		
Methyl amine	31.1	−6.5	−92.5	vapor	2204.0	988.0	190.0	6.5	0.1 (s)	
Methyl mercaptan	48.1	6.8	−121.0	vapor	1292.0	577.0	109.0	3.8	0.1	
Ozone	48.0	−112.0	−251.0	gas	—	—			760.0	0.17
Sulfur dioxide	64.1	−10.0	−72.7	vapor	2432.0	1140.0	225.0	8.4 (s)		

[a]Vapor pressures are referred to the following refrigerant temperatures:

Refrigerant system	Temperature (°C)
O_2 (l) ⇌ O_2 (g)	−183.0
CS_2 (s) ⇌ CS_2 (l)	−118.5
CO_2 (s) ⇌ CO_2 (g)	−78.5
NH_3 (l) ⇌ NH_3 (g)	−33.4
H_2O (s) ⇌ H_2O (l)	0.0

bPolymerizes to paraformaldehyde or o-trioxymethylene.

gaseous diffusion. A method of separating fissionable isotopes from nonfissionable isotopes; it is based on the fact that gases with different masses will diffuse through a porous barrier at different rates.

gaseous diffusion separator. An instrument used to separate a gaseous mixture into its components, as when a gas is an isotopic mixture and is allowed to diffuse through another (preferably heavier) gas. Separation of the various isotopes is based on the principle that the lighter isotopic molecules diffuse through the heavier gas more readily than the heavier isotopic molecules.

gaseous matter, methods of collection.

Method	Advantages and limitations
High Concentrations	
Displacement, using pipettes, bottles, etc.	Standard gas analysis methods and equipment available for most gases. High degree of accuracy.
Direct reading devices	Immediate and often continuous analysis for a limited number of gases. Commercial equipment, not always portable.
Modifications of methods for low concentrations listed below	Usually involves dilution to determinable concentrations. Only method for some gases.
Trace Concentrations	
Absorption in suitable media using gas-washing bottles, fritted or sintered disks, impingers, U tubes with solid absorbents, etc.	Variety of types, often must be used in series, choice of absorbent most critical. Most widely used method.
Freeze-out traps	Limited use owing to difficulties associated with use of liquid air, dry ice, etc. Usually a "last resort" method.
Adsorption, using silica gel, activated charcoal, etc.	Excellent method for most solvent vapors and other gases that will absorb. Care necessary in properly activating adsorbent.
Direct reading instruments	Most desirable, available commercially for a few gases, can give continuous readings and permanent records. Usually expensive.

gasoline and diesel engine emission factors.

Pollutant	Gasoline engine emission (lb/1000 gal)	Diesel engine emission (lb/1000 gal)
Aldehydes	4	10
Benzo[a]pyrene	0.3 g/100 gal	0.4 g/100 gal
Carbon monoxide	2910	60[a]
Hydrocarbons	524[b]	180
Oxides of nitrogen	113	222
Oxides of sulfur	9	40
Ammonia	2	n.a.[c]
Organic acids	4	31
Particulates	11	110

[a]Includes blowby emissions, but not evaporation losses.
[b]Includes 128 lb/100 gal blowby emissions.
[c]Not available.

gauss. The cgs electromagnetic unit of magnetic induction (flux density), equal to one maxwell per square centimeter. It has a value such that if a conductor 1 cm long moves through a magnetic field at a velocity of 1 cm/s, in a mutually perpendicular induction, the induced emf is one abvolt.

Gay–Lussac's law of combining volumes. If gases interact and form a gaseous product, the ratio of the volumes of the reacting gases and the volumes of the gaseous products can be expressed by small whole numbers.

Geiger counter. A particular kind of detector consisting of a gas-filled tube in which an electrical discharge occurs if it is triggered by a particle or photon produced by a radioactive disintegration. A hollow cylinder acts as the negative electrode and a thin wire down the center is positively charged. The tube is filled with a low-pressure gas, and a potential difference of about 1000 V is applied to the electrodes. When a charged particle or gamma ray enters the Geiger tube, it knocks out electrons from the gas atoms in its path and produces ions. These electrons are then attracted to the positively charged wire. As they move in the direction of the wire, they collide with more gas atoms to produce additional free electrons and positive ions.

Geiger region. In an ionization radiation detector, the operating voltage interval in which the charge collected per ionizing event is essentially independent of the number of primary ions produced in the initial ionizing event.

Geiger threshold. The lowest voltage applied to a counter tube for which all pulses produced in the counter tube are of substantially the same size, regardless of the size of the primary ionizing event.

Gelman sequential sampler. An instrument consisting of a metal box containing 12 microimpingers, each connected through a rotary valve to a vacuum pump. An electric timer programs the valve and pump in such a manner that it is possible to sample through each impinger in turn, for a preset time interval, i.e., the sample could be preset to collect 12 two-hour samples of SO_2 each day. The impingers are then returned to the laboratory for analysis of the samples, and clean impingers with fresh reagent are loaded into the sample for another series of sequential samples.

gene. The fundamental unit of inheritance which determines and controls hereditarily transmissible characteristics. Genes are arranged linearly at definite loci on chromosomes.

generator. A mechanism which converts mechanical energy into electrical energy.

generator, Van de Graaff. An electrostatic generator which employs a system of conveyor belts and spray points to charge an insulated electrode to a high potential.

genetic effect of radiation. Inheritable changes, chiefly mutations, produced by the absorption of ionizing radiations. On the basis of present knowledge, these effects are purely additive and there is no recovery.

geostrophic wind. (1) A wind blowing with a steady speed along lines of constant pressure (isobars). (2) The horizontal wind velocity for which the Coriolis acceleration exactly balances the horizontal pressure force.

geothermal energy. The energy of the heat of the earth's interior. The nether regions of the earth are very hot, but in most parts of the world the observed temperature gradient in the outer crust averages only about one degree centigrade for every hundred feet of depth. In certain regions of the earth much steeper temperature gradients occur, sometimes as much as a hundred times the normal. It is the heat in these regions that is termed "geothermal energy." Such thermal regions are usually, but not always, closely associated with volcanic activity and earthquakes. The vast majority of earthquakes studied by scientists occurred in closely defined belts or zones, mostly of comparatively narrow width. The most important of these earthquake zones, which also contains a great number of active and extinct volcanoes, more or less follows the periphery of the Pacific Ocean. Another important zone runs along the middle of the Atlantic Ocean, with an easterly branch passing through the Mediterranean and the Middle East into Tibet. Another important zone more or less follows the direction of the Great African Rift and the Red Sea. All these zones are interconnected. Geothermal areas generally tend to lie within the earthquake belts, though not necessarily close to volcanoes. A primitive classification of geothermal field types distinguishes three broad classes: (a) Hot water fields, which contain a water reservoir at temperatures ranging from 60 to 100°C. Such fields can be useful for space heating, agricultural, and various industrial purposes. (b) Wet steam fields, containing a pressurized water reservoir at temperatures exceeding 100°C. This is the commonest type of economically exploitable geothermal field. Such fields can be suitable for power generation, as well as for other purposes. (c) Dry steam fields, or those that yield dry or superheated steam at the wellhead, at pressures above atmospheric. The temperatures of superheated steam may vary from 0 to 50°C. Geologically, wet steam and dry steam fields are generally similar.

geothermal energy application. The generation of power has been the most important application of geothermal energy, and will remain so in the future. It is being used extensively for area heating, particularly in Iceland, and for raising "forced" vegetables and flowers in heated glasshouses. It is being used for heat-intensive industries, such as paper making in New Zealand, and for the recovery and processing of diatomite in New Zealand by geothermal means. The most pressing need in the coming years will be for freshwater. Geothermal desalination holds great promise of cheap and abundant supplies of freshwater at certain places. Some geothermal waters are known to contain small proportions of highly valuable mineral ingredients, and their extraction also holds considerable promise. The industrial and other potentials of geothermal energy suggest that great economic advantages could be gained from dual or multipurpose plants combining power production with one or more other applications.

geothermal fluids. The character of fluids discharged from geothermal wells varies considerably. In some areas only steam (slightly superheated) is produced; elsewhere wells yield a mixture of steam and water. In other cases only hot water is produced, either above or below 100°C; if the temperature is at least 100°C, sufficient pressure is maintained to suppress boiling and the formation of steam. In most cases the fluid produced contains incondensible gases (typically CO_2 and H_2S, with minor amounts of other gases), the proportion of gas sometimes being so high that economic utilization is not feasible. Where water is present, it contains a variety of chemicals in solution. Water contained in steam can also be regarded as an impurity, but in many cases the heat in the water can be used to good effect. The water–steam ratio may be quite high, and ratios of 4:1 to 8:1 (by weight) are not uncommon.

geothermal low-temperature hot-water field. A low-temperature hot-water field may sometimes occur in an environment similar to that in the diagram (p. 146) showing the basic model of a steam field. It can also occur in fields devoid of cap rock, as shown in the diagram, where the thermal gradient and depth of the previous aquifer are sufficient to maintain a convective circulation. The temperature in the upper part of the reservoir will not exceed the boiling point at atmospheric pressure, partly because water brought up convectively from below will lose pressure (and temperature) as it rises and partly because there may be mixing with cool groundwaters.

geothermal steam field. (1) A source of natural heat of great output. (2) An adequate water supply. (3) An "aquifer," or permeable reservoir rock. (4) A cap rock. The source

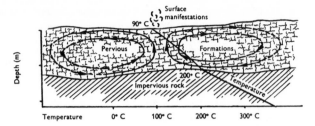

Basic model of a low-temperature hot-water field.

of heat is a magmatic intrusion into the earth's crust, having a temperature of 600–900°C, often at depths on the order of 7–15 km. In an active volcano, a magmatic intrusion reaches the surface through a large fault system. In compact, hard rocks, faulting may provide a channel for the upward flow of magma, while plastic rocks such as clay may flow by gravity into the fault space and seal it from above. Magmatic intrusions without present eruptions are common in acidic volcanoes and can also occur in basic volcanoes. Surface volcanic products like lavas and tuffs cool too quickly to originate a "commercial" geothermal field. The large structural depressions of the earth appear to be caused by the splitting of the crust at rates of several centimeters per year. There are good reasons for believing that grabens are the preferred site of magmatic intrusions, the chief reason being the fact that in a graben the overburden is less than in the adjoining positive structures, so that magma can intrude to a level where its energy is balanced by the weight of the overlying stratigraphic series. It appears that most of the water in an aquifer is of meteoric origin and that it is heated by conduction through a large impermeable base rock, even though some relatively small quantities of magmatic steam may penetrate this base rock through faults and fissures. As hot fluid is withdrawn from bores or from surface vents, the hydrological balance of the system is restored, or partially restored, by the inflow of new water. Many hydrothermal systems are dynamic, with water entering at

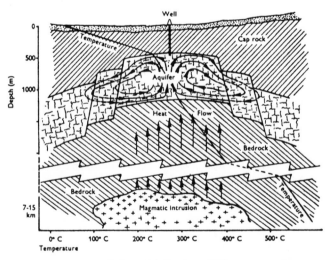

Basic model of a steam field.

some high level and leaving at some low level. A good productive geothermal well should produce at least 20 ton/hr of steam, many wells should produce a great deal more. A "wet" well may produce some hundreds of tons per hour of mixed fluid. Any permeable rock can serve as a good geothermal reservoir. A cap rock is a layer of rock of low permeability overlying the aquifer. All steam-producing fields have a cap rock. Some have been formed as original impervious rocks, others may have become impervious as a direct result of thermal activity. Hydrothermal alteration can be recognized by the bleaching of graywacke, and in places by the lack of vegetation.

germide. Any compound that kills disease-carrying microorganisms. These must be registered as pesticides with the EPA.

Gibbs' phase rule. $F = C + 2 - P$; F, the number of degrees of freedom of a system, is the number of variable factors (temperature, pressure, and concentration) of the components, which must be arbitrarily fixed so that the condition of the system may be perfectly defined. C, the number of the components of the system, is chosen equal to the smallest number of independently variable constituents by means of which the composition of each phase participating in the state of equilibrium can be expressed in the form of a chemical equation. The components must be chosen from among the constituents which are present when the system is in a state of true equilibrium and which take part in that equilibrium. As components are chosen, the smallest number of such constituents is necessary to express the composition of each phase participating in the equilibrium, zero and negative quantities of components being permissible. In any system the number of components is definite, but may alter with changes in the conditions of the experiment. A qualitative but not a quantitative freedom of selection of components is allowed, the choice being influenced by the suitability and simplicity of the application. P, the number of phases of the system, are the homogeneous, mechanically separable, and physically distinct portions of a heterogeneous system: The number of phases capable of existence varies greatly in different systems: there can never be more than one gas or vapor phase, since all gases are miscible in all proportions; a heterogeneous mixture of solid substances forms as many phases as there are substances present.

gilbert. The cgs electromagnetic unit of magnetomotive force. One gilbert (Gi) = $^{10}/_4\pi$ ampere-turns.

glasphalt. A material used in highway construction made of recycled glass that has been crushed and added to asphalt.

glazing. A transparent covering (e.g., glass or plastic) for a solar collector, window, or skylight.

gob pile. A fire which results when air can flow readily through the interstices of large piles of low-combustible refuse from coal mine preparation plants. Slow oxidation occurs, heat is liberated, and higher temperatures develop until ignition occurs.

Goetz particle spectrometer. A form of centrifuge which can be used for accurately determining particle size distribution. The air flows through a helical channel constructed on a rotating cone. Suspended particles are impelled outward from the axis of rotation under the influence of centrifugal force and collected on a metallic foil wrapped around the cone. Because of the systematically increasing centrifugal force exerted on the particles as they progress through the channel, the finer particles are collected farther downstream than the coarser ones.

gradient. Either the rate of change of a quantity (as temperature, pressure, etc.) or a diagram or curve representing this rate.

gradient wind. The wind that blows along curved isobars or contours with a velocity corresponding to the spacing of the isobars or lines of contour. The wind 2000 feet above the earth's surface is often referred to as the gradient wind.

Graham's law. The relative rates of diffusion of gases under the same conditions are inversely proportional to the square roots of the densities of those gases.

grain. A unit of weight equal to 65 milligrams or 2/1000 of an ounce.

grain loading. The rate at which particles are emitted from a pollution source—measurement is made by the numbers of grains per cubic foot of gas emitted.

gram-atomic-weight. A mass in grams numerically equal to the atomic weight of an element.

gram equivalent of a substance. The weight of a substance displacing or otherwise reacting with 1.008 g of hydrogen or combining with one-half of a gram-atomic-weight (8.00 g) of oxygen.

gram-molecular-weight (gram-mole). The mass in grams numerically equal to the molecular weight of the substance.

gram-rad. Unit of integral dose equal to 100 ergs.

gram-roentgen. A unit of integral dose; the real energy conversion when one roentgen is delivered to one gram of air (83.8 ergs).

graphite. An allotropic form of carbon found as a mineral or formed from coke. It has a very high sublimation temperature, around 4000°C, coupled with a low density (2.3 g/cm³). The heat required to vaporize graphite is 10,000 Btu/lb, about six times that required to ablate beryllium. Graphite is an outstanding material in respect to strength at very high temperatures (above 2000°C) and only four known compounds, all carbides, have melting points superior to it. Graphite is used to make crucibles for the manufacture of crucible steel and to melt brass and other nonferrous alloys. Such crucibles are composed of approximately 50 percent graphite, bonded with clay and sand. Finely pulverized graphite is used in foundaries to give the surface of molds a smooth finish. Graphite is also used in various types of lubricants.

grating. A plate on whose flat (or slightly curved) surface many parallel, equidistant lines have been engraved; it is used to measure the wavelength of a beam by measuring the angle of its diffraction.

gravitation. The force of attraction existing between all material bodies in the universe. The magnitude of the force between any two bodies is proportional to the product of the masses of the two bodies and inversely proportional to the square of the distance between them.

gravitational water. The water that continues to move down into deeper strata by gravitational force and exists in surface soil layers for only a very brief period after a rain.

gravity settling chamber. A collector that slows the gas from conveying velocities to settling velocities for a period of time sufficient to enable the heavier dust to settle under the influence of gravity into the dust hoppers, where it is periodically removed. Settling velocities range from 60 to 600 ft/min. The disadvantage of this type of collector is its very low collection efficiency for fine and moderately fine dusts. This characteristic limits the use of the gravity settling chamber to that of a precleaner. The diagram on p. 149 shows a typical gravity settling chamber.

gray body. A hypothetical body which absorbs some constant fraction, between zero and one, of all electromagnetic radiation incident upon it. This fraction is called the

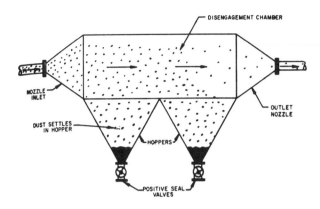

Gravity settling chamber.

absorptivity and it is independent of wavelength. As such, a gray body represents a surface of absorptive characteristics intermediate to those of a white body and a black body. No such substances are known in nature.

green belt. A buffer zone created by restricting development from certain land areas.

green flash. A predominantly green coloration of short duration often in the form of a flash, seen at the extreme upper edge of a luminary (sun, moon, or sometimes even a planet) when disappearing below or appearing above the horizon.

greenhouse effect. In the atmosphere carbon dioxide does not effect shortwave radiation. Upon striking the earth, shortwave radiation is transformed into longwave radiation or heat, which is reradiated. This heat is absorbed by the carbon dioxide molecules and transferred to the atmosphere. It is this heat transfer that is ultimately responsible for long-term climate and short-term weather changes. Because of the balance of incoming light and outgoing heat, the mean temperature of the earth remains at about 58°F. This phenomenon of carbon dioxide absorption of heat is called the greenhouse effect. A greenhouse permits most of the short wave solar radiation to pass through the glass roof and sides to be absorbed by the floor or ground and objects inside. These objects reradiate energy at a temperature of about 300°K, with a principal intensity around 10 μm. The glass absorbs the energy at these wavelengths and sends part of it back into the greenhouse, causing the inside of the greenhouse to remain warmer than the outside. The atmosphere similarly acts as a greenhouse, the gases absorbing selectively in somewhat the same way as the glass. This is the reason why the most efficient flat solar collectors provide a layer or two of glazing, transparent to sunlight but trapping the infrared heat energy absorbed by a black collector plate.

grease. In sewage, fats, waxes, free fatty acids, calcium and magnesium soaps, mineral oils, and other nonfatty materials. The type of solvents to be used for its extraction should be stated.

grit. (1) Solid particles of natural or man-made origin, retained on a 200 mesh British Standard sieve, i.e., 76 μm or greater in diameter. (2) The heavy mineral matter in water or sewage such as sand, gravel, and cinders.

ground. The state of a nucleus, atom, or molecule when it is at its lowest energy. All other states are termed excited.

ground cover.   Plants grown to keep soil from eroding.

ground level concentration.   The amount of solid, liquid, or gaseous material per unit volume of air, from zero to two meters above the ground.

ground level inversion.   An increase in temperature with increasing height of the atmospheric layers nearest to the ground (normally the temperature drops with increasing height).

ground state.   The arrangement of electrons in an atom in which they have the lowest energy and are therefore most stable.

groundwater.   The variable depth below the soil surface where there is a zone in which the soil or impermeable material is saturated with water. The upper level of this zone of saturation is called the water table. At times groundwater may be present in a position above the main zone of saturation; this is called perched groundwater and the upper surface of this layer is called the perched water table.

H

H-coal process. *Process:* The H-coal process can produce synthetic crude oil from coal. Coal is dried, pulverized (−40 mesh), and slurried with coal-derived oil. Slurry, mixed with hydrogen, is preheated and fed to an ebullient-bed reactor containing a cobalt–moly catalyst. Coal is hydrogenated and converted to liquid and gaseous products. Reactor conditions are about 850°F and 2700 psig. A constant catalytic activity is maintained by adding and withdrawing catalyst continuously. A slurry of unconverted coal and liquid product is let down from the reactor to an atmospheric-pressure flash drum. Flash vapors go to an atmospheric distillation tower. Bottom products are further processed in a vacuum tower to obtain vacuum distillate overhead and a vacuum bottom slurry product. The light hydrocarbon vapors in the gas leaving the reactor are removed in a recycled-gas scrubber. Ammonia and hydrogen sulfide are removed and gas is recycled to the H-coal reactor. Part of the heavy distillate product from the top of the vacuum distillation and the bottom of the atmospheric distillation are recycled as slurry oil (Hydrocarbon Research, Inc.).

h-particle. The positive hydrogen ion or proton resulting from the bombardment of a hydrogen atom by alpha rays or fast-moving positive ions.

habitat. The sum of environmental conditions in a specific place that is occupied by an organism, population, or community.

hadron. A joint name for baryons and mesons, that is, all particles which interact strongly with others. Elementary particles can be grouped into two broad classes depending on how they interact. Particles that exert strong nuclear forces are called hadrons (the strong ones). Protons and neutrons are hadrons.

hafnium (Hf). A metallic element with an atomic weight of 178.5; specific gravity, 13.10; melting point, 2000°C. It occurs in all the useful zirconium ores and is available as a byproduct.

hail. Precipitation in the form of small balls of irregularly shaped lumps of ice, invariably produced by convective clouds, nearly always cumulonimbus.

hair hygrometer. An instrument used to measure humidity, made with strands of human hair mounted under tension. The hairs contract when the humidity content of the air is low, and expand when it increases. One end of a sheaf of hairs is fixed, and the other is arranged so that it operates a set of levers. The moment the levers begin to move, their motion is magnified mechanically so that it causes a pointer or some other indicating device to move over a calibrated scale. The readings on this scale can be used in conjunction with tables of specific humidity and temperature measurements and can be converted into other humidity factors. This type of hygrometer is often employed in hygrography.

half-life, biological.   The time required for the body to eliminate one-half of an administered dose of any substance by regular process of elimination. The time is approximately the same for both stable and radioactive isotopes of a particular element.

half-life, radioactive.   The time taken for half of the nuclei in a radioactive element to disintegrate. The number N remaining in a given sample falls off exponentially with time, $N = N_0 e^{-\lambda t}$, where λ is the radioactive decay constant.

halite.   The mineral name for sodium chloride or common salt, NaCl. Colorless to gray, it occurs as cubic crystals. Its hardness is 2 to 2.5 on the Mohs' scale.

halogenated hydrocarbon.   A hydrocarbon having one or more hydrogen atoms replaced by a halogen atom.

halogenation.   The incorporation of one of the halogen elements, usually chlorine or bromine, into a chemical compound; e.g., benzene is treated with chlorine to form chlorobenzene, and ethylene is treated with bromine to form ethylene dibromide.

hammer mill.   A machine that breaks waste materials into smaller pieces by using a system of heavy, moving hammers.

hardness.   (1) The property of substances determined by their ability to abrade or indent one another. An arbitrary scale of hardness is based upon ten selected minerals. For metals the diameter of the indentation made by a hardened steel sphere (Brinnell) or the height of rebound of a small drop hammer (Shore Scleroscope) serve to measure hardness. (2) A measure of the capacity of water for precipitating soap; a characteristic of water which represents the total concentration of just the calcium and magnesium ions expressed as calcium carbonate.

harmonic.   Of a nonsinusoidal period quantity; a sinusoidal component of the periodic quantity having a frequency which is an integral multiple of the fundamental frequency. (2) An oscillation of a system, e.g., of a violin string or the air in an organ pipe, with a frequency which is a multiple of the fundamental (that is, the lowest) frequency of the system.

hazardous air pollutant.   A substance covered by the Air Quality Criteria, which may cause or contribute to illness or death: Asbestos, beryllium, mercury, and vinyl chloride are among a few.

hazardous waste.   A waste material which by its nature is inherently dangerous to handle or dispose of, such as old explosives, radioactive materials, some chemicals, and some biological wastes; usually produced in industrial operations.

haze.   A state of atmospheric obscurity due to the presence of fine dust particles in suspension; the visibility exceeds 1 km, but is less than 2 km. The particles are so small that they cannot be felt or individually seen with the naked eye. They diminish horizontal visibility and give the atmosphere an opalescent appearance. Many haze formations are caused by the presence of abundant condensation nuclei which may grow in size and become mist, fog, or cloud. (Heat haze is caused by volatilized essential oils of growing vegetation. Constituents of heat haze are examples of naturally occurring particulate materials in the atmosphere.) A distinction may be made between dry (0.1 μm) haze particles and damp (> 0.1 μm) haze particles. Dry haze particles produce a bluish color when the haze is viewed against a dark background; this same type of haze appears as a yellowish veil when viewed against a light background.

haze horizon. The top of a haze layer which is confined by a low-level temperature inversion in such a way as to give the appearance of the horizon when viewed from above against the sky. The true horizon is usually obscured by the haze in such instances.

header. A section of pipe which carries the main liquid flows at the top and bottom of a solar collector, and into which are fitted smaller pipes extending the length of the collector. The bottom header carries cool water to the collector, the top header carries sun-heated water away from it.

hearing. The process of perceiving sounds. Hearing starts with the occurrence of sound pressure fluctuations in air; these in turn vibrate the ear drum or tympanic membrane, which in turn activates a lever system of three very small bones, the auditory ossicles, situated in the air-filled cavity of the middle ear. The ossicles are grouped in such a way that the vibrations of the tympanic membrane drive the ossicle known as the malleus, which drives the second ossicle, the incus, and finally the third ossicle, the stapes, sets up vibrations of the coiled canals in the bone of the skull, known as the inner ear. These canals include the semicircular canals, concerned with balance, and the cochlea, in which is situated the end organ of hearing, or organ of Corti, where mechanical vibrations set up nerve impulses in the fibers of the auditory nerve. The information travels in this coded manner to the appropriate parts of the brain, to be perceived there as a sensation, probably stored as a memory, capable of arousing pleasure or annoyance, as well as having other secondary effects.

hearing defect (central deafness). A hearing impairment that cannot be explained by abnormality of the cochlea sense organ or auditory nerve. The cause is to be sought in any condition of the brain which may affect the nerve pathways from the auditory nerve to the temporal lobe of the brain where auditory sensations are represented. This category of deafness is sometimes described as of intracranial origin, that is, inside the space in the skull occupied by the brain and brain stem. Intracranial conditions which have deafness as a symptom include various types of tumors or abcess formations in the brain, interference with the blood supply to parts of the brain, such as caused by cerebral thrombosis or hemorrhage, or from injuries.

hearing defect (cochlea and auditory nerve). A hearing defect either in the cochlea itself or in the fibers in the auditory nerve; described collectively as sensory-neural deafness. This form of deafness or hearing loss is one suffered by the majority of adults in later life, for deterioration of hearing, especially for high frequencies, is apparently a characteristic of growing older. High frequencies are particularly affected, and the name presbycusis denotes the connection with age. The changes in this condition are progressive and may be detected from young adult life onward. In later life a well-recognized aspect consists of changes in the cochlea near the base, where the hair cells degenerate to varying degrees and the nerve fibers which would have carried their messages to the brain also degenerate. Certain drugs are known to be capable of damaging the cochlea, as are diseases not specifically otological in nature, such as meningitis resulting from any one of a number of different kinds of infective agents; in addition, certain virus diseases, such as mumps and measles, appear to be able to damage the organ of Corti.

hearing defect, sound-conducting pathway (conductive deafness). The conducting mechanism consists of the pinna, the external ear canal, the tympanic membrane, and the ossicles and their attachments, including the orifices and the oval and round windows. Defects which may involve the conduction mechanism may be easily anticipated. Since the passage of sound from the exterior to the oval window is a physical process involving the movement of air molecules and anatomical structures, any interference with these events will tend to reduce the transfer of sound energy into the cochlea. The obstruction may be in the meatus itself, due to a variety of things, such as a foreign body, water, or earwax. Infection of the middle ear can occur with the common cold and diseases such as measles and scarlet fever. Fluid collects in the middle ear, and the infective process may involve the formation of pus; the tympanic membrane itself may rupture, the discharge then running out of the ear canal, a condition known as suppurative otitis media. Chronic types of middle-ear infection may appear without a preceding acute stage. The result of these conditions is interference to varying degrees of the mobility of the ossicles, as well as damage to the eardrum and a reduction of the energy transfer to the oval window, and hearing is affected.

hearing effects of intense sound.   Any of the moving parts of the hearing mechanism can be damaged or even destroyed by intense sound. In the kind of hearing defect produced by prolonged exposure to intense sound, it is the hair cells which appear to be critically affected. In the human ear, whatever damage is produced in the inner ear is normally impossible to examine directly. The signs of harm are a reduction of hearing sensitivity.

hearing loss, occupational.   Certain kinds of work are sufficiently noisy to cause damage to hearing. As a result of work in a noisy situation, deterioration of hearing occurs at a rate mainly determined by the level of the exposure, but usually this goes initially unnoticed. Temporary dullness of hearing after exposure to the noise at work, with perhaps some noise in the ears (tinnitus) are the usual signs that damage is being done to hearing. Tinnitus takes various forms, and may be of a rushing or hissing nature or a more musical type of noise. Obvious tinnitus persists for periods lasting minutes or hours after a severe noise exposure, and is a sign of damage or potential damage. It is also found in ear diseases. Since noticeable dullness of hearing and tinnitus tend to disappear after some hours away from noise, they usually make little impression and are often ignored. A stage is reached, however, when the person begins to realize that his hearing is impaired, by which time little or nothing can be done to reverse the condition. The pattern of onset of occupational hearing loss and its subsequent development are characterized by a fairly predictable sequence of events. The first sign is a small depression in the audiogram between 3000 and 6000 Hz, commonly at 4000 Hz. If exposure to noise continues, the dip at 4000 Hz deepens, but still remains predominantly in the same frequency region. Continued exposure produces a definite slowing of the deterioration at the most affected frequency (about 4000 Hz), but there is a subsequent gradual extension to higher and lower frequencies, the precise pattern depending on the spectral characteristics of the noise exposure. At audiometric frequencies of 3000–4000 Hz, a permanent noise-induced threshold shift is almost at maximum after 10–15 years, irrespective of the eventual duration of exposure. For 2000 Hz and below, the rate of deterioration and the eventual degree of hearing loss is less the lower the frequency and is absolutely less in the case of higher frequencies. The deterioration, however, appears to continue indefinitely, or at least for observed durations up to 45 years.

heart disease.   All chronic respiratory diseases involve the heart, for stress on the heart and blood vessels is an inevitable result of a constricted or otherwise obstructed and injured respiratory tract. The cardiorespiratory system functions as a unit, one part making up for the occasional failure of the other. The heart must work harder to pump enough blood to compensate for any loss of oxygen due to respiratory disease. As a result the heart may show significant changes, sometimes doubling in size, as a secondary effect of lung affliction. The heart's burden is also increased by carbon monoxide, which can reduce the oxygen content of the blood. It is not known how much of a hazard is presented by small quantities of this gas, but such amounts may have a deleterious effect on the hearts of those already suffering from anemia or a cardiorespiratory disease.

heat. Energy transferred by a thermal process. Heat can be measured in terms of dynamic units of energy, like the erg, joule, etc., or in terms of thermal change in some substances, for example, the energy required per degree to raise the temperature of a unit mass of water at some temperature (calorie, Btu).

heat balance. The equilibrium which exists, on the average, between the solar radiation received by a planet and its atmosphere and that which it emits. That the equilibrium does exist in the mean is demonstrated by the observed long-term constancy of the earth's surface temperature. On the average, regions of the earth nearer the equator than about 35° latitude receive more energy from the sun than they are able to radiate, where a latitude higher than 35° receives less. The excess of heat is carried from lower latitude to higher latitudes by atmospheric and oceanic circulations and is reradiated there.

heat capacity. The quantity of heat required to increase the temperature of a system or substance by one degree. It is usually expressed in calories per degree centigrade. Molar heat capacity is the quantity of heat necessary to raise the temperature of one molecular weight of the substance by one degree.

heat conduction. The transfer of heat between particles in solid, liquid, and gaseous materials.

heat conductivity factor. The amount of heat (in kcal) which passes through a 1-m^2 cross section of a uniform material 1 m thick, with uniform heat transfer, when the temperature difference between the opposite surfaces is 1°C.

heat convection. The transfer of heat by the motion of liquids or gases as a result of a density change or force circulation.

heat equivalent. The quantity of heat necessary to change one gram of solid to a liquid with no temperature change; also called latent heat of fusion.

heat exchanger. (1) A device used for transferring heat from one substance to another, often through metal walls, usually to extract heat from a medium flowing between two surfaces. Regenerative cooling is one example. Sodium potassium is used as a heat transfer medium. (2) A device used for the efficient transfer of heat energy from one liquid to another, from air to liquid, or from liquid to air. In some solar systems, a heat exchanger may be required between the transfer medium circulated through the collector and the storage medium or between the storage and the distribution medium.

heat exchanger, double-wall. A heat exchanger providing a double separation between the transfer medium and the potable water supply consisting of tubing or a plate coil wrapped around and bonded to a tank. The potable water is heated as it circulates through the coil or tank. When this method is used, the tubing coil must be adequately insulated to reduce heat loss.

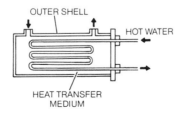

Double-wall heat exchanger.

heat exchanger, shell and double-tube. A type of heat exchanger similar to the shell and tube heat exchanger except that a secondary chamber is located within the shell to surround the potable water tube. The heated toxic liquid then circulates inside the shell but around this second tube. An intermediary nontoxic heat-transfer liquid is then located between the two tube circuits. As the toxic heat-transfer medium circulates through the shell, the intermediary liquid is heated, which in turn heats the potable water supply circulating through the innermost tube. This heat exchanger can be equipped with a sight glass to detect leaks by change in color (toxic liquids often contain a dye) or by a change in the liquid level in the intermediary chamber, which would indicate a failure in either the outer shell or intermediary tube lining.

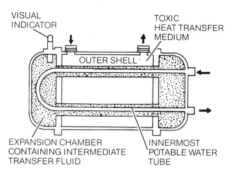

Shell and double-tube heat exchanger.

heat exchanger, shell and tube. A type of heat exchanger used to transfer heat from a circulating transfer medium to another medium used in storage or in distribution. Shell and tube heat exchangers consist of an outer casing or shell surrounding a bundle of tubes. The water to be heated is normally circulated in the tubes and the hot liquid is circulated in the shell. Tubes are usually metal, such as steel, copper, or stainless steel. A single shell and tube heat exchanger cannot be used for heat transfer from a toxic liquid to potable water because double separation is not provided and the toxic liquid may enter the potable water supply in case of tube failure.

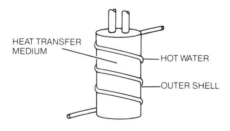

Shell and tube heat exchanger.

heat flow factor. The amount of heat in kcal which passes in 1 hr through 1 m^2 of a structure (e.g., a wall) of thickness d (in m) in still air when the temperature difference between the air spaces on the two sides (e.g., air in a room and outside air) is 1°C (heat flow normal to the surface).

heat island effect. The phenomenon of air circulation peculiar to cities in which warm air builds up in the center, rises, spreads out over the town, and, as it cools, sinks at the edge, while cooler air from the outskirts flows in toward the city center to repeat the flow pattern. In this way a self-contained circulation system is put in motion that can be broken only by relatively strong winds.

heat of adsorption. The heat given off when molecules are adsorbed.

heat of combustion of a substance. The amount of heat evolved by the combustion of one gram-molecular-weight of the substance.

heat of condensation. The amount of heat given up by one gram-molecule of a vapor condensing into a liquid at the same temperature and pressure as that of the vapor.

heat of formation. The heat evolved or absorbed when one gram-molecule of a compound is formed from its elements at constant volume.

heat of melting (heat of fusion). The quantity of heat absorbed when a unit mass of a solid is changed to a liquid at constant temperature and pressure.

heat of sublimation. The quantity of heat absorbed when a unit mass of solid is converted directly into vapor.

heat of vaporization. The quantity of heat required to convert one gram-molecule of a liquid at its boiling point at a given pressure completely into vapor at the same temperature and pressure.

heat pump. A device or machine that moves heat in the direction opposite to where it would usually go (not toward a colder area, but from a cold place to a warmer place). The heat pump makes cool areas cooler and warm areas warmer. In summer the heat pump sucks heat out of the hot inside air and exhausts it outside. In winter it pulls heat out of even subzero outside air and uses it to heat the house's interior. Most models use a Freon-type coolant, or antifreeze, to effect the movement of heat. In a space-heating installation, the heat pump can extract sun-derived heat from outside air and deliver it inside, acting as a kind of solar distribution pump.

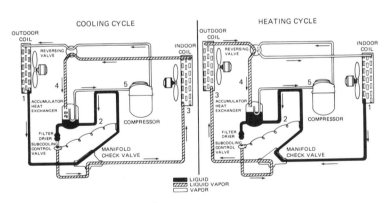

Heat pump.

heat quantity. The cgs unit of heat is the calorie, the quantity of heat necessary to change the temperature of one gram of water from 3.5 to 4.5°C (called a small calorie). If the temperature change involved is from 14.5 to 15.5°C, the unit is the normal calorie.

The mean calorie is 1/100 the quantity of heat necessary to raise one gram of water from 0 to 100°C. The large calorie is equal to 1000 small calories. The British thermal unit is the heat required to raise the temperature of one pound of water at its maximum density by 1°F; it is equal to about 252 calories.

heat ray. An electromagnetic wave which transfers heat between surfaces and their surroundings at different temperatures.

heat sink. A structure designed to absorb excess heat. For example, water pipes placed under an array of solar cells heated by sunlight pouring through Fresnel lenses; the water keeps the silicon solar cells cool enough to operate efficiently.

heat storage capacity. The amount of heat in kcal taken up in heating a structure, or given out in cooling it.

heat storage factor. The amount of heat in kcal required to heat 1 m^3 of a body by 1°C.

heat transfer. The transfer or exchange of heat by radiation, conduction, or convection within a substance and between the substance and its surroundings. Radiation represents the transfer of radiant energy from one region to another by electromagnetic waves with or without an intervening medium. Conduction, or diffusion of heat, implies the elastic impact of fluid molecules, without any net transfer of matter. Convection arises from the mixing of relatively large volumes of fluid due to fluid motion and may be caused by either local temperature inequalities (free convection) or an applied pressure gradient (forced convection).

heat transfer coefficient. The rate of heat transfer per unit area per unit temperature difference, a quantity having the dimensions of reciprocal length.

heat transfer factor. The amount of heat in kcal which is exchanged in 1 hr between a surface of 1 m^2 and the moving air, when the temperature difference between the air and the surface is 1°C.

heat unit. A unit quantity of heat, such as the calorie or British thermal unit; the heat required to raise a unit mass of water by one degree, within a specified temperature range.

heating season. The coldest months of the year, when pollution increases in some areas because people burn fossil fuels to keep warm.

heavy hydrogen. The hydrogen isotope having an atomic weight of 2 because of the addition of one neutron to the single proton in the ordinary hydrogen nucleus; also called deuterium.

heavy metal. A general name given to the ions of metallic elements such as copper, zinc, iron, chromium, and aluminum. Heavy metals are normally removed from waste water by forming an insoluble precipitate (usually a metallic hydroxide).

heavy water. Water in which the molecule consists of a combination of heavy hydrogen and oxygen. It is used as a moderator in a nuclear reactor.

heliographic. Referring to positions on the sun measured in latitudes from the sun's equator and in longitude from a reference meridian.

helionics. The conversion of solar heat to electric energy.

heliopyranometer. An instrument used to measure the sun's heat energy at any suitable site.

heliostat. A sun-tracking mirror that continually directs focused solar energy to a receiver; a mirror or array of mirrors, usually controlled by computer, placed as one of many heliostats which focus solar energy on a receiver or boiler on a tower, heating liquid in this boiler to very high temperatures under pressure.

helium (He). An inert, colorless, volatile, nonflammable, odorless, gas with an atomic weight of 4.003, a boiling point of $-268°C$, and a density of 0.187 47 g/liter at STP, found in natural gases and in the atmosphere.

hematite. A red iron oxide, Fe_2O_3, the chief ore of iron. It has a blood-red streak, like rouge or red paint, and may be soft and earthy or compact and hard. Hematite forms shiny, platy, steel-gray crystals, submetallic reddish-brown to black masses, red granules, and dull, red ocher. Hematite occurs as a primary mineral, as an oxidation product of weathering, and in sedimentary beds.

hemispherical. Referring to thermal radiation properties in all possible directions from a flat surface.

hemoglobin. The red, iron-containing protein pigment of the erythrocytes that transport oxygen and carbon dioxide and aid in the regulation of pH in blood.

henry. A unit of electrical inductance; the inductance of a closed circuit in which an electromotive force of one volt is produced when the electric current in the circuit varies uniformly at the rate of one ampere per second.

Henry's law. The mass of a slightly soluble gas that dissolves in a definite mass of a liquid at a given temperature is very nearly directly proportional to the partial pressure of that gas. This holds for gases which do not unite chemically with the solvent.

heptachlor. $C_{10}H_5Cl_7$, a compound generally used in pesticides. Heptachlor is a synthetic fat-soluble, but water-insoluble, chemical and because it is stable for periods of nine or more years, it is regarded as a persistent pesticide. Heptachlor is listed by the EPA's Carcinogen Assessment Group. In 1964, heptachlor epoxide was found in the body fat of the general population of the United States in amounts ranging from 0.10 to 0.24 ppm and in smaller, but detectable, amounts in the body fat of persons living in England. The oral LD_{50} of heptachlor for rats has been reported as 40 mg/kg and the LD_{50} for dermal exposure in rats has been established as approximately 195 mg/kg. A dose of one to three grams in a human can produce tremors, convulsions, kidney damage, respiratory collapse, and death. Subjects with liver impairment are particularly susceptible to adverse health effects. Heptachlor has been used as an experimental carcinogen in animals.

herbicide. Any chemical substance used to destroy plants. *Sources:* spillover from weed-control programs, herbicide industry. A general effect is leaf deformation and an example is cotton and grape leaves becoming deformed after treatment with one millionth of a gram or less, of 2,4-dichlorophenoxyacetic acid.

herbivore. An animal that feeds on plants.

hereditary. Pertaining to the transmission of characters and traits from parent to off-spring.

hertz (Hz). A unit of frequency equal to one cycle per second.

Hess's law of constant heat summation. The amount of heat generated by a chemical reaction is the same whether the reaction takes place in one step or in several steps, or all chemical reactions which start with the same original substances and end with the same final substances liberate the same amounts of heat, irrespective of the process by which the final state is reached.

heterocyclic compound. A cyclic compound which contains, in addition to carbon, one or more other atoms in the ring. The atoms which occur most frequently in heterocyclic rings are nitrogen, sulfur, and oxygen.

heterosphere. The part of the upper atmosphere where the relative proportions of oxygen, nitrogen, and other gases are variable, and wherein radiation particles and micrometeoroids are mixed with air particles.

heterotopes. Atoms having different atomic numbers, and therefore occupying different positions in the periodic system. They are separated by chemical means.

heterotroph. An organism which cannot synthesize its own food from inorganic materials and must therefore live either at the expense of autotrophs or upon decaying matter.

hexachlorobenzene. C_6Cl_6, a chlorinated hydrocarbon used as a fungicide. It can be acutely toxic when administered to animals via the oral route. The oral LD_{50} for rats is 3500 mg/kg. Accidental ingestion of hexachlorobenzene has proved fatal to humans as in a Turkish disaster in the 1950s when families during a food shortage ate seed grain that had been treated with the chemical. More than 3000 persons were poisoned and about 300 of the total, mainly infants and children, died. In some villages there were no children between the ages of two and five for a period of five years. Surviving children acquired darkly pigmented skin and a layer of dark hair over most of their bodies. Other victims developed photosensitive skin which was easily irritated and often rubbed off, leaving scar tissue. Other effects were enlarged livers, darkened urine, loss of appetite and weight, loss of muscle tissue and strength, and neuromuscular disorders that interfered with normal activities such as eating food with tableware or climbing stairs. Most of the survivors recovered eventually but for some the symptoms persisted into the 1970s.

hexachlorophene. $C_{13}H_6Cl_6O_2$, a halogenated aromatic hydrocarbon suspected of a rather wide range of adverse health effects, particularly in infants and fetuses. Although it was known that hexachlorophene was a toxic substance when administered orally, it was not known that toxic effects throughout the body could result from application of the germicidal substance to the skin. Around 1970, Swedish doctors became aware of a higher than average incidence of teratogenic effects among infants born to mothers who had used hexachlorophene soaps or cleansers during pregnancy. At about the same time, it was reported that 41 infants and young children had apparently died of hexachlorophene poisoning after a baby powder containing the chemical had been applied to skin inflamed with diaper rash. In 1973, several separate incidents involving death or tissue damage to babies exposed to hexachlorophene were reported. The cases included death after bathing infants for several days with hexachlorophene soap, skin burns associated with use of the soap, and infants born with ichthyosis, or "fish skin disease." A common complication in most cases was brain damage. More was learned about the potential toxic effects of hexachlorophene when manufacturers proposed that it would be an effective fungicide to protect fruit and other foods. Food and Drug Administration tests with laboratory animals showed that rats exposed to the chemical developed muscular weakness and paralysis due to an unpredicted mechanism of hexachlorophene causing destruction of the myelin sheaths protecting the nerve fibers of the brain and central nervous system. The myelin damage, called status spongiosum because of the spongy appearance of the nervous tissue, was also found in monkeys exposed to hexachlorophene and in post-mortem examinations of babies bathed with hexachlorophene soaps.

hexamethylphosphoric triamide. $C_6H_{18}N_3OP$, a colorless liquid possessing unique solvent properties and widely used in organic and organometallic reactions in laboratories. It also is used as a processing solvent in the production of Kevlar ™ aramid fiber. Hexamethyl-phosphoric triamide (HMPA) has been found to produce malignant tumors and nasal squamous cell carcinomas in rats exposed to it. However, there are no data available on the toxic effects of HMPA in humans. In addition to the evidence that hexamethylphosphoric triamide may be a carcinogen in rats, laboratory tests have produced toxic effects that include kidney disease, severe bronchiectasis and bronchopneumonia, and squamous metaplasia and fibrosis of the lungs. Oral doses administered to rats and cockerels resulted in testicular atrophy and aspermia. Rabbits developed weight loss, gastrointestinal disorders, and apparent nervous system dysfunction after application of HMPA to the skin. NIOSH has estimated that 5000 people are occupationally exposed to the chemical. The nasal tumors in rats occurred after repeated daily exposure to HMPA in concentrations ranging from 400–4000 ppb over a period of eight months. In some cases, the tumors originating in the lining of the nasal cavity spread into the brain. OSHA has not yet established standards for human exposure to hexamethylphosphoric triamide.

hiemal period. The winter interval which begins sometime in mid or late November in most temperate areas and lasts until the end of February; normally a period during which precipitation in the form of snow, frozen rain, or rain is plentiful. The average temperatures are usually freezing or somewhat above. Most of the organisms that are intolerant of these lower thermal values have hibernated, have encysted, or have been killed.

high-density polyethylene. A material used to make plastic bottles that produces toxic fumes when burned.

High Volume Air Sampler. A filtering device for particulate collection, used by the National Air Pollution Control Administration to determine the concentrations of suspended particulates in American urban and rural areas. A glass-fiber medium is used which can filter at the rate of 70 cubic feet per minute. The filter is over 99 percent efficient for the collection of particles as fine as 0.3 μm in diameter. High Volume Air Samplers generally run as long as 24 hours without an appreciable increase in airflow resistance.

histogram. A graphical representation of a frequency distribution. The range of the variable is divided into class intervals for which the frequency of occurrence is represented by a rectangular column; the height of the column is proportional to the frequency of observation within the interval.

hoarfrost. A deposit of ice having a crystalline appearance, generally assuming the form of scales, needles, feathers, or fans.

holding pond. A pond or reservoir usually made of earth built to store polluted runoff.

hole. A minute area in a semiconductor, with a positive charge equal in absolute value to the negative charge on an electron.

homocyclic. Containing rings of one kind of atom only (usually carbon atoms).

homogeneous reactor. A reactor in which the uranium fuel is mixed with a liquid moderator, which can be light or heavy water or a molten metal, instead of being formed into metallic rods.

homologous. Having a similar structure, proportion, proportional to each other, identical in nature, relation, or the like.

homologous turbulence. A turbulence in which the mean value of the squares and products of the velocity components and their derivatives differ only in scale from point to point.

homologous series. (1) A series of organic compounds, each of which differs from its contiguous members in the series by containing one group more, or one group less, of two hydrogen atoms attached to one carbon atom. (2) A series of similar compounds which conform to a general formula.

homopause. The top of the homosphere, or the level of transition between it and the heterosphere. The homopause probably lies between 80 and 90 kilometers above the earth's surface, where molecular oxygen begins to dissociate into atomic oxygen. The homopause is somewhat lower in the daytime than at night.

homosphere. The lower portion of a two-part division of the atmosphere according to the general homogeneity of atmospheric composition (as opposed to the heterosphere); the region in which there is no gross change in atmospheric composition, that is, all the atmosphere from the earth's surface to about 90 kilometers. The homosphere is about equivalent to the neutrosphere and includes the troposphere, stratosphere, and mesosphere; it also includes the ozonosphere and at least part of the chemosphere.

Hooker Chemical Company pollution. *The Los Angeles Times* reported on June 18, 1979, that internal memorandums and laboratory reports (the internal memorandums and laboratory reports were obtained by the Securities and Exchange Commission in connection with Occidental Petroleum's unsuccessful bid to take over the Mead Corporation in 1978) revealed that Occidental Chemical Co. had dumped five tons of toxic chemicals each year for several years near its plant in Lathrop, California. The documents further revealed that officials of Hooker Chemical Co., who managed the Occidental Chemical plant in California, had been aware as early as April 1975 that its discharges of pesticides, fertilizers, and nitrates were infiltrating drinking wells near the plant and posing serious health hazards to the local population. The dumping practices of Hooker Chemical Co. were under investigation by the Justice Department and the Environmental Protection Agency as a result of the investigation into the Niagara Falls Love Canal incident, where chemicals leached into local residents' backyards and basements and forced the evacuation of 239 homes.

In Michigan, the state filed a lawsuit against Hooker Chemical Co. seeking damages for cleanup operations costing $220 million. Investigations showed that Hooker Chemical Co. had been storing 55-gallon drums of pesticide waste in a site near its plant in Montague, Michigan since 1957. The drums eventually rotted, allowing a poisonous compound called C-56, used to make several insecticides, to leak into the soil and flow toward nearby White Lake. In Florida, where Hooker was convicted of polluting the air with fluoride in 1978, authorities discovered that the company's phosphorus plant in White Springs was violating emission standards. *The New York Times* reported that documents dealing with the White Springs plant indicated that although the emissions permit for one stack of the phosphorus plant allowed for a discharge of up to 34 pounds (15.5 kilograms) a day, on some days as much as 3000 pounds (1364 kilograms) of fluoride were released into the atmosphere.

The White Springs plant produced concentrated phosphorus compounds from phosphate rock mined in the area. The Hooker Chemical Company plant's release of fluoride and phosphate went directly into the Suwannee River. (See **Love Canal seepage.**)

Hook's law. (1) A description of the behavior of many elastic objects, it states that a distance an object is stretched is proportional to the force exerted on it, provided it is not stretched beyond the elastic limit. (2) Within the limits of perfect elasticity, strain is directly proportional to stress.

horizon. The great circle of the celestial sphere midway between the zenith and the nadir.

horsepower. The commercial unit of mechanical power or the rate of doing work. One horsepower (hp) is equal to the expenditure of 33,000 foot-pounds per minute, or 550 foot-pounds per second. The force or weight moved in pounds multiplied by the velocity in feet per minute gives the foot-pounds per minute. The product is divided by 33,000 to give the horsepower.

hot.   Slang for radioactive material.

hour angle.   The angular distance west of a celestial meridian or hour circle; the arc of the celestial equator, measured westward through 360°.

hour circle.   A great circle of the celestial sphere passing through the celestial poles and a celestial body on the vernal equinox. An hour circle moves with the body as the celestial sphere rotates, unlike the celestial meridian of a point which remains fixed.

humic acid.   One of the various organic acids derived from humus, the organic portion of the soil formed of partly decomposed animal and plant matter.

humidity.   The measure of the quantity of water vapor contained in a given portion of the atmosphere. This may be expressed directly as the number of grams of water vapor in a cubic meter, or in terms of the contribution of water-vapor pressure to the total pressure of all the atmospheric gases. The latter is called the vapor pressure. It may be stated in terms of specific humidity, relative humidity, temperature of the dew point, or possibly some other quantities.

humidity, absolute.   The mass of water vapor present in a unit volume of the atmosphere, usually measured as grams per cubic meter. It may also be expressed in terms of the actual pressure of the water vapor present.

humidity, relative.   The ratio of the actual amount of water vapor present in the atmosphere to the amount needed to completely saturate air at the given temperature.

humus.   Decomposed organic material.

humus soil.   Humus which is formed and located in the lower portion of the A_0 section of the soil profile is organic in origin, derived from plant and animal remains that combine with mineral parts of the soil in the A horizon to form the complex colloidal structures within the soil. The importance of humus layers is that it provides food for soil organisms as well as chemical elements such as carbon, nitrogen, phosphorus, calcium, iron, and manganese that are essential for proper plant growth. Humus soils also tend to impede the leaching of nutrient substances from the upper soil layers. The amount of humus in any particular soil will vary with the vegetation and climate. In general, prairie soils contain the greatest quantities of humus, forest soils somewhat lesser amounts, and desert soils the smallest amount.

hurricane.   A tropical storm with wind speeds of 75 miles per hour and more, spawned in the North Atlantic, the Caribbean, the Gulf of Mexico, or the eastern North Pacific. It is usually accompanied by torrential rains, giant waves, and high tides. In the Northern Hemisphere, tropical storms blow counterclockwise around a calm central area called the "eye." In the Southern Hemisphere, hurricane winds blow clockwise. The maximum speed of hurricane winds can range anywhere from 100 to 150 miles per hour; gusts have frequently been measured at 200 miles per hour. The area lashed by the wind and rain of an average-size hurricane may be from 200 to 500 square miles. The storm does not remain stationary, but travels along a path several thousand miles in length before it spends itself. Usually hurricanes move from lower to higher latitudes.

Huygens wavelets.   The assemblage of secondary waves asserted by Huygens to be set up at each instant at all points on the advancing surface of a wave or phase front. Many phenomena of wave optics can be explained on this assumption (the Huygens principle) of the continual creation of new wavelets and subsequent destruction or constructive interferences between the wavelets to set up the next imagined state of the advancing wave front.

HYDRANE. *Process:* Pulverized coal is fed to the coal-hydrogenation reactor, where it is contacted with a concurrent stream of hot gas (about 50 percent methane and 50 percent hydrogen). Pretreatment of even-caking coals is not necessary in this dilute-phase mixture; however, the walls of the reactor must be maintained at about 1500°F to prevent coal from sticking to them. About 20 percent of the coal is converted to methane in this action. Char, which is rendered noncaking in the coal hydrogenator, is fed to a fluid-bed hydrogenation reactor, where it is contacted with hydrogen to form methane. Methane, along with excess hydrogen, flows from this second reactor to the first reactor to further react with fresh coal feed. Char from the fluid-bed hydrogenator flows to a hydrogen generator, where the hydrogen required for the process is generated. Steam and oxygen are fed to the hydrogen generator and coal ash is removed from it. Raw product gas from the coal hydrogenator is treated for sulfur (H_2S) removal. No oxygen is introduced into the gasification system; the only carbon oxides in the gas are those which can be formed from the oxygen of the coal and this quantity is small. Acid-gas removal and final methanation systems are greatly simplified (U.S. Bureau of Mines).

hydrapulper. A machine used in the papermaking process, which cooks used paper until it forms a thick soup of wastepaper fibers called pulp. The machine also cleans the wastepaper by removing ink and other materials.

hydrate. A compound formed by the intimate union of water with a molecule of some other substance, the molecular structure of which is represented as actually containing water, a hydroxide, such as calcium hydrate, from which water may be separated by a simple readjustment of the molecular structure.

hydrated lime. The same as slacked lime, made by adding water to quick lime and used in the manufacture of soap for making hydrated grease.

hydration. (1) The attachment of water molecules to particles of solute. (2) The solvation process in which water is the solvent. (3) The addition of hydrogen and oxygen atoms to a substance in the proportion in which they occur in water. (4) The chemical addition of water to the minerals of a rock to form new minerals, chiefly hydrous silicates and hydrous oxides. Carbonation frequently occurs together with hydration. Thus orthoclase feldspar, a mineral abundant in granite, is decomposed and converted largely to kaolin, the principal mineral in common clay. The potassium and excess silica are released at the same time.

hydraulic radius (streams). The current velocity of any stream will be governed by the steepness of the basic gradient or the ratio of vertical drop to the length of the stream and the hydraulic radius. The hydraulic radius is obtained by dividing the area of the cross section of stream by the wetted perimeter. The hydraulic radius will be proportional to the stream velocity.

hydrazine. H_2NNH_2, a colorless fuming liquid at room temperature and a powerful reducing agent that can self-ignite on contact with soil, asbestos, cloth, or wood. Hydrazine is a powerful explosive often used as a rocket fuel. When heated to decomposition, it can produce highly toxic fumes of nitrogen compounds. However, because of its sensitivity, it is more likely to explode when exposed to heat unless it is in a carefully controlled reaction. Compounds of hydrazine, such as phenylhydrazine, are also dangerous and highly toxic. Workers exposed to phenylhydrazine often develop a severe form of dermatitis and may experience systemic effects such as anemia and general weakness, gastrointestinal disorders, and kidney damage. Hydrazine and its compounds tend to destroy red blood cells, converting the hemoglobin molecules to methemoglobin. Hydrazine is an experimental carcinogen of the lungs, nervous system, liver, kidney, bone marrow, breast, and subcutaneous tissue. In 1980 it was listed by the EPA's Carcinogen Assessment Group and accepted by OSHA. The oral LD_{50} for rats is 60 mg/kg and the LC_{50} by inhalation for rats has been established at 570 ppm for four hours.

hydrocarbon. A compound which consists solely of carbon and hydrogen. *Acetylene hydrocarbon.* Acetylene hydrocarbons are those which contain at least one pair of triple-bonded carbon atoms in their structure, satisfying the general formula C_nH_{2n-2}; also known as alkynes. These are unsaturated hydrocarbons. *Aromatic hydrocarbon.* Aromatic hydrocarbons are characterized by a molecular structure involving one or more six-carbon-atom rings and having properties similar to those of benzene, which is the simplest member of this group. *Branched-chain hydrocarbon.* Branched-chain hydrocarbons are nonaromatic hydrocarbons in which not all the carbon atoms of the molecule are in a single chain. The simplest is isobutane. *Olefin hydrocarbon.* Olefin hydrocarbons contain at least one part of double-bonded carbon atom in their structure, with a single double bond satisfying the formula C_nH_{2n}. These are unsaturated hydrocarbons. *Paraffin hydrocarbon.* Paraffin hydrocarbons are those in which the proportion of hydrogen to carbon is such as to satisfy the general formula C_nH_{2n+2}, such as methane, CH_4, or ethane, C_2H_6. These are also called saturated hydrocarbons. *Straight-chain hydrocarbon.* Straight-chain hydrocarbons are those in which all the carbon atoms of the molecule are in a single unbranched chain; they are designated also as normal hydrocarbons. *Unsaturated hydrocarbon.* Unsaturated hydrocarbons have fewer hydrogen atoms or equivalent groups than the corresponding saturated compounds. *Polycyclic hydrocarbon.* Polycyclic hydrocarbons are molecules having more than one ring in their structure. These may be aromatic rings, cycloalkane rings, cycloalkene rings, heterocyclic rings (with other atoms besides carbon in the ring), and combinations of such rings.

hydrocarbon, aromatic. A hydrocarbon which has alternating single and double covalent bonds in six-membered carbon rings.

hydrocarbon chain. A hydrocarbon series such as the paraffin series in which the hydrocarbon atoms are bound to the carbon atoms in a chain form, or with the succeeding carbon atoms aligned along a straight line with hydrogen atoms on either side.

hydrocarbon cracking. The thermal decomposition, with or without a catalyst, of petroleum or heavy petroleum fractions, with production of lower-boiling-point materials useful as motor fuels, domestic fuel oil, or other needed products; generally, a thermal decomposition process, where ammonia may be cracked into nitrogen and hydrogen, and methane may be cracked into carbon and hydrogen or other hydrocarbons.

hydrocarbon, naphthene. A saturated hydrocarbon similar to paraffins in its properties, but the carbon atoms in each molecule are arranged in a closed ring. Therefore naphthenes are also known as cycloparaffins. They occur in naphthas and higher-boiling-point fractions. Naphthenes can be changed into higher octane compounds.

hydrocarbon, olefin. An unsaturated hydrocarbon. If one hydrogen atom can be added to an olefin, it is a monolefin; if two can be added, it is a diolefin. Olefins are almost entirely formed by cracking. Diolefins are also present, but are so chemically unstable that they react with one another to form a gum and must therefore be removed from gasoline or prevented from overacting. The monolefins have names corresponding to those of the paraffins, having the endings -ene or -ylene, ethane (ethylene) and propane (propylene). Diolefins also have names corresponding to paraffins, but end in -diene, as in butadiene.

hydrocarbon, paraffinic. A saturated hydrocarbon containing the maximum number of hydrogen atoms, chemically inactive under most conditions but chemically reactive when broken apart. The carbon atoms form chains. Those without branches (straight-chain structures) are normal paraffins; those with branches are isoparaffins. Paraffins with four or less carbon atoms are gaseous at ordinary temperatures; those with five to fifteen are liquids; those with over fifteen are waxes. The lighter gaseous paraffins (methane, ethane) are components of natural gas. The heavier gaseous paraffins (propane, butane, and isobutane) are components of liquefied petroleum gas (LPG).

hydrocarbon, ring compound. A hydrocarbon molecule in which the carbon atoms are arranged in a ring form, to which one hydrogen atom is connected to each carbon atom in the radical position. This is also known as a "cyclic compound."

hydrocarbon, saturated. A hydrocarbon molecule which is unable to absorb more hydrogen atoms, such as the paraffin series hydrocarbons.

hydrocarbon series. A group of hydrocarbon compounds which follow the same general law of molecular arrangement and composition. The arrangement follows the ring or chain order and the composition covers the proportion of carbon atoms to hydrogen atoms.

hydrocarbon, unsaturated. A hydrocarbon compound which has the ability to absorb additional hydrogen than naturally existing in the compound. The olefins or ethylenes are unsaturated hydrocarbons and form new compounds when additional hydrogen is supplied.

hydrochloric acid. A common acid made from hydrogen and chlorine; also called muriatic acid.

hydrocracking unit. The process for cracking heavy hydrocarbons to light products in the presence of high partial pressures of hydrogen and a special catalyst. Hydrocracked products are considered to be essentially sulfur-free.

hydrogen (H). A colorless, tasteless, and odorless gaseous element with an atomic weight of 1.0080 and a boiling point of -252.8°C. Hydrogen has the simplest atom of all the elements. It is found in most of the substances that form living matter; combined with oxygen it forms 90 percent of the energy in the sun and stars depend on it. It makes up 15 percent (by weight) of petroleum and all products secured from petroleum. It plays an important part in the combustion of gasoline and all other petroleum fuels. There are two molecular species of hydrogen, termed *ortho* and *para*. The two atoms in *ortho*-hydrogen spin in the same direction; the two atoms in *para*-hydrogen spin in opposite directions. The stable form of hydrogen at room temperature contains about 75 percent *ortho*-hydrogen, and liquid hydrogen at its boiling point contains over 99.8 percent of the *para* form; unless catalysts are included in the liquefaction circuit, the liquid hydrogen will be 75 percent *ortho*, a percentage that can be appreciably reduced by catalysts introduced during manufacture. The *ortho* constituent transforms to *para* in storage, releasing excess energy which boils off the liquid hydrogen. The excess energy arises from 338 cal/mol of the *ortho* during conversion, which is greater than the latent heat of liquid hydrogen, namely, 216 cal/mol. Space is filled with hydrogen atoms at densities varying from one in 20 cubic centimeters to more than two per cubic centimeter. About nine-tenths of the universe is hydrogen.

hydrogen equivalent. The number of replaceable hydrogen atoms in one molecule, or the number of atoms of hydrogen with which one molecule could react.

hydrogen ion concentration. The concentration of hydrogen ions in solution when the concentration is expressed as gram-ionic-weights per liter. A convenient form of expressing the hydrogen ion concentration is in terms of the negative logarithm of this concentration. The negative logarithm of the hydrogen ion concentration is called the pH. The significance of pH is still in dispute. Water at 25°C has a concentration of H ions of 10^{-7} and a concentration of OH ions of 10^{-7} moles per liter. Thus the pH of water is 7 at 24°C. Greater accuracy is obtained if one substitutes the thermodynamic activity of the ion for its concentration.

hydrogen peroxide. H_2O_2, a colored liquid with a molecular weight of 34.02; specific gravity, 1.4631; melting point, $-1.7°C$; boiling point, 152.1°C. When forced through a platinum mesh, it decomposes into hot steam and oxygen. Manganese dioxide is used as a catalyst for its decomposition.

hydrogen sulfide. H_2S, a compound produced in large quantities in the processing of petroleum, in the coking of coals, and in the operating of paper pulp mills, using the Kraft process. Hydrogen sulfide is involved in the distillation of tar, gas refining, the manufacturing of viscose rayon, and in certain chemical processes. Other sources include inadequately treated sewage dumps and other environments in which anaerobic bacteria can function. Hydrogen sulfide is objectionable because of its distinct and unpleasant odor even at very low concentrations. It causes odor nuisances when present in the air at concentrations 10 to 100 times smaller than the lowest concentration of sulfur dioxide detectable by smell. People can smell hydrogen sulfide in concentrations of approximately 0.035 to 0.10 ppm. Silver and copper tarnish rapidly in the presence of hydrogen sulfide. House paints that contain lead compounds darken rapidly in the presence of even low concentrations of hydrogen sulfide by forming black lead sulfide. It is unusual to find levels in the atmosphere that are high enough to cause damage to vegetation and irritation of the eyes and respiratory system. Higher concentrations can result from accidental industrial discharges from point of source, rather than from community-wide emissions. Many industries have taken extensive measures to remove hydrogen sulfide before it becomes a nuisance. It is standard practice in the petroleum industry to recover hydrogen sulfide for the production of sulfur, which is sometimes used later for manufacturing sulfuric acid.

hydrogenation. The chemical addition of hydrogen to a material. Hydrogenation may be either nondestructive or destructive. In the former, hydrogen is added to the molecules only if and where unsaturation with respect to hydrogen exists. In the latter, operations are carried out under conditions which result in the rupture of some of the hydrocarbon chains (cracking), the hydrogen adding on where the chain breaks have occurred.

hydrologic cycle. (1) The major pathway of the hydrologic cycle is an interchange between the earth's surface and the atmosphere via precipitation and evaporation. The relative and absolute amounts of precipitation and evaporation dictate a good deal of the structure and functions of ecosystems. (2) All water is locked into a recycling process called the hydrologic cycle. Solar energy, especially in the tropics, evaporates water from the ocean surface, filling the air mass above with large quantities of water vapor. When these warm, moist maritime masses conflict with cool, dry air over large land areas, some of the water vapor precipitates out. The reason that temperate latitudes produce frequent storms is that contrasting warm and cool air masses mix between the high and low latitudes.

hydrology. The science dealing with the properties, distribution, and circulation of water.

hydrolysis. (1) A double decomposition reaction involving the splitting of water into its ions and the formation of a weak acid or base or both. (2) A chemical reaction in which water acts upon another substance to form one or more entirely new substances. An example of hydrolysis is the conversion of starch to glucose by water in the presence of a suitable catalyst. (3) The splitting of a compound into parts by the addition of water between certain of its bonds, the hydroxyl group being incorporated in one fragment, and the hydrogen atom in the other.

hydrometeor. Any product of condensation or sublimation of atmospheric water vapor, such as rain, fog, or frost.

hydrometer. A device which floats upright in liquids at a depth which depends upon the density (or specific gravity) of the liquid; a scale on the hydrometer allows one to read the specific gravity of the liquid directly.

hydrometer, specific gravity. A hydrometer which indicates the specific gravity, or the relation of the weight of the given liquid per unit volume to the weight of a unit volume of water. It is used for all liquids, oils included.

hydrometer, thermometric. A hydrometer with any scale which is provided with a thermometer to take the temperature of the liquid at the same time that the density is measured. This is of importance, as the density of a liquid varies with the temperature.

hydrometer, Twaddell. A hydrometer used for liquids heavier than water and marked with the Twaddell scale, which when multiplied by 0.005 gives the specific gravity.

hydronic. A solar heating system in which the heat transfer fluid is a liquid, usually water, or a combination of an antifreeze mixture in the collectors and water in the storage and heating systems.

hydroponics. The science of growing food plants in water containing chemical nutrients and using lamps to simulate sunlight.

hydrosphere. The total water on earth, including the oceans; the inland lakes, swamps, rivers, and creeks; groundwater or water that has soaked into the ground and occupies openings in the lithosphere; water vapor in the atmosphere; water enclosed in sediments; and masses of continental ice like those of Antarctica and Greenland.

hydrostatic equilibrium. The state of a fluid whose surfaces of constant pressure and constant mass (or density) coincide and are horizontal throughout. Complete balance exists between the force of gravity and the pressure. The relation between the pressure and the geometric height is given by the hydrostatic equation.

hydrostatic pressure. The pressure caused by, or corresponding to, the weight of a column of water at rest.

hydrothermal. Pertaining to, or resulting from, the activity of hot aqueous solutions originating from magma or other sources deep in the earth.

hydrous. Containing water chemically combined, as in hydrates and hydroxides.

hydroxide. A compound containing the hydroxyl group OH, a group in which one oxygen atom is combined with one hydrogen atom.

hydroxyl. A monovalent group consisting of a hydrogen atom and an oxygen atom linked together; it may arise as a dissociation product in the combustion of rocket propellants.

hydroxylamine. NH_2OH, a member of a family of nitrogen compounds that includes nitrobenzene and the anilines. At room temperature, hydroxylamine exists in the form of white crystals, melting at 34°C to a colorless liquid. It is a moderate irritant to the skin, eyes, and mucous membranes and can cause alterations in red blood cells, producing methemoglobinemia. The LD_{50} in rats via the intraperitoneal route is 59 mg/kg and subcutaneous administration to rats has a reported LD_{50} of 29 mg/kg. Hydroxylamine also is a teratogen and was included in 1980 by OSHA on its list of suspected carcinogens. The compound is regarded as a dangerous explosion hazard in the presence of heat or flame and can react violently with sodium, zinc, and compounds of barium, copper, lead, potassium, and phosphorus.

HYGAS. *Process*: Coal is crushed, dried, pretreated (if necessary), and slurried with light oil. Coal slurry is fed to a three-stage gasifier. The first stage vaporizes light oil. Hydrogasification is carried out in the remaining two stages. The first stage of the hydrogasifier acts as a co-current transport reactor. Fresh lignite reacts with the hot effluent from the second-stage hydrogasifer. The latter is a fluid-bed gasifier in which char from the first stage reacts with steam and H_2. The required hydrogen can be provided from second-stage char by one of three alternate processes. (a) *Electrogasifier*. Heat for the reaction of steam with char to form H_2 and CO is supplied by an electric current passing through a fluid bed. (b) *Oxygen gasifer*. Heat for the reaction of steam with char to form H_2 and CO is supplied by combustion of a portion of the char with oxygen in a fluid bed. (c) *Steam-iron process*. Hydrogen is produced by reaction of steam with elemental iron. The iron oxide formed is contacted with producer gas obtained by combustion of char with air. The producer gas reduces the iron oxide to iron to complete the cycle. Raw gas from the HYGAS reactor is quenched to knock out oil. The net product oil is removed and slurry oil is recycled. CO_2 and H_2S are removed in purification stage 1 (Institute of Gas Technology).

hygrometer. An instrument which measures the water vapor content of the atmosphere. There are six basically different means of transduction used in measuring this quantity and an equal number of hygrometer types. These are (a) the psychrometer which utilizes the thermodynamic method; (b) the class of instruments which depend upon a change of physical dimensions due to absorption of moisture (hair hygrometer, torsion hygrometer, goldbeater's skin hygrometer, carbon-film hygrometer element); (c) those which depend upon moisture concentration (dew point hygrometer); (d) the class of instruments which depend upon a change of electrical or chemical properties due to moisture absorption (absorption hygrometer, electrical hygrometer, carbon-film hygrometer element); (e) the class of instruments which depend upon the diffusion of water vapor through a porous membrane (diffusion hygrometer); and (f) the class of instruments which depend upon measurements of water-vapor absorption spectra.

hygroscopic. Having the property of absorbing or becoming coated with moisture; changing form with changes of moisture content; absorbing and retaining moisture from the atmosphere.

hygroscopic water. A thin film of water surrounding soil particles, 15 to 20 molecules thick. The adhesive and cohesive forces are of such values that this water is unavailable to plants and animals.

hyperbaric. Pertaining to breathing atmosphere pressure above sea level.

hyperbarism. The phenomenon of disturbances in the body resulting from an excess of ambient pressure over that within the body fluids, tissues, and cavities.

hyperon. (1) Any particle with a mass intermediate to those of the neutron and the deuteron. (2) A particle, heavier than a nucleon, and tending to transform itself into one within a split second.

hypertonic. Having a higher osmotic pressure than the solution with which it is compared; having a higher concentration of solute and a lower concentration of solvent (water) molecules.

hyperventilation. Overbreathing. A respiratory-minute volume, or pulmonary ventilation, that is greater than normal. Hyperventilation often results in an abnormal loss of carbon dioxide from the lungs and blood, which may lead to dizziness, confusion, and muscular cramps.

hyperventilation syndrome. The blurring of vision, (feeling of) tingling of the extremities, faintness, and dizziness, which may progress to unconsciousness and convulsions, caused by reduction of the normal carbon dioxide tension of the human body, due to increased pulmonary ventilation.

hypocapnia. A deficiency of carbon dioxide in the blood and body tissues, which may result in dizziness, confusion, and muscular cramps.

hypoxia. An oxygen want or deficiency; any state wherein a physiologically inadequate amount of oxygen is available to, or utilized by, tissue without respect to cause or degree.

hysteresis. A time lag exhibited by a body in reacting to changes in the forces affecting it; any of several effects resembling a kind of internal friction, accompanied by the generation of heat within the substance affected. Magnetic hysteresis occurs when a ferromagnetic substance is subjected to a varying magnetic intensity; electric hysteresis occurs when a dielectric is subjected to a varying electric intensity; elastic hysteresis is the internal friction in an elastic solid subjected to varying stress.

hytherograph. A specialized type of climograph, which records temperature and precipitation. The amount of precipitation is plotted along the horizontal axis (abscissa) and thermal values are plotted along the vertical axis (ordinate). In a graph of this type it is possible to place a number of hytherographs on one graph for as many geographical locations.

Hz (hertz). A unit of frequency used in electronic measurement equal to one cycle or one wavelength of electromagnetic energy per second.

I

ice crystal. The structure of the ice crystal has great regularity and symmetry. Because this highly organized structure cannot accommodate other atoms or molecules without very severe local strain, salt and practically every other solute in the water is rejected by the advancing surface of a growing ice crystal. This means that the impurities accumulate just ahead of the crystal surface, with the purity of the ice crystal being very high. The ice crystal belongs to the hexagonal system. The hexagonal crystal axes are the three a axes in the horizontal plane, of equal length, with angles of 120° between the positive ends. The vertical axis is designated as the c axis. Ice crystals can be further identified as belonging to the ditrigonal pyramidal class.

ice point. The true freezing point of water; the temperature at which a mixture of air-saturated pure water and pure ice may exist in equilibrium at a pressure of one standard atmosphere.

illuminance. The total luminous flux received on a unit area of a given real or imaginary surface, expressed in such units as the foot-candle, lux, or phot. Illuminance is analogous to irradiance, but is to be distinguished from the latter in that illuminance refers only to light and contains the luminous efficiency weighting factor necessitated by the nonlinear wavelength response of the human eye.

illumination. Illumination on any surface is measured by the luminous flux incident on a unit area. The units in use are: the lux, one lumen per square meter; the phot, one lumen per square centimeter; and the lumen per square foot. Since at a unit distance from a point source of unit intensity the illumination is unity, unit illumination may be defined as that produced by a unit source at a unit distance, hence the meter-candle or candle-meter, which is equal to the lux, and the foot-candle equivalent to one lumen per square foot.

immission. The occurrence of solid, liquid, and gaseous substances polluting the air, primarily close to the ground, usually equivalent to ground-level concentration.

immunity. The power which a living organism possesses to resist and overcome infection.

impaction. A forcible contact of particles of matter; synonymous with impingement.

impactor. A sampling device which employs the principle of impingement. The "cascade impactor" refers to a specific instrument which employs several impactions in series to collect successively smaller sizes of particles.

impedance. The rate at which a substance absorbs and transmits sound.

impingement. The act of bringing matter forcibly into contact. As used in air sampling, impingement refers to a process for the collection of particulate matter in which the gas being sampled is directed forcibly against a surface. *Dry impingement.* The process of impingement in the gas stream, where particulate matter is retained upon the surface against which the stream is directed. The collecting surface may be treated with a film of adhesive. *Wet impingment.* The process of impingement in a liquid body and where that liquid retains the particular matter.

impingement samples. The standard impinger is used extensively for sampling air pollution. Its normal sampling rate is one cubic foot per minute (cfm) and the suction source is usually an electrically operated pump. The collecting medium is frequently water or water–alcohol mixtures. Air is drawn into the liquid medium at high velocities and at a fixed distance from an impingement plate, so that maximum wetting and collecting of particles is attained. The sample so collected can be analyzed both gravimetrically or chemically. In addition, particle size determination and dust counts can be made, although dust counts are not often of as much importance in air pollution as they are in industrial hygiene work.

impingement separator. A device used for the collection of medium-sized particles, employing impaction by permitting the gas to impinge on a plate, vane, or scroll. In essence, impingement separators utilize the inertia of particles in the dust-laden gas stream, which is directed through a virtual obstacle course. As the gas is deflected around each obstruction, the inertia of the particles causes them to maintain their original course, thus impinging on the obstacle's surface. The baffle chamber is the simplest form of impingement separator. It effectively collects particles 20 μm in size. Baffle chambers are designed so that the gas flows around zigzag plates or a staggered pattern of variously shaped obstacles. Material collected on these obstacles is flushed off by water. Baffle chambers are frequently used to remove suspended particles carried over from boiling liquids.

impinger. A sampling instrument employing impingement for the collection of particulate matter. Commonly this term is applied to specific instruments such as the midget and standard impinger. *Midget impinger.* A specific instrument employing wet impingement, using a liquid volume of 1 ml and a gas flow of 0.1 cfm. *Standard impinger.* A specific instrument employing wet impingement, using a liquid volume of 75 ml and a gas flow of 1 cfm. *Dry impinger.* A commercial type of impinger, usually consisting of a series of progressively smaller-sized jets impinging on standard microscope slides. This results in progressively higher jet speeds and the collection of progressively smaller-sized particles.

implementation plan. An outline of steps needed to meet environmental quality standards by a set time.

impoundment. A body of water confined by a dam, dike, floodgate, or other barrier.

incandescence. The emission of light due to high temperatures of the emitting material. Any other emission of light is called luminescence.

incineration. The disposal of solid, liquid, or gaseous wastes by burning.

incineration emission factors (in lb/ton of refuse burned).

Pollutant	Municipal multiple chamber	Industrial and commercial	
		Single chamber	Multiple chamber
Aldehydes	1.1	5–64	0.3
Benzo[a]pyrene	6,000 μg/ton	100,000 μg/ton	500,000 μg/ton
Carbon monoxide	0.7	20–200	0.5
Hydrocarbons	1.4	20–50	0.3
Oxides of nitrogen	2.1	1.6	2
Oxides of sulfur	1.9	n.a.[c]	1.8
Ammonia	0.3	n.a.[c]	n.a.[c]
Organic acids	0.6	n.a.[c]	n.a.[c]
Particulate	6[a]; 12[b]	20–25	4

Pollutant	Flue-fed apartment incinerator	Domestic single chamber	
		Without auxiliary gas burning	With auxiliary gas burning
Aldehydes	5	6	2
Benzo[a]pyrene	n.a.[c]	n.a.[c]	n.a.[c]
Carbon monoxide	n.a.[c]	300	n.a.[c]
Hydrocarbons	40	100	1.5
Oxides of nitrogen	0.1	1.5	2
Oxides of sulfur	0.5	2.0	2
Ammonia	0.4	0.4	Negligible
Organic acids	22	13	4
Particulate	26	39	6

[a]For incinerator with spray chamber.
[b]For incinerator without spray chamber.
[c]Not available.

incinerator. Equipment in which solid, semisolid, liquid, or gaseous combustible wastes are ignited and burned, leaving solid residues containing little or no combustible material. *Flue-fed incinerator.* An incinerator for multiple-occupancy units in which refuse is charged through openings on each floor into a flue, depositing the refuse into a combustion chamber below. *Multiple-chamber incinerator.* A two-stage mechanism involving the drying of the fuel, ignition and combustion of fixed carbon with gasification and partial combustion of the volatile components proceeding in the ignition chamber or primary stage. The gas-phase combustion is completed in the second stage, which consists of both a mixing chamber and a combustion chamber. Secondary air is admitted to the mixing chamber to aid in combustion. Multiple-chamber incinerators consist of two or more refractory lined chambers, interconnected by gas passage ports or ducts and designed in such a manner as to provide for complete combustion of the material to be burned. Depending upon the arrangement of the chambers, multiple-chamber incincerators are designed as in-line or retort types.

index of refraction. For any substance, the ratio of the velocity of light in a vacuum to its velocity in the substance; also the ratio of the sine of the angle of incidence to the sine of the angle of refraction. In general, the index of refraction for any substance varies with the wavelength of the refracted light.

index, purification. A criterion by which may be measured the degree of oxidation, reduction, or nitrification accomplished in sewage treatment.

index, sludge. The volume in milliliters occupied by an aerated mixed liquor containing one gram of dry solids after settling 30 minutes; commonly referred to as the Mohlman index. The Donaldson index, which is commonly used, is obtained by dividing 100 by the Mohlman index.

indicator. In biology, an organism, species, or community that shows the presence of certain environmental conditions.

individual perception threshold (IPT). The lowest concentration of a particular odor at which a subject indicated both an initial positive response and a repeated response.

indoor air pollution. Current research indicates that present energy conservation practices may be a health hazard in houses and offices. In the interest of conservation, energy contractors are building "tighter" offices and homes that reduce the flow of fresh air into a structure, thereby reducing the amount of energy needed to heat and cool it. In an office in an old-fashioned building with windows that can open and let air through, the breeze incurred may effect a complete exchange of air every hour or two. However, the energy-efficient seal windows with heavy insulation may exchange air only once in every ten hours. The tight office buildings trap a variety of noxious fumes including formaldehyde (used in glue for wood-laminated desks), radon (a radioactive substance present in concrete buildings), carbon monoxide, and particulates (from tobacco smoke), as well as other gases emitted from wallpaper adhesives and synthetic carpeting. It is known that exposure to formaldehyde, carbon monoxide, and other chemicals and gases can produce headaches, fatigue, and eye and throat irritations. Smoking in offices is also believed to increase health risks significantly. Recently the American Society of Heating, Refrigerating, and Air Conditioning Engineers started drafting standards to prescribe limits on the presence of formaldehyde and other substances in office air. The standards are to include subjective tests to determine the air quality in buildings. Other efforts are being made to improve indoor air quality by improving ventilation.

inductance. The change in magnetic field due to the variation of a current in a conducting circuit causes an induced counterelectromotive force in the circuit itself. This phenomenon is known as self-induction. If an electromotive force is induced in a neighboring circuit, the term "mutual induction" is used. Inductance is measured by the electromotive force produced in a conductor by a unit rate of variation of the current. Units of inductance are the centimeter (absolute electromagnetic) and the henry, which is equal to 10^9 centimeters of inductance. The henry is that inductance in which an induced electromotive force of one volt is produced when the inducing current is changed at the rate of one ampere per second.

induction. Any change in the intensity or direction of a magnetic field causes an electromotive force in any conductor in the field. The induced electromotive force generates an induced current if the conductor forms a closed circuit.

industrial waste (agricultural runoff). A great deal of phosphorus is being added to the environment in the fertilization of farmlands. When a highly soluble fertilizer is used, 10–25 percent is leached away into the surface runoff before the plants are able to use it. Another agricultural source of phosphorus is animal waste. Livestock, increasingly grouped in feedlots, contribute over seventy million pounds of phosphorus per year, almost as much as detergents.

industrial waste (fruit and vegetable processing). The processing of canned and frozen produce uses many millions of gallons of water. The use of water in processing starts with the cleaning, grading, and sorting process. Chemical peeling, the dipping of fruit and vegetables into a hot lye solution to dissolve the skin, produces a wash water that is hot and very alkaline, containing not only minerals, but large amounts of dissolved organic matter from the skins. In the freezing process most foods have to be blanched, that is, briefly dipped into hot water or steam to denature enzymes that might cause them to deteriorate after freezing. This blanching is also hot and contains much dissolved organic matter. When the steam used in the water-removal phase of making juice concentrate is condensed to avoid air pollution, a liquid waste with various dissolved materials remains. Although these processes have greatly increased pollution, progress is being made in recycling and waste treatment.

industrial waste (petroleum refining). Petroleum is a rich mixture of organic compounds; the refining process simply sorts these out, refines, and purifies them for many general and specialized uses. Crude oil is emulsified with water to separate the salts and other impurities which, when settled out, are released along with the water. Then, by a complex series of fractionations and cracking, various refining products ranging from heavy grease to high octane fuel are obtained. A vast quantity of water is required for this process, most of which is recycled.

industrial waste (phosphorus). Of all the minerals cycling in the environment, the one which most often limits the growth of plants and animals is phosphorus. Too little phosphorus results in a low productivity of aquatic systems; too much leads to the population explosion of plants and microorganisms. The concentration of phosphorus in aqueous environments is normally low, about 0.02 parts per million (ppm), because most phosphates are insoluble in water. If the input of phosphorus into aquatic systems is increased much beyond this figure, it is either organically fixed by organisms and ultimately deposited in sediments, or precipitated directly. Either way, excess phosphorus is incorporated into sediments eventually becoming of critical importance.

industrial waste (pulp and paper). Pulping is the process of breaking down wood into its component fibers, which are then formed into paper. The cellulose wood fibers are held together with lignin and a complex sugar or hemicellulose which can be dissolved under high pressures and temperatures with bisulfite and either sulfurous acid or sulfur dioxide. This treatment yields 50 percent of the dry weight of the wood as cellulose fibers and a waste sulfite liquor containing the remaining 50 percent in the form of lignosulfonate and various sugars hydrolyzed from the hemicellulose. Until recently, millions of tons of spent sulfite liquor were dumped into the environment. Today it is gradually being recycled. Lignosulfonates, because of their dispersed and sequestering characteristics, are being used in other industries as additives to drilling mud in oil wells and as stabilizers of unpaved road surfaces in the summer. Over 40 percent of the domestic vanillin supply (artificial vanilla flavoring) is synthesized from lignosulfonate, as are glacial acetic and formic acids. The sugar xylose, which is quite abundant in pulping paper, is very sweet but has few calories, making it a potential substitute for cyclamates. The complex sugars from hemicellulose can be converted to ethanol (grain alcohol).

inert gas. (1) A gas that does not react with other substances under ordinary conditions; also called noble or rare gas. (2) Any one of six gases—helium, neon, argon, krypton, xenon, and radon—all of whose shells of planetary electrons contain stable numbers of electrons so that the atoms are almost completely chemically inactive.

inertia. The resistance offered by a body to a change of its state of rest or motion, a fundamental property of matter.

inertial collector, Anderson Air Sampler. An inertial collector originally developed for sampling airborne microorganisms consisting of a nest of six sieves stacked one above the

the other. The top sieve has a great number of relatively large holes, while the other plates have an equal number of holes of progressively smaller diameters. Beneath each perforated plate is a Petri dish containing agar. The air, which is sampled at the rate of one cubic foot per minute, passes sequentially through the six stages from top to bottom, impacting the relatively large particles on the dish beneath the first stage, while the finer particles are collected in the subsequent stages. The Anderson Air Sampler can be used for analyzing inert dust by simply substituting an adhesive material for the agar.

inertial collector, cascade. When a high velocity jet of dust-laden air is directed onto a flat surface, the particles impact on the surface, lose their kinetic energy, and are thus collected. The cascade impactor utilizes this principle for collecting dust in four separate size fractions. Since the air flows through a series of slits of progressively smaller areas, the velocity of the gas stream passing through the slit correspondingly increases at each stage. As the efficiency of impaction is directly proportional to the particle velocity, the cascade impactor collects larger particles on the initial stages and relatively finer particles on the later stages. The collecting surfaces are glass slides which can be removed and placed in the microscope for counting and sizing.

inertial collector, wet impinger. Essentially a single-stage impactor, with gas flowing through a glass tube terminating in a fine nozzle submerged in water. The aerosol particles impinge on the bottom of the impinging flask and are trapped in the liquid, which is usually water.

inertial separator. A type of equipment that includes all dry-type collectors utilizing the relatively greater inertias of particles to effect particulate gas separation. Two types of equipment utilize this principle: cyclonic separators, which produce a continuous centrifugal force as a means of exerting the greater inertial effects of the particle; and simple inertial or impaction separators, which employ incremental changes of direction of the carrier gas stream to exert the greater inertial effects of the particle. Included in this category, besides cyclonic and impingement separators, are gravity settling chambers and high-velocity gas reversal chambers.

influent. The sewage, water, and other liquids, either raw or partly treated, flowing into a reservoir, basin, or treatment plant, or any part thereof.

infiltration. The action of water moving through small openings in the earth as it seeps down into the groundwater.

infrared gas analyzer. An infrared spectrometer with two equivalent calcium fluoride infrared window sources, one for the comparison cell and the other for the sample cell. It functions on the principle that the sample air will absorb infrared radiation at a rate different than that of the gas in the comparison cell. Thus, with proper instrumentation, the CO concentration in the sample gas may be measured.

infrared radiation. The electromagnetic radiation of wavelengths between 7500 Å, the limit of the visible spectrum at the red end, and those of centimetric radio waves. Infrared radiation is investigated by thermal detectors, sensitized photographic emulsions, and techniques similar to those used for very short radio waves. The main subdivisions of the infrared range are: (a) very near infrared, 0.75–2.5 µm; (b) near infrared, 2.5–25 µm; (c) far infrared, 25–300 µm; and (d) microwave range, 300–10,000 µm.

infrared, visible light, ultraviolet. Higher in frequency than the radio waves are the electromagnetic radiations called the visible light spectrum. Within this range are all the colors of the rainbow; starting with red, which has the lowest frequency, light progresses through orange, yellow, green, blue and indigo, and violet, which has the highest frequency. Frequencies just below visible red light cannot be seen; these are the infrared or heat radiations. At the other end of the light spectrum are frequencies just above visible violet light, the ultraviolet radiations, whose natural source is the sun. The infrared region consists of wavelengths from 0.8 to 1000 μm and contains most of the heat energy from the sun.

infrasonic. Pertaining to frequencies below the range of human hearing; frequencies of less than about 15 hertz.

inhibitor. A chemical included in the liquid used in solar collectors to prevent corrosion of fluid-carrying pipes.

inoculum. Bacteria placed in compost to start biological action.

insolation. The rate at which the total solar energy, direct plus sky radiation, is received on a horizontal surface. This would be the same as the solar constant only if the surface were outside the atmosphere and the sun were directly overhead as its mean distance from the earth. The insolation received at the surface of the earth depends upon the solar constant, the distance from the sun, the inclination of the sun's rays, and the amount depleted while passing through the atmosphere.

instantaneous sampling. Obtaining a sample of an atmosphere in a very short period of time, such that this sampling time is significant in comparison with the duration of the operation or the period being sampled; also known as grab sampling.

integrated pest management. Combining the best of all useful techniques—biological, chemical, cultural, physical, and mechanical—into a custom-made pest control system.

intensity. The amount of energy per unit time passing through a unit area perpendicular to the line of propagation at the point in question.

intensity level (sound-energy flux density level). The intensity level, in decibels, of sound is 10 times the base 10 logarithm of the ratio of the intensity of this sound to the reference intensity. The reference intensity shall be stated explicitly. A common reference sound intensity is 10^{-16} watt per centimeter in a specified direction.

interceptor sewers. The collection system that connects main and trunk sewers with the wastewater treatment plant. In a combined sewer system, interceptor sewers allow some untreated wastes to flow directly into the receiving streams so the plant won't be overloaded.

intermittent sampling. Sampling successively for limited periods of time throughout an operation or for a predetermined period of time. The duration of the sampling periods and the intervals in between is not necessarily regular and is not specified.

internal-combustion engine. An engine in which both the heat energy and the ensuing mechanical energy are produced inside the engine proper.

interstate carrier water supply. A source of water for planes, buses, trains, and ships operating in more than one state. These sources are regulated by the federal government.

interstate waters. Defined by law as: (1) waters that flow across or form a part of state or international boundaries; (2) the Great Lakes; and (3) coastal waters.

intrinsic brightness. The luminous intensity measured in a given direction per unit of apparent (projected) area when viewed from that direction.

inversion. When the surface air is cooler than a layer above it, and it cannot rise and mix, the atmosphere is said to be particularly stable. The phenomenon of cool surface air trapped by a layer of warmer air is known as an inversion. There are several kinds of inversions, the most important of which are those formed by the descent of a layer of air within a high-pressure air mass, and those formed by the radiation of the earth's heat into space. The first kind, called a subsidence inversion, occurs when a layer of air within a high-pressure air mass sinks down upon an area and in so doing is compressed by the high-pressure area above the heated layer, while at the ground level the air is unchanged. Radiation inversion is a normal nighttime formation, occurring when the ground surface radiates heat out into space and quickly cools on clear nights. In doing so, it cools the surface air. This kind of inversion breaks up as the morning sun heats the ground and reestablishes the moving currents of rising warm air. The intensity and duration of an inversion is affected by the season of the year. Fall and winter generally have the longest-lasting and the greatest number of inversions.

inversion height. A term commonly used to denote the height of a temperature inversion, which is a layer of atmosphere in which the temperature increases with height (this layer may be any distance above the ground, amounting to a discontinuity in the temperature lapse rate); an abbreviation for "inversion of temperature gradient." The temperature of air generally gets lower with increasing height, but occasionally the reverse is the case, and when the temperature increases with height there is said to be an inversion.

invert. The floor, bottom, or lowest portion of the internal cross section of a closed conduit.

inverter. The electrical power supply which converts dc into ac power.

iodine number. A number expressing the percentage (i.e., grams per 100 grams) of iodine absorbed by a substance. It is a measure of the proportion of unsaturated linkages present and is usually determined in the analysis of oils and fats.

ion. (1) An atom, or group of atoms, that is not electrically neutral, but carries instead a positive or negative electric charge. Positive ions are formed when neutral atoms or molecules lose valence electrons; negative ions are those which have gained electrons. When the electrically neutral atom gains or loses an electron in a chemical bond, its electrical balance is destroyed and the atom becomes an electrically charged ion. (2) In atmospheric electricity, any of several types of electrically charged submicroscopic particles normally found in the atmosphere. Atmospheric ions are of two principal types, small ions and intermediate ions, which have occasionally been reported. The formation of small ions depends upon cosmic rays and radioactive emanations. Each of these consists of very energetic particles which ionize neutral air molecules by knocking out one or more planetary electrons. The resulting free electron and positively charged molecule (or atom) attach themselves to one or, at most, a small number of neutral air molecules, forming new small ions. In the presence of Aitken nuclei, some of the small ions will attach themselves to the nuclei, thereby creating new large ions. The two classes of ions differ widely in mobility. Only the highly mobile small ions contribute significantly to the electrical conductivity of air under most conditions. The intermediate and large ions are important in certain space-charge effects, but are too sluggish to significantly contribute to electrical conductivity. The formation of ions is offset by processes such as recombination (reactions).

ion exchange. A process used primarily for the concentration and recovery of valuable constituents, such as hexavalent chromium cyanide, copper, and other metals used in plating and other metal-finishing operations. Ion exchange is also used to purify and reclaim process solutions for reuse and to concentrate constituents for destruction. Ion exchange is particularly applicable where high-quality water is required as part of the processing operation, such as the final rinse on a chromium plating line. Ion exchange operates as a fixed bed with water passing through and the ion of concern is replaced by the ion present on the exchange material.

ion pair. The pair of ions, one positively and the other negatively charged, formed by the ionization of an initially neutral gas atom when it collides with a high-energy particle.

ionic bond. The bond between atoms which creates ions; bonds in which electrons are completely transferred from one combining atom to another. An example of ionic bonding is ordinary table salt, NaCl, where sodium's single outer electron is transferred to chlorine's outer shell.

ionization. The process by which an atom becomes electrically charged. Normally the electrons and protons of an atom balance each other, making the atom electrically neutral. But the orbiting electrons can be pulled off. Then the atom that has lost them becomes positively charged, and the atom or atoms that acquire them become negatively charged. An atom that has lost one or more electrons is called a negative ion. Radioactivity is dangerous because the rays can ionize the air through which they pass and any substance exposed to them. They can also penetrate the human body and continue ionizing atoms within it, thus initiating a reaction that can end in damage to critical molecules of human cells. Radioactive decay cannot be shortened in time or altered by pressure or temperature, so the danger of exposure exists as long as the decaying process exists, that is, until a stable element is produced and all the excess energy has been released. The time lapse varies from isotope to isotope, but is constant for each one, and is designated by what is termed the half-life of a substance. The half-life is the period of time required by the radioactivity of a substance to drop to one-half of its original value, that is, the time necessary for one-half of the atoms of a radioactive substance to disintegrate. After a period of time corresponding to approximately 10 times the half-life, the radioactivity of a substance drops to about 0.1 percent of its original value.

ionization chamber. An apparatus used to study the production of small ions in the atmosphere by cosmic-ray and radioactive bombardment of air molecules. The chamber is an airtight container, usually cylindrical in shape and 25 to 50 liters in volume. An insulated electrode is centrally located in the chamber. During operation, a potential is applied between the electrode and the chamber wall. The ions produced in the chamber are collected by the electrode system and measured by an electrometer.

ionization energy. The energy required to remove an electron from an atom. The stability of an electron is described in terms of its ionization energy.

ionization gauge. A vacuum gauge with a means of ionizing gas molecules and correlating the number and type of ions produced with the pressure of the gas. Various types of ionization gauges are distinguished according to the method of producing the ionization. Some common types are: the hot-cathode ionization gauge, the cold-cathode ionization gauge, and the radioactive ionization gauge.

ionization potential. The work (expressed in electron volts) required to remove a given electron from its atomic orbit and place it at rest at an infinite distance. It is customary to list values in electron volts (eV): 1 eV = 23,053 cal/mol.

ionizing radiation. Any electromagnetic or particulate radiation capable of producing ions, directly or indirectly, in its passage through matter.

ionosphere. The atmospheric shell characterized by a high ion density. Its base is at about 70 or 80 kilometers above ground and extends to an indefinite height. The ionosphere is classified into layers. Each layer, except for the D layer, is supposedly characterized by a more or less regular maximum of electron density. The D layer exists only in the daytime. It is not strictly a layer, since it does not exhibit a peak of electron and ion density, and it starts at about 70 to 80 kilometers above ground, merging then with the bottom of the E layer. The lowest clearly defined layer is the E layer, occurring between 100 and 120 kilometers above ground. The F_1 and F_2 layers occur in a general region between 150 and 300 kilometers above ground, the F_2 layer being always present and having the higher electron density. From the upper portion of the stratosphere, the ionosphere rises into the soundless void of space. In a sense, the ionosphere is a kind of celestial electronic web that acts as a reflector of radio waves. The ionosphere is extremely sensitive to radiations from the sun, especially during periods when sunspots are more numerous than usual. The ionosphere gets its name from the countless ions it contains. The sun is also mainly responsible for the physical characteristics of the ionosphere. One of these is its temperature, which ranges from -27 to $4000°F$. Scientists have determined that this high maximum temperature is the result of the ionization process that is continually going on, triggered by solar radiation. The altitude, width, and surface levelness of the ionosphere are other characteristics influenced in some degree by the sun's radiant waves.

iron (Fe). A metallic element. The permissible criterion for iron in filterable form for public water use is 0.3 mg/liter. In general, iron is one of the less toxic pollutants. Criteria for waters designated for the preservation of aquatic species do not specifically define upper limits for iron.

iron dextran. A sterile, dark-brown, colloidal solvent which has been used for therapeutic purposes, although iron salts may cause degenerative changes in human tissues. The dangerous dosage level for humans has been established as 150 mg/kg. Children are particularly sensitive and at least 30 children have died as a result of ingesting iron salts. Injection of iron dextran has caused fever, rapid heart beat, skin irritation, back pains, swollen lymph glands, and anaphylactic shock. Death may occur as much as one week after ingestion of iron salts. Iron dextran has been used as an experimental carcinogen in animals and is a suspected carcinogen in humans. It has been included in the EPA's Carcinogen Assessment Group list of substances for screening as a potential occupational carcinogen.

isallotherm. A line connecting points between which an equal temperature variation is observed within a definite interval.

isentropic-condensation temperature. The temperature at which saturation would barely be reached if the air were cooled adiabatically without the removal or addition of moisture. It differs from the dew point temperature in that it is defined in terms of an adiabatic, instead of a constant pressure, process. The isentropic-condensation temperature is always higher than the dew point temperature, because in an adiabatic process, owing to expansion, the water vapor has an increasingly larger volume to saturate.

isobar. (1) A line drawn on a map or chart through places having the same atmospheric pressure at a given time and at a standard level (e.g., at sea level). (2) For chemistry, elements of the same atomic mass but of different atomic numbers. The sum of their nucleons is the same, but there are more protons in one element than in the other.

isoconic line. A line joining points of equal dust precipitation per unit area and time.

isocyanate. An insulating material commonly applied to the roofs of buildings, including schools and hospitals. It was associated in 1980 with complaints of illness among Suffolk County, New York, school children who experienced skin rashes, nausea, headaches, and respiratory disorders after isocyanate was applied to a school roof by spraying. Isocyanate reacts violently with alcohols. The chemical may contain fluorotrichloromethane, which can cause narcosis and anesthesia in high concentrations and may decompose to emit toxic fumes of fluorides and chlorides when heated.

isodose chart. A chart showing the distribution of radiation in a medium by means of lines or surfaces drawn through points receiving equal doses. Isodose charts have been determined for beams of x rays traversing the body, for radium applicators used for intracavitary or interstitial therapy, and for working areas where x rays or radioactive isotopes are employed.

isodrosotherm. A line joining points of equal dew points.

isodynamic line. A line connecting points on the earth's surface which have the same total magnetic intensity.

isogonic line. A line on a chart of the earth's surface connecting points having equal magnetic declinations.

isokinetic. A line in a given surface connecting points with equal wind speed (British term for isotach).

isokinetic sampling. A technique for collecting airborne particulate matter, in which the collector is so designed that the airstream entering it has a velocity equal to that of the air passing around and outside the collector. In principle, an isokinetic sampling devise has a collection efficiency of unity for all sizes of particulates in the sampled air.

isomer. One of several nuclides having the same number of neutrons and protons, but capable of existing for a measurable time in different quantum states with different energies and radioactive properties. Commonly, the isomer of higher energy decays to one of lower energy by the process of isomeric transition.

isomerism. The existence of molecules having the same number and kinds of atoms, but in different configurations.

isopropyl oil. A byproduct of isopropyl alcohol manufacture used in the manufacture of isopropylene. It consists of a mixture of trimeric and tetrameric polypropylene plus small amounts of acetone, alkylbenzene, benzene, ethanol, heptane, hexane, isopropyl alcohol, isopropyl ether, polyaromatic ring compounds, and toluene. It is a recognized cause of cancer of the nasal sinuses, throat, and lungs, and was cited as a significant occupational cancer agent in the 1980 Report to the President by the Toxic Substances Strategy Committee.

isosafrole. $C_{10}H_{10}O_2$, a component of pesticides and a suspected form of toxic residue from the degradation by oxidation of other pesticide compounds. It is soluble in most organic solvents, but insoluble in water. It is particularly dangerous when heated because of the production of toxic fumes. The oral LD_{50} of isosafrole in rats is 2000 mg/kg. It is an experimental carcinogen and tumorogen when administered via oral and subcutaneous routes. Isosafrole is included in a list of substances scheduled for scientific review to determine whether it meets the OSHA definition of a potential occupational carcinogen.

isostasy. The condition of balance, or flotational equilibrium, between continents and ocean basins. At some depth beneath areas of equal weight, but unequal density, the pressures must be substantially the same. Any potential disturbance of isostatic balance taking place at the surface (such as a decrease in weight due to erosion of lands or an increase in weight due to the accumulation of a thick ice sheet, like that on Antarctica) is presumably met by slow flowage in the earth's plastic interior in order to keep the masses in equilibrium.

isotach. A line joining points of equal wind speed; also isokinetic.

isotherm. A line on a chart connecting all points of equal or constant temperature.

isothermal. When a gas passes through a series of pressure and volume variations without a change in temperature, the changes are called isothermal. A line on a pressure—volume diagram representing these changes is called an isothermal line.

isothermal atmosphere. An atmosphere in hydrostatic equilibrium in which the temperature is constant with height and the pressure decreases exponentially upward. In an isothermal atmosphere there is no finite level at which pressure vanishes.

isothermal equilibrium. The state of an atmosphere at rest, uninfluenced by an external agency, in which the conduction of heat from one part to another has, after a sufficient length of time, produced a uniform temperature throughout its entire mass.

isotope. A nuclide having the same atomic number as another nuclide, hence constituting the same element, but differing in its mass number. Isotopes of a given element have the same number of nuclear protons, but different numbers of neutrons. Naturally occurring chemical elements are usually mixtures of isotopes, so that observed (noninteger) atomic weights are the average values for the mixture. The nucleus of every atom belonging to the same element always contains the same number of protons, but the number of neutrons the atomic nucleus contains may vary. The most common form of an element is called the ordinary isotope. In hydrogen, the ordinary isotope with only a proton in its nucleus is called protium. Hydrogen's other isotopes are stable deuterium, or heavy hydrogen, with one proton and one neutron, and radioactive tritium, with one proton and two neutrons. All known elements have two or more isotopes.

isotropic radiation. A diffuse radiation which has exactly the same density in all directions.

J

Japan lacquer. A varnish that imparts a hard and bright coating to metals and woods. Natural Japan lacquer, which contains the irritating chemical urushiol, frequently causes dermatitis. (Urushiol is derived from catechol and is the main constituent of the irritating substance of poison ivy.) Synthetic Japan lacquer, made by heating linseed oil with lead compounds and diluting with solvents, is also an irritant.

jet engine. Broadly, any engine that ejects a jet or stream of gas or fluid, obtaining all or most of its thrust by reaction to the ejection; specifically, an aircraft engine that derives all or most of its thrust by reaction to its ejection of combustion products (or heated air) in a jet and that obtains oxygen from the atmosphere for the combustion of its fuel.

jet engine combustion. Air enters through the front of the engine and goes to a compressor, where it is increasingly compressed and forced into combustion chambers that are arranged in circles around the engine. Fuel is sprayed into the front end of the combustion chamber in a steady stream so that it ignites and burns continuously. The burning air–fuel mixture expands and pushes toward the rear (on the way, it hits turbine-wheel blades and forces them to rotate, thus driving the compressor). As the expanded mixture moves toward the tailpipe, the areaway narrows and the stream of burning air–fuel mixture is compressed into the exceedingly strong jet stream that shoots out of the rear of the plane.

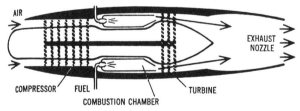

Jet engine combustion.

jet engine noise suppressors. Noise suppressors fitted to jet engines take on various forms, but they generally operate by modifying the flow of gases from the engine by directing them through some form of nozzle. The noise reduction is accompanied by a reduction in efficiency with increased operating costs.

jet engine noise (turbojet). The turbojet engine, as a whole, can be regarded as a tube whose axis is approximately parallel to the fore-and-aft axis of the airplane. The end of the engine pointing forward houses a compressor which takes in air and compresses it. Following this, fuel is burned in the combustion chamber, or chambers, and the resultant hot gases are expanded through a turbine which drives the compressor. After leaving the turbine, the hot gases are discharged as the jet exhaust to the atmosphere, where they produce part of the characteristic noise of the jet engine. The jet engine normally employs a number of rows of blades, or stages, in both the compressor and turbine. The noise output of jet engines comes from a number of sources, including the noise of the jet itself, the noise of the compressors, and the noise of the turbines. The characteristic roar of a jet engine is produced by the violent mixing of exhaust gases with the air into which they are discharged. The noise is influenced by a number of factors and is markedly dependent on the velocity of the actual jet relative to the surrounding air. The sound powers are proportional to the eight power of the jet velocity if the engine is restrained in a stationary position, that is, delivering its so-called static thrust.

jet fuels. Petroleum distillate products, similar in composition and properties to kerosene, are used alone or in combination with gasoline as jet fuels. Like kerosene, ingestion of jet fuels irritates the stomach and intestines and causes nausea and vomiting.

jet stream. A meandering river of high-velocity winds (60 and more miles per hour) blowing within the normal atmospheric wind patterns at high altitudes. Coursing through the atmosphere at tremendous speeds, these mighty currents of air sometimes exist as a single undulating ribbon encircling the globe. More often, jet streams are broken up into segments that range from 1000 to 3000 miles in length and from 100 to 400 miles in width. At times two or more jet streams may flow through the atmosphere on parallel courses separated by a considerable number of miles. The velocity of jet streams has been measured at as little as 50 miles an hour and as much as 150 miles an hour or more. The jet streams are strongest at altitudes ranging from 25,000 to about 40,000 feet.

joule. A unit of energy or work in the mks system; the work done when the point of application of one newton of force is displaced a distance of one meter in the direction of the force. $1 \text{ J} = 10^7$ ergs $= 1 \text{ W s}$.

joule, absolute. (1) A unit of energy equal to 0.2390 calories or 10^7 ergs; the energy dissipated in one second by a current of one ampere flowing across a potential difference of one volt. The international joule equals 1.001 65 absolute joules. (2) A unit of energy or work in the mks system of units; the work done when a force of one newton produces a displacement of one meter in the direction of the force.

Joule–Thomson effect. The cooling which occurs when a highly compressed gas is allowed to expand in such a way that no external work is done. This cooling is inversely proportional to the square of the absolute temperature.

juniper berries. The dried ripe fruit of *Juniperus communis* L. *pinacea,* used in medicines, liqueurs, and fumigating chemicals. The oil of the juniper berry contains such compounds as pinene, camphene, and terpineol, and, if ingested, it causes kidney irritations similar to those caused by turpentine.

jute. A plant fiber either of the East Indian plants *Corchorus olitorius* or *C. capsularis;* it is used to make burlap and twine. The dust that results during handling and processing is highly flammable.

K

kaolinite. Hydrous aluminum silicate, a soft and usually light-colored mineral that occurs in minute particles; also called china clay. It is the main constituent of many clays and shales. Kaolinite is also an important constituent of many soils and is used in making paper, china, brick, tile, and crockery.

Kelvin scale. A scale of temperature measured in degrees celsius from absolute zero ($-273.15°C$).

Kepone. $C_{10}Cl_{10}O$, a chlorinated polycyclic ketone used as an insecticide and a fungicide. Kepone has been identified as a potential carcinogen and, when inhaled or otherwise absorbed into body tissues, can cause damage to the central nervous system, characterized by tremors, ataxia, hyperexcitability, hyperactivity, and muscle spasms. Other symptoms and signs may include liver and kidney damage, testicular atrophy and low sperm count, sterility, estrogenic hormone changes, and breast enlargement. The oral LD_{50} in rats has been reported as 96 mg/kg and 250 mg/kg by skin exposure. The James River in Virginia became contaminated with Kepone by discharges from a manufacturing plant (Life Sciences Plant of the Allied Chemical Company) in Hopewell between 1966 and 1975. Virginia ordered the plant closed and banned fishing in the James River, which had been a dumping ground for thousands of pounds of the chemical. The river now remains partially closed to fishing, while scientists are still trying to determine if Kepone is a carcinogen, a cancer-causing substance, whose lethal effects may lie undetected for years. The Hopewell crisis erupted when quantities of Kepone were discovered in a Life Science supervisor's blood. Eventually some two dozen employees showed signs of illness. Ailments included severe trembling, vision problems, and sterility. During the last five years health officials have found, among other things, that the human body transforms Kepone into Kepone alcohol, which it excretes quickly. Workers who were exposed to Kepone during the years of operation of the Hopewell plant have been treated with cholestyramine, a medication originally intended to lower cholesterol, but also found to be capable of binding Kepone molecules and doubling the rate of excretion of the chemical from the body. In addition to previously reported health effects of Kepone, the chemical workers also experienced stuttering, loss of memory, and, in some cases, hallucinations. But whether Kepone causes cancer in humans remains unresolved. The EPA has estimated the cost of cleaning Kepone from the James River by dredging at $1 billion, with an additional expense of as much as $6 billion to clean the dredged material. Although the James River still has not been cleaned, there is a possibility under study to let nature run its course. Presently Kepone is being buried deeper and deeper in the river's sediment. Above ground Kepone has mostly been disposed of in special burial sites.

kerosene. A fuel oil fraction of petroleum distilled between 150 and 300°C. It is a mixture of various petroleum hydrocarbons that have 10 to 16 carbon atoms. The oral LD_{50} in the rat is 28 g/kg. Inhalation of high concentrations of kerosene vapor causes headaches and stupor. Ingestion causes irritation of the stomach and intestines and accompanying nausea and vomiting.

Kerr effect. When plane-polarized light is incident on the pole of an electromagnet, polished so as to act like a mirror, the plane of polarization of the reflected light is not the same when the magnet is "on" as when it is "off." It was found that the direction of rotation was opposite to that of the currents exciting the pole from which the light was reflected.

ketones. Organic compounds that contain the chemical group =CO, with acetone—dimethyl ketone—being the most familiar member. Some members have a narcotic or anesthetic effect, but none have been shown to have a high chronic toxicity.

kiln. A furnace in which the heating operations do not involve fusion. Kilns are most frequently used for calcining, and free access of air is usually permitted. Raw materials may be heated by combustion of the solid fuel with which they are mixed, but more usually they are heated by gas or the waste heat from other furnaces.

kilo. A prefix meaning 1000, used in the metric system. A kilogram is 1000 grams.

kinematic viscosity. A coefficient defined as the ratio of the dynamic viscosity of a fluid to its density. The kinematic viscosity of most gases increases with increasing temperatures and decreasing pressures. For dry air at 0°C, the kinematic viscosity is about 0.13 square centimeter per second.

kinematics. The branch of mechanics dealing with the description of the motion of bodies or fluids without reference to the forces producing the motion.

kinetic energy. The energy which a body possesses as a consequence of its motion, defined as one-half the product of its mass m and the square of its speed v, $\frac{1}{2} m v^2$. The kinetic energy per unit volume of a fluid parcel is thus $\frac{1}{2}\rho v^2$, where ρ is the density and v is the speed of the parcel.

kinetic theory. A theory that holds that a material body is not one continuous uninterrupted extent, but is made up of minute, invisible particles, molecules, that are in constant motion, oscillating, bumping into each other, and bouncing back. Increasing the temperature causes a rise in the speed of the molecules, thus increasing the kinetic energy of the molecules.

kinetic theory of gases. A theory in which gases are considered to be made up of minute, perfectly elastic particles which are ceaselessly moving about with high velocities, colliding with each other and with the walls of the containing vessel. The pressure exerted by a gas is due to the combined effect of the impacts of the moving molecules upon the walls of the containing vessel, the magnitude of the pressure being dependent upon the kinetic energy of the molecules and their number.

Kirchhoff's law of radiation. The relation between the powers of emission and the powers of absorption for rays of the same wavelength is constant for all bodies at the same temperature. First, a substance, when excited by some means or other, possesses a certain power of emission; it tends to emit definite rays, whose wavelengths depend upon the nature of the substance and the temperature. Second, the substance exerts a definite absorptive power, which is a maximum for the rays it tends to emit. Third, at a given temperature the ratio between the emissive and absorptive powers for a given wavelength is the same for all bodies, and is equal to the emissive power of a perfect black body.

konimeter. An instrument used for determining the dust content of a sample of air; also called a conimeter. One form of the instrument consists of a tapered metal tube through which a sample of air is drawn and allowed to impinge upon a glass slide covered with a viscous substance. The particles caught are counted and measured with the aid of a microscope. Because this method does not reveal particles of diameters less than 0.8 μm, and since the dust in steel foundaries is generally below this limit, the konimeter is favored for use in mines, while Owen's jet test is mainly used in foundaries.

Koppers–Totzek. *Process:* Gasifier can be operated on all types of coal without pretreatment. Coal is dried and pulverized (70 percent through 200 mesh). A homogeneous mixture of oxygen and pulverized coal is introduced to the gasifier through coaxial burners at each end. The gasifier is a refractory-lined, horizontal, cylindrical vessel with conical ends. Oxygen, steam, and coal react at about atmospheric pressure and at about 3300°F. Fixed carbon and volatile matter are gasified to produce offgas containing carbon monoxide and hydrogen. Coal ash is converted into molten slag. Some of this drops into a water-quench tank, the remainder being carried by the gas. Low-pressure steam is circulated around the burners and refractory to protect them from excessive temperatures. Gas leaving the gasifier is quenched with water to solidify entrained molten ash and prevent it from solidifying on the walls of the waste-heat boiler. After passing through the waste-heat boiler, the gas is scrubbed to removed entrained solids. Scrubbed gas is compressed to 450 psig and purified to remove hydrogen sulfide and a controlled quantity of carbon dioxide. Purified gas is shifted and then methanated. Methanated gas is purified to remove remaining CO_2 and dehydrated to produce pipeline-quality gas (Heinrich Koppers G.m.b.h. of Essen).

Kraft pulp processing emission factors.
Emission Factors for Kraft Pulp Processing (lb/ton dry pulp produced)

Source	Gaseous pollutants			Particulate pollutants	Type of control
	Hydrogen sulfide	Methyl mercaptan	Dimethyl sulfide		
Digester blow system	0.1–0.7	0.9–5.3	0.9–3.8	Negligible	Untreated
Smelt tank	n.a.[a]	n.a.[a]	n.a.[a]	20	Uncontrolled
				5	Water spray
				1–2	Mesh demister
Lime kiln	1	Negligible	Negligible	18.7	Scrubber (approx. 80% efficient)
Recovery furnace[b]	3.6	5	3	150	Primary stack gas scrubber
	3.6–7.0	n.a.[a]	n.a.[a]	7–16	Electrostatic precipitator
	0.7	n.a.[a]	n.a.[a]	12–25	Venturi scrubber
Multiple-effect evaporator	1.2	0.004	n.a.	Negligible	Untreated
	0–0.5	0.003–0.030	Negligible	Negligible	Black liquor oxidation
Oxidation towers	n.a.[a]	n.a.[a]	0.1		Black liquor oxidation

[a]Not available.
[b]Gaseous sulfurous emissions are greatly dependent on the oxygen content of the flue gases and furnace operating conditions.

krypton (kr). A colorless inert gas with atomic weight, 83.8; melting point, −157°C; and boiling point, −152°C. [85]Kr, with a half-life of 11 years, decays to stable [85]Rb by the emission of 0.67-MeV particles. Krypton in large concentrations is an asphyxiant.

kytoon. A captive balloon used to maintain meteorological equipment aloft at approximately a constant height. The kytoon is streamlined and combines the aerodynamic properties of a balloon and a kite.

L

labile. A term used synonymously with "unstable" in reference to systems of two phases that undergo spontaneous change as soon as the transition temperature is passed.

lagoon. A shallow pond where sunlight, bacterial action, and oxygen work to purify waste water.

lagooning. The holding of waste water in man-made ponds or lakes for the removal of suspended solids and insoluble oils. Lagoons are also used as retention ponds after chemical clarification to polish the effluent and to safeguard against upsets in the clarifier, for stabilization of organic matter by biological oxidation, for storage of sludge, which is hauled out intermittently, and for cooling water.

lambert. A unit of luminance equal to $1/\pi$ candles per square centimeter. It is equal to the uniform luminance of a perfect diffusing surface emitting or reflecting light at the rate of one lumen per square centimeter.

Lambert's law. The radiant intensity (flux per unit solid angle) emitted in any direction from a unit radiating surface varies as the cosine of the angle between the normal to the surface and the direction of the radiation. The radiance (or luminance) of a radiating surface is, therefore, independent of direction. Lambert's law is not obeyed exactly by most real surfaces, but an ideal black body emits according to this law. This law is also satisfied (by definition) by the distribution of radiation reflected by a perfectly diffuse reflector. In accordance with Lambert's law, an incandescent spherical black body, when viewed from a distance, appears to be simply a uniformly illuminated disk. This law does not take into account any effects that may alter the radiation after it leaves the source.

land pollution. The misuse of land in a way which makes it unfit for man's future needs, such as the construction of buildings or growth of foods or other materials which are used in daily life and which could cause either dangerous toxic contamination of the air and water resources or give them a disagreeable appearance, taste, or odor.

landslide. The bodily sliding of large masses of earth and rock down steep slopes, breaking into discrete fragments. Conditions especially favorable to landslides are: (a) steep slopes (sea cliffs, riverbanks, road cuts, fault scarps, and glaciated valleys); (b) weak, slippery, impermeable material (clay, shale, volcanic ash or tuff, serpentine, soapstone); (c) jointed or much broken rock on top of the formation; (d) an inclination (dip) of sedimentary beds toward the open side; and (e) heavy rains or melting snows.

langley. A unit of energy per unit area, equal to one gram-calorie per square centimeter, commonly employed in radiation theory. The langley is almost always used in conjunction with some time unit, to express a flux density.

Langmuir probe. A small metallic conductor or pair of conductors inserted within a plasma in order to sample the plasma current.

lapse rate. The rate at which temperature decreases or lapses with altitude.

large calorie. The amount of heat energy needed to raise the temperature of one kilogram of water by one degree centigrade.

large ion. An atmospheric ion of relatively large mass and low mobility which is produced by the attachment of a small ion to an Aitken nucleus. The ion density of large ions varies widely, depending upon the degree of atmospheric pollution. Representative low-altitude values might be 10^3 ions per cubic centimeter in clean country air, 10^4 ions per cubic centimeter in an industrial area, and 10^2 ions per cubic centimeter over the oceans.

laser. (1) An acronym for light amplification by stimulated emission of radiation. Basically, lasers are devices that produce intense, highly directional light beams of an exceptionally pure color. Stimulated emission depends upon the medium of excited atoms, the column that contains those atoms, and the source of stimulation. A laser is a pencil-shaped device for amplifying the effect of such excited atoms. The light from a laser is relatively powerful because the atoms are stimulated to emit photons much faster than they would spontaneously. Laser light is usually of a single frequency, or pure color, because of the resonance nature of the stimulated emission process. (The atoms emit and respond to a frequency or wavelength of light only if they are in resonance with it.) A laser light is coherent, meaning that the photons are in phase (in step) with one another. (2) A tubelike device in which two mirrors are placed facing each other at either end. Atoms inside the tube are then stimulated, usually by an intense light to produce light waves that are emitted through an opening in one of the mirrors. The laser beam is very narrow and intense, with all the excited atoms contributing photons of the same size and moving in the same direction, whereas ordinary light waves move in many directions and are of many frequencies. These coherent light waves are so nearly parallel that the laser beam can travel long distances without spreading out and fading.

latent heat. The thermal equivalent of the energy expended in melting a unit mass of a solid or vaporizing a unit mass of a liquid, or, conversely, the thermal equivalent of energy set free in the process of solidification or liquefaction.

latent heat of vaporization. The quantity of heat necessary to change one gram of liquid to vapor with no change in temperature, the heat is measured in terms of calories per gram. When one pound of water is vaporized, the result is measured in terms of Btu's per pound.

lateral sewers. Pipes running underneath city streets that collect sewage.

lateritic-type soil. One of a group of soils that have very thin organic and organic–mineral strata overlying leached reddish soils. The parent rock material is often rich in iron oxides or hydrates of aluminum and has only sparse amounts of silica. True lateritic soils are not found in the continental United States; they develop in humid, tropical environments.

law of conservation of electric charge. The total electric charge of the universe remains constant; the total electric charge entering into a reaction must be equal to the total electric charge that comes out. Therefore, if a positive charge appears during an interaction, then a balancing negative charge must also appear. Conversely, if a positive charge disappears, then a negative charge must also disappear, resulting in a net charge of zero.

law of conservation of mass–energy. Neither mass nor energy can be created or destroyed; mass and energy are equivalent and interchangeable; the total mass–energy of the universe remains constant. In subatomic interactions, the total mass–energy that enters into the action must equal the total mass–energy that comes out, i.e., when a neutron decays, its mass of 940 MeV is converted into a proton, an electron, and a neutrino. The total mass of these newly created subatomic particles plus their energy of motion must equal the original neutron's 940 MeV.

law of conservation of momentum. The total momentum of the universe remains constant; momentum cannot be lost or gained, momentum can only be shared or exchanged. Momentum is the tendency of a moving particle to keep moving and its resistance to being stopped; a thrown ball, a speeding bullet, or a moving automobile all have momentum. Momentum is calculated by multiplying the mass of the moving object by its velocity. In subatomic interactions, the total momentum of the particles involved before a collision must equal the total momentum after the interaction. If one particle gains momentum, then the other must lose an equal amount.

LC$_{50}$. The median lethal concentration, a standard measure of toxicity. It tells how much of a substance is needed to kill half of a group of experimental organisms.

LD$_{50}$. A dose that is lethal to 50 percent of a group of test animals.

leachate. Materials that pollute water as it seeps through solid waste.

leaching. A process in metal recycling where "tin" cans help remove copper from copper ore. Water with sulfuric acid passes through copper ore. This chemical solution is then poured into vats full of shredded steel cans. Because the iron in the cans is a more chemically active metal than copper, some of the iron dissolves and takes the place of copper in the chemical solution. Next, the copper forms a deposit on the surface of the remaining shreds. Later, the copper manufacturer shakes and washes the shredded cans to remove the copper for refining.

lead (Pb). A metallic element. (1) Lead occurs naturally in soil, rocks, water, and food; only a small fraction of it contaminates the air. The estimated average daily intake of lead in food is substantially greater than the intake from air pollution, and only 5 to 10 percent of the lead ingested with food is retained by the body. As an air pollutant lead is present in the form of particles so small that as much as 50 percent of what is inhaled may be retained. Lead is emitted into the air from smelters, from the combustion of certain fuels, and from dust and sprays, with the predominant amount coming from automobile exhausts. Of the more than 160,000 metric tons of lead emitted into the nation's air annually, about 90 percent comes from automobile exhaust. The second principal source of airborne lead is from industrial plants such as nonferrous smelters. Most primary smelters are located in sparsely populated areas, but even these smelters can pose a definite health hazard to people living near them. In addition to air pollution there are other sources of lead exposure. Lead is found in paint, inks, water supply and distribution systems, pesticides, and fresh and processed foods. (2) Lead, like most other elements, is present in the natural environment in very low concentrations. Although some lead comes from industrial sources, the internal combustion engine has been the major source of lead in the atmosphere. Lead has been added to most gasolines in the form of tetraethyl lead mixed with ethylene dibromide and ethylene dichloride, plus a marker dye. Since oral intake of lead every day is 15 times the amount of lead breathed, airborne lead is believed by some not to be a significant hazard. Little lead is permanently retained by the body, but lead is toxic while it is in its temporary state. (3) Lead enters the human body principally through ingestion and inhalation with subsequent absorption into the blood stream and distribution to all body tissues. Exposure to airborne lead can occur directly by breathing or indirectly by eating lead-contaminated food, water, or nonfood materials including dust and soil. Lead accumulates in the body throughout life, to a large extent immobilized in the bone. A significant amount is in the blood and soft tissues. Lead has its most pronounced effects on the blood-forming, nervous, and kidney systems, but may also harm the reproductive, endocrine, hepatic (liver), cardiovascular, immunologic, and gastrointestinal systems. Exposure to high lead levels may have severe and sometimes fatal consequences such as brain disease, colic, palsy, and anemia. (4) A permissible criterion of 0.05 mg/liter and complete absence as a desirable criterion are the Federal Water Pollution Control Agency Committee on Water Quality Criteria recommendations for public waters. Lead may arise as a contaminant of groundwaters both from natural sources and in the form of various industrial and mining effluents. A major problem with lead pollution is that the element is a cumulative poison. There is considerable variation in the toxicity of lead, depending upon the form of the element. Drinking water supplies for animals should not contain concentrations of lead exceeding 0.5 mg/liter.

lead equivalent. The thickness of lead affording the same reduction in radiation dose rates under specific conditions as the material in question.

Legionnaires' disease. The so-called Legionnaires' disease which killed 29 persons associated with an American Legion convention in Philadelphia in July, 1976, subsequently afflicted 48 persons in 19 states, according to the National Center of Disease Control. Federal health officials investigating the cases said 11 deaths had been attributed to the pneumonia-like ailment since August 1, 1976. Two small clusters of the disease appeared later, five cases confirmed in Burlington, Vermont, and nine cases in Columbus, Ohio. Other isolated cases occurred at various intervals in widely scattered locations. Researchers determined that the disease was caused by a previously unidentified bacterium unlike any other known to science. Experts at the Center for Disease Control said the antibiotic erythromycin, which had been administered to some victims, might be an effective drug in the treatment of Legionnaires' disease. The Center for Disease Control in Atlanta announced August 11, 1980, that researchers had been successful in isolating Legionnaires' disease bacteria from water. Previously, the bacterium had been isolated only in the tissue of the diseased victims. It was felt that it was a breakthrough that increased the odds for controlling epidemics of the illness. The bacterium was found in water taken from a cooling tower atop the Indiana University student union hotel in Bloomington and from a nearby creek. An investigation was undertaken there after it was reported on May 28, 1980, that three persons had died and 18 others had sickened. The isolation technique concentrated on other mediums because other outbreaks of the disease did not involve air conditioners.

Lenz's law. When an electromotive force is induced in a conductor by any change in the relation between the conductor and the magnetic field, the direction of the electromotive force is such as to produce a current whose magnetic field will oppose the change.

lepton. A subatomic particle. The lepton family particles have the lightest mass of all subatomic particles. They include the electron, along with the muons and neutrinos. The muon is similar to the electron, except that it is unstable and two hundred times heavier. Neutrinos conserve and carry off energy released in many interactions between particles or in beta decay. Neutrinos have no charge, almost no mass, and travel with the speed of light. Neutrinos are emitted during nuclear fusion on the sun and carry some of the sun's heat to the earth. As the neutrinos are created in the sun's core, they move right through the mass of the sun as if it were empty and out into space. To a neutrino, matter practically does not exist. Detecting neutrinos was accomplished in 1958 by two physicists, Clyde L. Cowan and Frederick Reines, working with the nuclear reactor at the Atomic Energy Commission's Savannah River plant. The positron is a positively charged elementary particle otherwise similar to an electron; it is the antiparticle of an electron.

The Leptons

Name	Symbol	Spin	Mass (MeV)	Mean lifetime (s)
e-Neutrino	v_e	one-half	0	Stable
μ-Neutrino	v_μ	one-half	0	Stable
Positron	e^+	one-half	0.5	Stable
Electron	e^-	one-half	0.5	Stable
Muon	μ^+	one-half	106	2×10^{-6}
	μ^-	one-half	106	2×10^{-6}

leukemia. A disease in which there is great overproduction of white blood cells, or a relative overproduction of immature white cells, and great enlargement of the spleen. The disease is variable, at times running a more chronic course in adults than in children. It is almost always fatal.

LiBr (lithium bromide). A compound mixed with water and used in absorption chillers supplied with heat energy from solar collectors.

life cycle. The stages an organism passes through during its existence.

lift. In a sanitary landfill, a compacted layer of solid waste and the top layer of cover material.

light. An electromagnetic radiation that stimulates the human eye. Visible light has wavelengths ranging from 7800 Å (0.000 078 cm) on the red side down to 3900 Å on the violet side.

light, velocity of. Light travels with velocity of $2.997\ 925 \times 10^8$ m/s; 186,283 miles/s; 0.1618 nautical miles/μs, or it takes 6.180 μs to cover one nautical mile.

lightning. A sudden flash of light caused by electrical discharge produced by thunderstorms.

limestone. An important building stone and a source of crushed stone for concrete. Limestone is quarried and mined extensively for the manufacture of cement, as a flux for blast furnaces, and for the manufacture of quicklime, hydrated lime, soda ash, and other products that are used on a large scale in the chemical industry. The raw materials most commonly used in making cement are mixtures of limestone and clay, which are heated to form a clinker. A small amount of gypsum is added and the clinker is then finely ground for making concrete.

limiting factor. A condition whose absence, or excessive concentration, exerts some restraining influence upon a population through incompatibility with species requirements or tolerance.

limnology. The study of the physical, chemical, meteorological, and biological aspects of fresh water.

limonite. The yellow, brown, or black hydrous oxide of iron. It is a noncrystalline weathering product of various iron minerals and is responsible for the yellow or brown color of many soils. Its forms include compact masses, nodules, porous bog iron ore, earthy yellow ocher, and rusty stains. Limonite is a minor source of iron.

lindane. $C_6H_6Cl_6$, also known as 1,2,3,4,5,6-hexachlorocyclohexane, an organochloride pesticide that also occurs in several isomeric forms. Like other halogenated hydrocarbons, lindane is relatively stable, fat-soluble, and produces central nervous system effects such as convulsions in warm-blooded animals. Reported LD_{50} oral toxicity levels have ranged from 5 mg/kg in cattle to 500 mg/kg in rats. An oral lethal dose of 188 mg/kg was reported for a human child. A single dose of 45 mg in an adult male, the equivalent of 0.65 mg/kg, resulted in convulsions, but in other reported studies subjects tolerated daily doses of 40 mg for 10 to 14 days. A dose of 30 mg three times a day for a week produced no adverse effects, but a daily dose of 180 mg resulted in dizziness and diarrhea. Four children who drank a home-made beverage containing sugar contaminated with lindane experienced vomiting and convulsions, but recovered without special treatment. But a five-year-old girl who ingested a solution containing 4.5 g of lindane developed breathing difficulty, cyanosis, and convulsions, and died despite emergency treatment for poisoning. Postmortem examination revealed pulmonary edema, fatty invasion of the liver, dilation of the heart, and necrosis of the circulatory system in several vital organs. Adverse effects observed in humans exposed to lindane parallel those found in laboratory animals administered lindane in toxicity experiments, except that tumors found in the animal subjects have not been detected in humans who have experienced chronic or acute exposure.

line of force. In the description of an electric or magnetic field, a line such that its direction at every point is the same as the direction of the force which would act on a small positive charge (or pole) placed at that point. A line of force is defined as starting from a positive charge (or pole) and ending on a negative charge (or pole). The line (of force) is also used as a unit of magnetic flux, equivalent to the maxwell.

linear accelerator. A straight-line particle accelerator. There are two kinds, the standing-wave and the traveling-wave type. In the standing-wave or drift tube linear accelerator the accelerating field exists across the gaps of a series of tubes. The moving particle just drifts along in a tube section at a constant rate of speed, but when it comes to a gap between two tube segments the particle is given a boosting "kick." Each tube segment is longer than the segment before it to compensate for the higher speed of the accelerated particle. This allows the particle to spend the same amount of time passing through each segment although its speed is steadily increased. In the traveling-wave accelerator the accelerating field moves down the tube with the charged particle in the form of an electromagnetic wave. This type of accelerator can deliver a powerful beam of electrons, since electrons moving in a straight-line path do not radiate bremsstrahlung as they do in a circular one. As the electromagnetic wave travels from one end of the accelerator to the other, it carries the electron along with it. The electromagnetic wave is controlled so that it stays with the particle and increases its energy. By the time the electrons have come to the end of the tubes, they are "bullets" moving close to the speed of light. Their speed is limited by the length of the straight-line tube and the energy and number of transmitters. Linear accelerators deliver power in very short bursts, because the tubes would burn out if they were to deliver their power for more than a few microseconds at a time.

light attenuation. A reduction in the intensity of light due to the action of the transmitting medium (absorption and scattering).

light scattering. The reflection or scattering of light in all directions when it enters a body of matter, however transparent. This is due to the interposition of particles of varying size in the light stream, from microscopic specks down to electrons and the deflection of light quanta resulting from the encounters with these small obstacles.

liquefaction. Any process by which a gas is converted to the liquid state.

liquid. A state of matter in which the molecules are relatively free to change their positions with respect to each other, but are restricted by cohesive forces so as to maintain a relatively fixed volume.

liquid fluorine. The element fluorine in its liquid form, having a specific gravity of 1.108, a melting point of $-219.6°C$, and a boiling point of $-188°C$.

liquid helium. Helium liquefies at a temperature lower than any other known substance. Helium II is a mixture of two fluids, one a normal component and the other a superfluid component, and the fraction of the latter increases until it reaches unity at absolute zero.

liquid hydrogen. The lightest (specific gravity 0.07) and the coldest (boiling point $-252.8°C$) of all known fuels. Its melting point is $-259.2°C$. It has a very low density, one quarter that of kerosene, and the highest heat of combustion of any chemical fuel, its heat per unit weight being 2.7 times that of kerosene, and 1.8 times that of pentaborane. When oxidized by liquid oxygen, liquid hydrogen develops a practical specific impulse ranging between 317 and 364 s, depending upon the mixture and pressure ratios.

liquid nitrogen. The element nitrogen in its liquid form, having a boiling point of $-195.8°C$ and becoming a solid at $-210.0°C$; the density of the liquid is 50.8 lb/ft³. It is used in cold radiation shields and in the manufacture of liquid hydrogen.

liquid oxygen (LOX). Oxygen supercooled and kept under pressure so that it remains a liquid. Oxygen solidifies at $-218.8°C$, boils at $-183°C$ at sea-level pressure, has a specific gravity of 1.14 and its heat of vaporization is 91.6 Btu/lb.

liquor, mixed. A mixture of activated sludge and sewage in an aeration tank undergoing activated sludge treatment.

liquor, supernatant. (1) The liquor overlying deposited solids. (2) The liquid in a sludge digestion tank which lies between the sludge at the bottom and the floating scum at the top.

liter. The volume of one kilogram of water at $4°C$, used in the metric system.

lithium (Li). The lightest metallic element, soft and silvery white in color, with atomic weight, 6.940; atomic number, 3; specific gravity, 0.534; melting point, $180°C$; and boiling point, $1330°C$. Lithium has a very high specific heat.

lithosphere. The land portion of the earth's surface.

live room. A room that is characterized by an unusually small amount of sound absorption.

loudness. The intensity of an auditory sensation, in terms of which sounds may be ordered on a scale extending from soft to loud. Loudness depends primarily upon the sound pressure of the stimulus, but it also depends upon the frequency and waveform of the stimulus.

loudness contour. A curve which shows the related values of the sound pressure level and frequency required to produce a given loudness sensation for the typical listener.

loudness level. The loudness level of a sound, in phons, is numerically equal to the median sound pressure level, in decibels, relative to 0.0002 microbar, of a free progressive wave with a frequency of 1000 cycles per second presented to listeners facing the source, which in a number of trials is judged by the listeners to be equally loud.

Love Canal seepage. The Hooker Chemicals and Plastics Corporation was accused of dumping toxic chemicals in the Love Canal area in Niagara Falls, New York. These chemicals had been dumped for many years and they began seeping into homes and a school playground. The Love Canal area was declared a health hazard in 1978 and New York State purchased the houses of residents wanting to move. By April 12, 1978, 235 of the 239 families in the area had moved away. A former official of the chemical company stated that the Board of Education was warned that a school which had been built on land donated by the company had a seepage problem. But he said residents of the area had not been warned for fear of causing legal problems for the Board. At a hearing of a senate panel in April 1978 the Senate Judiciary Committee was told that more than 40 million tons of toxic substances were being generated every year with no knowledge of how it was being disposed. The Environmental Protection Agency issued regulations on April 19, 1979, banning the manufacture of polychlorinated biphenyls, or PCB's, and phasing out most uses of the chemicals. The manufacture of the toxic chemicals, used primarily as insulating fluids and coolants in electrical equipment, had already been stopped by the Toxic Substances Control Act. (See **Hooker Chemical Company pollution.**)

lumen. The unit of luminous flux, equal to the luminous flux through a unit solid angle (steradian) from a uniform point source of one candle, or to the flux on a unit surface, all points of which are at a unit distance from a uniform point source of one candle.

luminance. In photometry, a measure of the intrinsic luminous intensity emitted by a source in a given direction; the illuminance produced by light from a source upon a unit surface area oriented normal to the line of sight at any distance from the source, divided by the solid angle subtended by the source at the receiving surface. It is assumed that the medium between the source and the receiver is perfectly transparent; therefore luminance is independent of extinction between the source and the receiver. The source may or may not be self-luminous. Luminance is a measure only of light; the comparable term for electromagnetic radiation in general is radiance.

luminescence. Any emission of light below the temperatures required for incandescence. Frequently, a body becomes luminescent because of previous exposure to some form of radiation. If the body is luminous during the time it is being irradiated, the phenomenon is called fluorescence. In phosphorescence the emission of light continues after irradiation.

luminous energy. The energy of visible radiation, weighted in accordance with the wavelength dependence of the response of the human eye.

luminous flux. The total visible energy emitted by a source per unit time. The unit of flux, the lumen, is the flux emitted in a unit solid angle (steradian) by a point source having a luminous intensity of one candle. A uniform point source having an intensity of one candle thus emits 4π lumens.

luminous intensity. The property of a source of luminous flux which may be measured by the luminous flux emitted per unit solid angle; also called the candle power. The accepted unit of luminous intensity is the international candle. The Hefner unit, which is equivalent to 0.9 international candle, is the intensity of a lamp of a specified design burning amyl acetate (the Hefner lamp). The mean horizontal candle power is the average intensity measured in a horizontal plane passing through the source. The mean spherical candle power is the average candle power measured in all directions and is equal to the total luminous flux in lumens divided by 4π.

lung cancer. A cancer characterized by an abnormal, disorderly new cell growth originating in the bronchial mucous membrane; it is usually fatal. Lung cancer, like emphysema, cannot be attributed to any single cause; however, it seems very likely that atmospheric contaminants, together with other factors, especially cigarette smoking, contribute to the development of this disease. Many scientists believe that those atmospheric pollutants which paralyze ciliary action in the respiratory tract may play a role in the development of cancer, even though they are not carcinogens themselves. The paralysis of the cilia would permit cancer-causing substances to remain in contact with the sensitive bronchial cells over a longer period of time.

Lurgi. *Process*: The Lurgi process is presently limited to noncaking coals. Crushed and dried coal is fed to a moving-bed gasifier, where gasification of coal takes place at 350 to 450 psi. Devolatization occurs initially and is then accompanied by gasification in the temperature range of 1150 to 1400°F. The residence time is about one hour. Steam is the source of hydrogen and the combustion of a portion of the char with oxygen supplies the heat required. A revolving grate at the base of the reactor supports the fuel bed, removes the ash, and introduces the steam and oxygen mixtures. Crude gas leaving the gasifier at temperatures between 700 and 1100°F (depending on the type of coal) contains tar, oil, naphtha, phenols, ammonia, plus coal and ash dust. Quenching with oil removes tar and oil. Part of the gas passes through a shift converter. The catalyst used in the shift also promotes the desulfurization of light oils. Gas coming out of the shift converter is washed for the removal of naphtha and unsaturated hydrocarbons. The CO_2, H_2S, and COS are removed. The gas is methanated and pipeline gas is produced by final CO_2 removal and dehydration. Substitution of air for oxygen to the gasifier will produce a low-Btu raw gas (Lurgi Mineraloltechnik G.m.b.h.).

Lyman-alpha radiation. A hydrogen series of radiation lines in the extreme ultraviolet region of the spectrum from 1216 to 512 Å. The earth's atmosphere stops this radiation, but it has been observed in the laboratory and detected and measured by instruments in space satellites and probes. A cloud of hydrogen surrounds the earth and gives rise to the observed glow, the intensity of which is enhanced in certain active plages of the sun's chromosphere. The emission line at a wavelength of 1216 Å is the most important ionizing component in the D region of the ionosphere.

M

macrometeorology. The study of the largest-scale aspects of the atmosphere, such as the general circulation, weather types, and the Grosswetterlagen. There is a wide gap between this and the relatively small scale of mesometeorology. The gap is bridged by those atmospheric characteristics referred to as of cyclonic scale. (Grosswetterlag is defined as the mean pressure distribution for a time interval during which the positions of the stationary cyclones and anticyclones remain unchanged.)

magmatic sediment. Sediments of magmatic origin are not extensive. They represent dissolved substances that were transported from within the earth by the heated waters associated with magmas. Much of the material reaches the surface in hot springs, which may make deposits on land, as in Yellowstone Park, or may discharge on the floor of the sea, adding their dissolved load to the seawater.

magnesite. Magnesium carbonate, a compound forming white or variously colored compact masses or crystalline deposits. Much of it was formed through the alteration of dolomite or serpentine by magnesium-bearing solutions. Its principal use is in making refractory brick for open-hearth steel furnaces, cement kilns, and other furnaces.

magnesium (Mg). A silver-white metallic element; atomic weight, 24.32; specific gravity, 1.70; melting point, 650°C; boiling point, 1100°C. Magnesium is easily tarnished by moisture. When heated in air, it ignites at a comparatively low temperature and burns brilliantly. Although magnesium occurs in great abundance in combination with other minerals, these compounds cannot be used as a source of metallic magnesium. The metal is produced by passing a direct current through fused magnesium chloride.

magnetic field. The region in the neighborhood of a magnet, or of a conductor carrying an electric current, in which magnetic forces can be detected.

magnetic flux. The surface integral of the magnetic field intensity normal to the surface. The magnetic flux through any closed figure is the product of the area of the figure and the average component of the magnetic induction normal to the area. The unit of magnetic flux, the maxwell, is the flux through a square centimeter normal to a field of one gauss. The line is also a unit of flux, it is equivalent to the maxwell.

magnetic moment of a magnet. The moment measured by the torque experienced when at right angles to a uniform field of unit intensity. The value of the magnetic moment is given by the product of the magnetic pole strength and the distance between the poles.

A unit magnetic moment is that possessed by a magnet formed by two poles one centimeter apart, of opposite signs and of unit strength.

magnetic storm. A disturbance of the terrestrial magnetic field caused by a stream of corpuscles travelling at about 4 million mph through interplanetary space, covering the distance from the sun to the earth in about one day. The storms are caused by solar flares and sunspots.

magnetite. Black iron oxide, a dark, heavy magnetic mineral that is present in small amounts in most igneous rocks and in black sands. Magnetite is an ore of iron.

magnetohydrodynamic (MHD) steam power plant. The diagram shows a magnetohydro-dynamic (MHD) steam power plant. The efficiency of such an installation could reach 60 percent or more, considerably above the overall plant efficiency of 41 percent for the best steam turbine electrical plant. The use of fossil fuels in an MDH generator results in levels of air pollution far lower than conventional fossil fuel electrical power plants and substantially less thermal pollution than in nuclear plants. In place of the solid conductors of the usual turbine-driven generator, MHD devices substitute a gas raised to 2000–2500°C. The gas is the combustion product of fossil fuel "seeded" with an alkali metal (about 1 percent) to promote ionization. Ionized gas is passed through a magnetic field to generate a current. Gas-cleaning equipment satisfies both the need for air pollution control and the economic requirement for seed recovery. The success of electrostatic precipitation in this application has been demonstrated on a pilot scale.

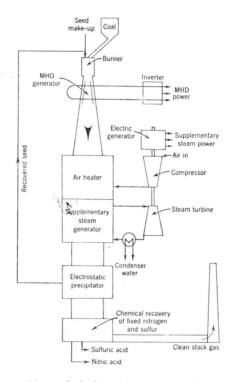

Magnetohydrodynamic steam power plant.

magnetohydrodynamics (MHD). A simple MHD generator resembles a piece of pipe in which electrodes are embedded. The pipe is placed between the poles of a powerful electromagnet. When hot gas is forced through the pipe at a great velocity, the magnetic link of force generates current that is tapped by the electrodes and fed into a tower distribution system. Its advantages include an ability to withstand very high temperatures, its lack of moving parts, and at least a 50 percent efficiency.

magnifying power. For an optical instrument, the ratio of the angle subtended by the image of the object seen through the instrument to the angle subtended by the object when seen by the unaided eye. In the case of the microscope or simple magnifier, the object as viewed by the unaided eye is supposed to be at a distance of 25 cm (10 in.).

makeup carbon. Fresh, granular activated carbon which must be added to a column system after a regeneration cycle, or when deemed necessary, to bring the total amount of carbon to specification.

manganese (Mn). A grayish-white, polyvalent, nonmagnetic metallic element that in trace amounts is essential to life processes, but in larger than necessary levels can produce adverse health effects. An important industrial use of manganese is in the production of iron alloys such as stainless steel. In water supplies, manganese contributes a brownish color to laundered fabrics and impairs the taste of coffee, tea, and other beverages. Because of these effects and the possibility of physiological disorders in humans consuming excessive amounts of the water, the EPA has recommended a maximum concentration of 50 micrograms of manganese per liter of water. Laboratory studies and clinical reports indicate an intraperitoneal LD_{50} of 53 mg/kg for mice, and an inhaled dose of 11 mg/m^3 in humans can result in toxic central nervous system symptoms. Manganese is considered a known mutagen and a suspected carcinogen. Chronic exposure resulting in central nervous system disorders can occur in less than three months, although in some cases symptoms developed after one to three years of exposure to heavy concentrations of manganese dust or fumes. Removal of the patient from the site of exposure may or may not be followed by a remission of symptoms. Permanent disability from exposure to manganese compounds may involve symptoms that simulate Parkinson's disease, multiple sclerosis, or amyotrophic lateral sclerosis.

manometer. An instrument used for measuring pressure differences, usually consisting of a U-shaped tube containing a liquid. The height of the liquid in one end of the tube moves proportionally with changes in pressure upon the liquid in the other end. The manometer is applied to a tube-type differential pressure gauge. A mercury barometer is a type of manometer.

marsh. A wet, soft, low-lying land that provides a niche for many plants and animals. It can be destroyed by dredging and filling.

maser. An acronym for microwave amplification by stimulated emission of radiation. The maser produces concentrated radio microwaves and it is based on the concept that excited atoms not only radiate by themselves, but can be stimulated to emit radiations by an incoming wave of proper energy. The maser can receive a weak radio signal and emit a signal at exactly the same frequency, but 1000 times more powerful than the original. Masers have increased the sensitivity of radars, helping to locate objects more accurately by bouncing radio signals off them. By amplifying weak signals from outer space received by radio telescopes, masers have also made it easier to listen and map the distant invisible galaxies.

masking. (1) The process by which the threshold of one sound is raised owing to the presence of another. (2) The increase, expressed in decibels, of the threshold of hearing of masked sound due to the presence of masking sound.

mass. The amount of matter in a body, not affected by changes in gravitational pull, but measured by the extent to which it resists efforts made to start or stop it as motion. The gram is 1/1000 the quantity of matter in the International Prototype Kilogram; it is one of the three fundamental units of the cgs system. The British standard of mass is the pound, of which a standard is preserved by the government. The U.S. standard mass is the avoirdupois pound defined as 1/2.20462 kilogram.

mass action, law of. At a constant temperature, the product of the active masses on one side of a chemical equation divided by the product of the active masses on the other side of the chemical equation is a constant, regardless of the amounts of each substance present at the beginning of the action. At constant temperatures, the rate of the reaction is proportional to the concentration of each kind of substance taking part in the reaction.

mass concentration. A concentration expressed in terms of mass of substance per unit volume of gas or liquid.

mass median size. For samples of particulate matter, the median measurement of particle size related to the diameter size.

mass, relativistic. The increased mass associated with a particle when its velocity is increased. The increase in mass becomes appreciable only at velocities approaching the velocity of light, 3×10^{10} cm/s.

material damage. Air pollution damage to materials of inanimate bodies. This may be brought about through: (a) abrasion; (b) deposition and removal, where there might not be much damage except for the appearance in deposition, but where damage occurs in removal or cleaning operations; (c) direct chemical attack such as the tarnishing of silver by H_2S; (d) indirect chemical attack, which is deterioration caused by chemical reactions after absorption of a pollutant; and (e) electrochemical corrosion.

matter (gas). The molecules in a gas are generally far apart and independent of each other. They are free to move about at random, and diffuse to fill their container regardless of its shape. As the molecules dart about, they collide with each other and bang against the container walls, producing gas pressure.

matter (liquid). The molecules in liquids are generally farther apart than those in solids. They cluster together in a random manner, yet are free to move in relation to each other. This allows liquids to conform to the shape of their container. Liquids, although not as dense as solids, still have a definite volume.

matter (plasma). A form of intensely hot gas. Gases start to become plasma at 5000°F. The intense heat breaks apart some of the gas molecules, and the atoms which compose the "broken" molecules dart off independently. These independent atoms are sometimes stripped of one or more of their electrons, creating negative and positive ions. Darting about furiously in the chaos of a plasma gas are molecules, atoms, and positive and negative ions. Plasma is the most common form of matter in the universe.

matter (solid). The molecules forming solids are very close together and vibrate in fixed positions. The arrangement of the molecules in a substance determines the characteristic form of the solid, whether it is hard and rigid like a diamond or soft and crumbly like sulfur.

maximum allowable concentration (MAC). The concentration of a pollutant considered harmless to healthy adults during their working hours, assuming that they breathe uncontaminated air for the remainder of the time.

maximum emission concentration. Standards for the maximum air pollutant emission from stationary or moving sources. These include opacity standards, such as the Ringelmann chart gradations often applied with time specification, gravimetric emission standards (expressed as the weight of emitted pollutant per volume or unit weight of carrier gas), volumetric emission standards (percentage of gaseous pollutant by volume in the emitted gas or as parts per volume in specific parts of effluent gas, e.g., where SO_2 is limited to 0.2 percent by volume in emitted gases or zinc mineral is limited to one part by volume per 1000 parts of effluent gas).

maxwell. The cgs electromagnetic unit of magnetic flux; the flux through a square centimeter normal to a field at one centimeter from a unit magnetic pole.

Maxwell's rule. Every part of an electric circuit is acted upon by a force tending to move it in such a direction as to enclose the maximum amount of magnetic flux.

mechanical equivalent of heat. The quantity of energy which when transformed into heat is equivalent to a unit quantity of heat; 4.18×10^7 erg = 1 cal (20°C).

mechanical turbulence. The erratic movement of air influenced by local obstructions.

mega. A prefix meaning one million (10^6).

megacycle. One million cycles; one thousand kilocycles. The term is often used as the equivalent of one million cycles per second.

mel. A unit of pitch. The pitch of any sound judged by listeners to be n times that of a one-mel tone is n mels. 1000 mels is the pitch of a 1000-Hz tone at a sensation level of 40 decibels.

membrane filter. A controlled pore filter composed of cellulose esters. Membrane filters can be manufactured with uniformly controlled pore sizes. Nylon mesh may be used for reinforcement. Types commonly used for air sampling have a pore size of about 0.45 to 0.8 µm. The pores constitute 80 to 85 percent of the filter volume. Because of electrostatic forces and the formation of a precoat of collected particles on the surface, these filters can collect particles down to about 0.1 µm in diameter. They are particularly suited to particle sizing, since most of the collected particles are retained on or near the surface.

mercury (Hg). A liquid metallic element. Mercury in naturally occurring stable compounds causes no problems, but its liberation in the form of soluble salts from a group of mercury compounds with industrial or agricultural uses cause some concern. Inorganic mercury compounds are used in plastics, industrial chlorine, and the electronics industry. There has been mercury poisoning derived from occupational exposure; accumulation of inorganic mercury in the kidneys affects reabsorption and secretion of sugar, protein, and salts; accumulation in the brain causes a loss of coordination. There are two types of organic mercury; aryl salts of mercury, which break down into inorganic mercury in the body, or the alkyl salts of mercury, particularly methyl mercury, which are able to diffuse easily through membranes and spread throughout the body. Mercury is a fungicide, a potent killer of fungi, therefore organic compounds of mercury, especially methyl mercury, have been widely used by pulp mills to keep fungi, bacteria, and algae which thrive on wood pulp from clogging up the machinery. In the United States, methyl mercury fungicides are widely used in agriculture, especially in wheat-growing areas.

mercury barometer. A barometer consisting of a glass tube a little more than 30 inches long, with the top end sealed. The tube, filled with mercury, stands in a vertical position with its open end submerged in a pool of mercury. The height of the column of mercury inside the tube varies according to the pressure of the atmosphere on the pool of mercury at the base. When high-pressure systems move through the atmosphere, they cause the mercury to rise in the tube, low-pressure systems cause it to fall. A high barometer reading is a sign of good weather; a low barometer reading indicates that unpleasant weather exists already or is approaching.

mercury vapor. The liberation of mercury vapor occurs when mercury salt comes into contact with organic matter in the soil or with the organic vapors given off plants. The extent and rate at which metallic mercury injures plants vary with the concentration of the vapor in the air, and this in turn varies with temperature. Vapors from soil treated with mercury compounds injure plants more quickly at higher temperatures. In a recent study, flower buds of roses in all stages of development were affected by the vapors emanating from soil treated with bichloride. The peduncles of very young buds turned yellowish and then black; half-mature buds turned brown, and the corollas abscised from the receptacle without opening; the older buds continued to unfold, but the petals lacked the pink color; stamens were also injured, turning nearly black in half-mature buds.

mesh size. The particle size of granular activated carbon as determined by the U.S. sieve series. Particle size distribution within a mesh series is given in the specification of the particular carbon.

meson. Greek for "an intermediate one," an unstable particle with integral spin, having a mass between those of the proton and electron, tending to transform itself into a photon or lepton. All mesons are radioactive and have very short half-lives. There are K mesons, or kaons, and pi mesons, or pions. Kaons are the heavier of the two and can appear when protons or neutrons are bombarded. Kaons decay to the smaller pions. Pions are believed to be the particle of exchange in nuclear force, as they are exchanged between protons and neutrons, they "cement" these nucleons together and keep the nucleus stable. Once inside the nucleus, pions exist for only a few millionths of a second, decaying immediately to muons and neutrinos.

mesosphere. Above the stratosphere is the stratopause, a transitional zone below the mesosphere. The mesosphere is located 20 to 52 miles above the earth's surface. Experiments have indicated that there is a gradual increase in temperature in the mesosphere, rising from about $-42°C$ at its lower limit to about $0°C$, followed by a thermal reversal at higher altitude, bringing about a gradual drop in values.

metabolism. The sum of all physical and chemical processes by which living organized substances are produced and maintained and by which energy is made available for the uses of organisms.

metallic bond. A bond forming the crystalline solids called metals. Metal atoms have only a few electrons in their outer shells and these electrons are easily detached, leaving behind positive ions. The detached or free electrons flow through the structure of the metal in the form of a negatively charged "electron gas," which reacts with the positively charged metal ions. It is the attraction between the negative "electron gas" and the positive metallic ions that forms the metallic bond that holds a metal together in its solid state. This migrating "electric gas" also makes metals excellent conductors of heat and electricity.

metastable atom. An atom with an electron excited to an energy level where simple radiation is forbidden and thus the atom is momentarily stable.

metastable compound. A chemical compound of comparative stability which becomes unstable under a particular set of conditions.

metastable state. An excited state of a nucleus which returns to its ground state by the emission of a gamma ray over a measurable half-life.

methane. CH_4, a light, odorless, flammable gas. It is the chief constituent of natural gas and is often produced by the partial decay of plants in swamps. It is the first member of the gas family composed of one carbon and four hydrogen atoms.

methanol. CH_3OH, methyl alcohol, a colorless toxic liquid with a molecular weight of 32.03, a specific gravity of 0.791, a freezing point of $-97.8°C$, and a boiling point of $64.1°C$.

methanol synthesis. *Process:* Synthesis gas $(CO + H_2 + CO_2)$ is converted to methanol in the catalytic converter. Two chemical reactions generalize the formation of methanol: $CO + 2H_2 \rightarrow CH_3OH$, and $CO_2 + 3H_2 \rightarrow CH_3OH + H_2O$. Synthesis gas can be produced by any of a number of processes such as those used in the production of intermediate-Btu fuel gas. The raw gas composition depends on many operating factors such as temperature, pressure, excess steam, etc. For methanol synthesis, methane in the synthesis gas is undesirable. A gasifier operated at high pressure (1000–1500 psia) and high temperatures (2200–2700°F) will produce the most suitable synthesis gas. Raw gas from the gasifier is passed through a water–gas shift and purification to produce synthesis gas of the correct composition. Methanol synthesis catalysts are vulnerable to sulfur poisoning. The removal of sulfur compounds from the synthesis gas is essential. Methanol is produced in processes operating at high pressures (about 4500 psig), intermediate pressures (2000–2500 psig), or low pressures (750–1500 psig). High-pressure processes use zinc chromium oxide catalysts. Low- and intermediate-pressure processes use highly active Cu–Zn–Cr catalysts. Compressed synthesis gas is combined with recycle gas and passed through the methanol catalytic converter, which operates at about 500°F. Crude methanol obtained from the converted gas is condensed and separated from the unreacted recycle gas. Dissolved gas and some high-pressure recycle gas is purged to control the concentration of inert gases. Crude methanol is purified by distillation to remove components such as dimethyl ether, methyl formate, water, and higher alcohols.

methyl alcohol. CH_3OH, methanol, commonly called wood alcohol. It is made by combining carbon monoxide and hydrogen under high pressure and in the presence of catalysts. Methyl alcohol is extremely poisonous.

methyl chloromethyl ether. C_2H_5OCl, a clear, colorless, corrosive liquid that emits vapors that smell of hydrogen chloride and formaldehyde. It is used in ion exchange resins, the manufacture of polymers, bactericides, and drugs to lower cholesterol, and in the production of gelatin and sugar. The chemical has been found to cause oat-cell carcinoma, a form of lung cancer, in humans. It has also been observed to produce tumors and cancers in mice. The adverse effects of methyl chloromethyl ether can occur as a result of skin absorption or inhalation. Commercial products may also be contaminated with bischloromethyl ether, also a known carcinogen. The oral LD_{50} is 55 ppm for seven hours. NIOSH has recommended exposure limits of not more than 0.1 percent by weight or volume.

methyl iodide. CH_3I, a colorless, light-sensitive liquid used as a narcotic and anesthetic. It is regarded as a potential disaster hazard that may decompose when heated, producing highly toxic iodide fumes. The boiling point is 42.5°C. The oral LD_{50} of methyl iodide for rats has been established as 220 mg/kg; the subcutaneous LD_{50} is 110 mg/kg, and the lowest lethal concentration for rats via the inhalation route is 3790 ppm for a period of 15 minutes. Methyl iodide has been found to produce tumors in experimental animals and has been listed by the EPA's Carcinogen Assessment Group as a potential occupational carcinogen accepted for review by OSHA.

methyl methanesulfonate. $C_2H_6O_3S$, an ester related in toxic effects to aldehydes, ethers, and ketones. The LD_{50} for methyl methanesulfonate has been established as 125 mg/kg when administered subcutaneously. It has been observed to have carcinogenic, tumor-genic, or mutagenic effects in animals when administered by subcutaneous, intraperitoneal, or oral routes. The compound was added in 1980 to the EPA's list of suspected occupational carcinogens.

N-methyl-N-nitro-N-nitrosoguanidine. An experimental mutagen and carcinogen listed by the EPA's Carcinogen Assessment Group and accepted by OSHA as a possible occupational carcinogen.

3-methylcholanthrene. $C_{21}H_{16}$, a compound that occurs in the form of yellow crystals and a potential fire hazard in the presence of oxidizing materials. 3-Methylcholanthrene is a powerful irritant and has been associated with the development of cancers through exposure of skin surfaces as well as by subcutaneous, oral, and implantation routes. It has been identified by the EPA Carcinogen Assessment Group as a potential occupational carcinogen and accepted by OSHA for review.

methylene blue number. The amount in milligrams of methylene blue absorbed by one gram of carbon in equilibrium with a solution of methylene blue having a concentration of 1.0 mg/liter.

4,4'-methylenebis (2-chloraniline). $C_{13}H_{12}N_2Cl_2$, a form of aniline regarded as an industrial hazard. The compound is also known as MOCA and 3,3'-dichloro-4,4'-diamino diphenyl methane. It is a neoplastic agent and a carcinogen with a threshold limit value of 0 ppm. It has been listed by EPA's Carcinogen Assessment Group and accepted by OSHA for review as a potential occupational carcinogen.

metric system. A system of measurement developed in France during the 1790s, ten and multiples of ten are used exclusively. The meter, defined as 1/10,000,000 of the distance from the equator to the North Pole, was its original standard. Since then, the meter has been redefined in terms of the wavelength of light emitted by ^{86}Kr. Universally adopted by scientists, the metric system is also used commercially in most countries. The major approximate equivalents are:

Length	1 cm	=	0.4 in.
	1 m	=	1.1 yd
	1 km	=	0.6 mile
Volume	1 liter	=	1.06 quart
Weight	1 kg	=	2.2 lb

metric units. The cgs system which uses the centimeter (cm), the gram (g), and the second (s) as units of length, mass, and time is gradually being ousted by the mks system, which uses the meter (m), the kilogram (kg), and the second (s). Other units are linked to these; thus velocity is measured in units of meters per second (m/s). Some units have their own names, like the unit of force, one newton (1 N = 1 m kg/s^2), the force needed to accelerate 1 kg of mass from 0 to 1 m/s in 1 s. (One kilogram at rest exerts on its support a force of about 10.2 N at medium latitude.) Other units are the energy unit, one joule (1 J = 1 N m), and the common unit of power, one watt (1 W = 1 J/s). For decimal multiples and submultiples, see decimal notation. Electrical units like the coulomb (for charge), the ampere = one coulomb per second (for current), and the volt (for potential and hence often called voltage) are also linked to the mks system; but temperature is not, nor is illumination.

mgd. Millions of gallons per day, a measurement of water flow.

mica. Any of a group of minerals that crystallize in thin, easily separated layers. There are two main varieties of mica; muscovite, hydrous potassium aluminum silicate, which is white or colorless, and biotite, black mica, which resembles muscovite except for its color. Other micas are phlogopite, a bronze colored magnesium mica; and lepidolite, a lilac-colored lithium-bearing mica. Micas can be separated into extremely thin sheets. Muscovite, in high-grade cleavage sheets, is used extensively for electrical insulation for irons and toasters. Mica in sheet form is used for glazing, and in flake or powder form for filler, heat insulation, decoration, and tire powder. Sheet mica is obtained from pegmatites in association with large crystals of feldspar and quartz.

micro. A prefix meaning 1/1,000,000 (10^{-6}) used in the metric system. A microgram is 1/1,000,000 of a gram.

microbar. A unit of pressure commonly used in acoustics; one microbar is equal to one dyne per square centimeter.

microbe. A tiny plant or animal. Some microbes that cause disease are found in sewage.

microcurie. The amount of radioisotope undergoing 3.7×10^4 disintegrations per second.

micrometeorology. The portion of meteorology which deals with the observation and explanation of the smallest-scale physical and dynamic occurrences within the atmosphere. So far, studies within the field are confined to the surface boundary layers of the atmosphere, that is, from the earth's surface to an altitude where the effects of the immediate underlying surface upon air motion and composition become negligible. More confining limits have been suggested. To date, the bulk of the work in this field has centered around the evaluation of low-level atmospheric turbulence, diffusion, and heat transfer. Small-scale temperature, evaporation, and radiation studies are included; so far these have been made primarily in the field of microclimatology.

micron. A unit of length equal to one-millionth of a meter, or one-thousandth of a millimeter.

microphone. An electroacoustical transducer that responds to sound waves and delivers essentially equivalent electric waves.

mil. One-thousandth of an inch; a unit of angular measurement equal to 1/6400 of a circle.

milli. A prefix meaning 1/1000 (10^{-3}) used in the metric system. A milligram is 1/1000 of a gram.

millibar. A unit of pressure equal to 1000 dynes per square centimeter, or 1/1000 of a bar; a unit used by meteorologists to measure air pressure (34 millibars equals one inch of mercury).

milligrams per liter. A weight per volume designation used in water and waste analysis.

Mirex. $C_{10}Cl_{12}$, a chemical similar in formulation to Kepone, and a pesticide that was used to control the spread of fire ants in the southern United States until it was banned in 1978 because of its observed effects on mammals. The oral LD_{50} for Mirex has been reported as 306 mg/kg. It is considered a persistent chlorinated hydrocarbon that is toxic to nontarget species of animals and it has cumulative effects. In addition to its application in fire-ant control in the southern states, Mirex has been employed in Hawaii to prevent mealy bug wilt in pineapples. In laboratory studies, Mirex was found to be a potential carcinogen and teratogen. In 1980, the EPA issued a conditional registration for a substitute fire-ant pesticide, called Amdro, for use on lawns, pastures, and range lands, but clearance for use on agricultural crop lands was delayed pending further studies. The fire ant, which entered the United States from Brazil, has spread through more than 230 million acres of the south, as far north as the Carolinas, and has become a serious threat not only to the soil it inhabits, but to humans and domestic animals who become victims of its sting. The sting of the fire ant causes pain, swelling, and edema and may require steroid medications to relieve the symptoms.

mist.   A state of atmospheric obscurity produced by suspended water droplets (visibility exceeds 1 km, but is less than 2 km); according to international definitions, a hydrometeor consisting of an aggregate of microscopic and more or less hygroscopic water droplets suspended in the atmosphere. It generally produces a thin grayish veil over the landscape. It reduces visibility to a lesser extent than does fog. The relative humidity with mist is often less than 95 percent. It is intermediate in all respects between haze (particularly damp haze) and fog. In popular usage in the United States, mist is the same as drizzle.

Mitomycin C.   An experimental carcinogen and also an antineoplastic substance. It is a member of the mitomycin complex produced by *Streptomyces caespitosus* from soil collected at Jochi-machi, Japan. It has the chemical formula $C_{15}H_{18}N_4O_5$ and occurs as blue-violet crystals. The intravenous LD_{50} in mice is 5 mg/kg. It has been included among other drugs such as chlorambucil and diethylstilbestrol for OSHA review as a possible substance which may by itself or in combination with other chemicals cause an increase in the incidence of benign or malignant tumors or metabolize into one or more potential occupational carcinogens.

mixing liquor.   Activated sludge and water containing organic matter being treated in an aeration tank.

mixing depth.   The expanse in which warm air rises and mixes with cooler air; the stretch between a sunlit meadow and the height at which cooling air meets its equal in temperature. It delineates the upward boundary for pollution dispersion. Seasons change the mixing depth. During the summer daylight hours, the depth usually extends several thousands of feet. During the winter, when the sun gives less heat, it may measure as little as a few hundred feet. The mixing depth also varies with the time of day. At night, the air close to the earth is cooled by contact with it, while the air higher up stays relatively warm. If surface cooling is great enough, the unchanged higher air may be even warmer than the air near the earth; then the colder, heavier air at the surface has no place to go and the mixing depth is minimal.

mixing height.   The height at which stack effluent begins mixing with the atmosphere.

mksa system.   A system of units based on the meter, kilogram, second, and ampere.

mobile source.   A moving producer of air pollution, mainly forms of transportation—cars, motorcycles and planes.

moderator.   A material used for slowing down neutrons in an atomic pile or reactor, usually graphite or "heavy water" (deuterium oxide). The moderator is a material that surrounds the fuel rods and reduces the speed and energy of the free neutrons. Slow-moving neutrons have a better chance than fast ones of hitting the nucleus of a fuel atom and are more efficient in starting and sustaining a chain reaction. Heavy hydrogen and graphite are good moderators.

modulus of elasticity.   The stress required to produce a unit strain, which may be a change of length (Young's modulus), a twist or shear (modulus of rigidity or modulus of torsion), or a change of volume (bulk modulus) expressed in dynes per square centimeter.

Mohs' scale.   A scale of hardness for minerals ranging from 1 for the softest to 10 for the hardest.

molar (M).   The strength of a solution expressed in terms of moles per liter of the dissolved substance. A bottle labeled "5M NaOH" contains a five-mole solution of sodium hydroxide.

molasses number. A number calculated from the ratio of the optical densities of the filtrate of a molasses solution treated with a standard activated carbon and the activated carbon in question.

mole. The amount of substance containing the same number of atoms as 12 grams of pure ^{12}C. The gram-mole or gram-molecule is the mass in grams numerically equal to the molecular weight.

molecular effusion. The passage of gas through a single opening in a plane wall of negligible thickness, where the largest dimension of the hole is smaller than the mean free path.

molecular flow. The flow of gas through a duct under conditions such that the mean free path is greater than the largest dimension of a transverse section of the duct.

molecular flux. The net number of gas molecules crossing a velocity component in a unit interval of time, those having a velocity component in the same direction as the normal to the surface at the point of crossing being counted as positive and those having a velocity component in the opposite direction being counted as negative.

molecular volume. The volume occupied by one mole; numerically equal to the molecular weight divided by the density.

molecular weight. The sum of the atomic weights of all the atoms in a molecule.

molecule. The smallest part of a substance that can exist separately and still retain its chemical properties and characteristic composition; the smallest combination of atoms that will form a given chemical compound.

molybdenite. Molybdenum sulfide, the chief ore of molybdenum (a metal used in steel), a very soft, heavy (specific gravity 4.75), blue-gray mineral that forms shiny, flexible scales or flat, hexagonal crystals with platy cleavage. It resembles graphite, but differs in color and specific gravity.

molybdenum (Mo). A silvery white metal or a gray-black powder with atomic weight, 95.95; specific gravity, 10.2; melting point, 2620°C; boiling point, 4600°C; with nearly twice the Young's modulus of nickel, iron, or cobalt and a density only 10−15 percent greater. One of the five most abundant refractor metals in the earth's crust.

molybdenum disulfide. MoS_2, black, dry, nonvolatile lubricant suitable for use between metals at extremely low temperatures and pressures. It has molecular weight, 160.12; specific gravity, 4.80; and melting point, 1185°C.

momentum. The product of the mass of a body and its velocity; its cgs unit is g cm/s.

monaural hearing. The perception of sound stimulation by a single ear.

monitoring. The instrumentation and procedures for the continuous measurement of air pollutants and the application of regulatory or control measures when an established standard has been exceeded.

monitoring systems. An automatic system placed in a smokestack to measure and record the amounts of air pollutants being discharged into the atmosphere. The data can either be stored for later examination or be transmitted to a central unit.

monophagous species. Animals able to utilize only one type of food. The geographic distribution is limited to regions where the specific food type (plant or animal) is located. If the food type is removed or destroyed by some catastrophic action, the species is destined by sheer lack of nutriments to disappear from the area if encysted or dormant stages of the life cycle are not present in the immediate environment.

monostatic reflector. The characteristic of a reflector which reflects energy only along the line of the incident ray, e.g., a corner reflector.

monsoon. Winds that consistently blow onshore during the summer (bringing humid, rainy weather) and offshore during the winter (heralding cool, dry weather). Monsoons are typical of tropical Southeast Asia.

mortality. The number of individuals dying per unit of time; also called death rate. The minimum possible mortality equals the population loss under ideal or nonlimiting conditions. The actual observed mortality is the rate of loss of individuals under given environmental conditions where predators, accidents, competition, and other factors share in causing death.

moving bed. An application with Filtrasorb granular carbons in which a single carbon column offers the efficiency of several columns in series. This is accomplished by the removal of spent carbon from one end of the carbon bed and the addition of fresh carbon at the other end, with little or no interruption in the process.

muck soil. Earth made from decaying plant materials.

mucus. The sticky fluid covering the airway lining of the respiratory system.

mulch. A layer of material (wood chips, straw, leaves) placed around plants to hold moisture, prevent weed growth, and enrich soil.

multiple-chamber incinerator. A two-stage incinerator consisting of the following basic components: (a) a primary chamber wherein preheating and combustion take place, (b) a secondary chamber for the combustion and expansion of gases, (c) a chamber for settling fly ash, and (d) a stack which discharges the gases to the atmosphere, or in certain cases where state or municipal regulations require it, a scrubbing system. The combustion process in a two-stage or multiple-chamber incinerator consists first of a primary chamber into which the solid waste material is charged. In this chamber, drying, ignition, and

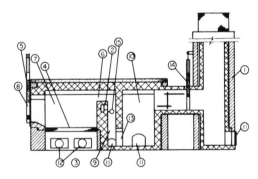

SIDE ELEVATION

1. STACK	6. FLAME PORT	11. CLEANOUT DOORS
2. SECONDARY AIR PORTS	7. IGNITION CHAMBER	12. UNDERFIRE AIR PORTS
3. ASH PIT CLEANOUT DOORS	8. OVERFIRE AIR PORTS	13. CURTAIN WALL PORT
4. GRATES	9. MIXING CHAMBER	14. DAMPER
5. CHARGING DOOR	10. COMBUSTION CHAMBER	15. GAS BURNERS

Multiple-chamber incinerator.

combustion of the solid refuse occur. As the burning proceeds, the moisture and volatile components of the fuel are vaporized and partially oxidized in passing from the ignition chamber through the flame port, which connects the ignition chamber to the mixing chamber. From the flame port, the products of combustion and volatile components of the refuse flow through the mixing chamber, where secondary air is introduced. In a third chamber known as the final combustion chamber, there is also a refractory wall or curtain wall between the mixing chamber and the final combustion chamber. Fly ash and other solid particulate matter are collected in the combustion chamber by impingement on these walls or by gravity settling. Finally, the gases are discharged through a stack or combination of gas coolers, which are watery spray chambers, and an induced draft system.

multiple-hearth furnace. A method for sludge disposal in which liquid–solid waste sludges are fed by conveyor to the top of the incinerator. The incinerator is a vertical cylindrical enclosure with a number of horizontal hearths, each having an opening to the hearth below. The waste sludge is moved to lower hearths by the plowing action of air-cooled metal rabble arms. The waste sludge is dried on the first two hearths and burned on the next three, and the ash is cooled on the lower hearth. The ash is removed for burial or reuse.

multiple use. The harmonious use of land for more than one purpose, i.e., grazing of livestock, wildlife production, recreation, watershed, and timber production, not necessarily the combination of uses that will yield the highest economic return or greatest unit output.

mutagen. Any substance that causes changes in the genetic structure in subsequent generations.

mutation. A change in the characteristics of an organism by an alteration of the usual hereditary pattern.

mutualism. The cooperation of two organisms in order to survive. The algal–fungal association, or lichens, which are common in nature are examples of mutualism. The alga depends upon the fungus for water and protection and the fungus, in turn, receives food produced photosynthetically by the alga.

N

nano. A prefix meaning multiplied by 10^{-9}.

naphthylamine. A toxic nitrogen compound of the aromatic amine group. Naphthylamine commonly occurs in either alpha or beta isomers, with the formula $C_{10}H_7NH_2$. Both the α- and β-naphthylamines are found as crystals that are white to pink in color. The oral LD_{50} of both naphthylamines is about 700 mg/kg in rats. α-Naphthylamine has been associated with the occurrence of bladder cancer. Both forms are skin and bladder irritants and have been identified as a cause of liver and kidney damage. α-Naphthylamine is a suspected human carcinogen and neoplastic agent. β-Naphthylamine is regarded as a highly toxic compound in solid, liquid, or vapor states. It can be absorbed through the skin, as well as through the gastrointestinal tract and respiratory system. Tumors and cancer have resulted from continued or long exposure to even small amounts of the chemical. It has been identified as a carcinogen in humans when absorbed through the digestive tract or the skin. In addition, β-naphthylamine produces toxic fumes when exposed to heat or flame.

β-naphthylamine. A compound developed as an antioxidant in rubber, greases, and oils, and in the synthesis of dyes and other antioxidants. β-Naphthylamine was found to be a human-bladder carcinogen in the 1940s and was replaced by phenyl-β-naphthylamine, which was believed safe until studies in the 1970s showed that the phenyl (PBNA) form is metabolized to the original β-naphthylamine (BNA) molecule in the human body. In addition, commercial PBNA is contaminated with 20 to 30 ppm of BNA. Thus phenyl-β-naphthylamine is a precursor of a known human carcinogen and should be handled in the same manner as a human carcinogen. Production in the United States of phenyl-β-naphthylamine has declined considerably from a level of nearly five million pounds in 1973. The majority of the estimated 15,000 workers exposed to the chemical are employed in the rubber industry. Finished rubber products may contain as much as one percent phenyl-β-naphthylamine. Evidence that the phenyl form of BNA is metabolized to the carcinogenic form has been substantiated by experiments with dogs and human volunteers, who excreted BNA in the urine after ingesting PBNA. Reports that PBNA may be a carcinogenic precursor in humans has been challenged by epidemiological studies which found no significant increase in the incidence of bladder cancer among rubber workers whose contact with PBNA began after it had replaced the BNA form as an antioxidant. The findings of the epidemiology studies have been questioned in turn by investigators who claim that the PBNA form may not have been in use long enough to produce the human data risk factors necessary to show a carcinogenic effect. NIOSH has recommended that industrial hygiene practices be followed to minimize exposure to phenyl-β-naphthy-lamine in the workplace.

narrow band. A narrow band of transmitted waves with neither the critical nor the cutoff frequencies of the filter being zero or infinite.

natality. The production of new generations of organisms by sexual or asexual repro-ductions. The reproductive potential of a population is based on natality and mortality, intraspecific competition for food, shelter, mates, predation, and parasitism. All of these factors enter the total picture and often radically change population growth patterns. The maximum, or potential, natality refers to the greatest number of organisms that can possibly be produced by a parent organism over a unit period of time. A normal, or realized, natality rate is an observed population birth rate occurring under normal circumstances for the species.

National Air Sampling Network (NASN). The network of air sampling stations established in the United States in urban and nonurban communities. The objectives of this network include the determination of the extent and nature of air pollution, as well as the determination of relationships between air pollution and socioeconomic, geographic, topographic, meteorological, and other factors. Initially, operations of the network included the collection and analysis of suspended particulate matter only. Activities have been expanded to include the collection and analysis of precipitation samples and the determination of atmospheric contamination levels of specific gaseous pollutants.

natural gas. A combustible gas which is recovered from an underground source. Not all natural gases occur in association with oil accumulations, combustible natural gas also occurs in association with coal. Natural gas contains 50 to 99 percent methane (CH_4) by volume and differing amounts of higher paraffins, together with some inert gases, mainly carbon dioxide and nitrogen. The calorific value may vary between 700 and 1500 Btu/ft^3 at standard temperature and pressure. The term "wet gas" is used to describe a natural gas which contains pentane (C_5H_{12}) (boiling point 27.9°C) and heavier paraffins with still higher boiling points at ordinary atmospheric pressure. A "dry gas" is that which contains only a slight amount of these hydrocarbons. Natural gasoline is obtained from wet gas by such methods as compression and cooling, absorption, adsorption, and low-temperature separation. "Liquefied petroleum gas" is the fraction of wet gas, mainly butane and isobutane (C_4H_{10}, normal boiling points −0.5 and −11.7°C).

natural gas combustion emission factors (in lb/10^6 ft^3 of natural gas burned).

Pollutants	Power plants	Industrial boilers	Domestic and commercial heating units
Aldehydes	1	2	Negligible
Benzo[a]pyrene	n.a.	20,000/10^6 ft^3	130,000/10^6 ft^3
Carbon monoxide	Negligible	0.4	0.4
Hydrocarbons	Negligible	Negligible	Negligible
Oxides of nitrogen	390	214	116
Oxides of sulfur	0.4	0.4	0.4
Ammonia	n.a.	0.3	0.3
Organic acids	n.a.	62	62
Particulates	15	18	19

natural selection. The process of survival of the fittest, by which organisms that adapt to their environment survive and those that don't disappear.

necrosis. The death of cells that can discolor areas on a plant or kill the entire plant.

negative charge. One of the two kinds of electric charge, the other being positive.

neoplasm. A new growth of cells which is more or less unrestrained and not governed by the usual limitations of natural growth. It is benign if there is some degree of growth restraint and no spread to distant parts. It is malignant if the growth invades the tissues of the host, spreads to distant parts, or both.

neper. (1) A dimensionless unit for stating the ratio of two power values; the number of nepers is one-half the Napierian logarithm of the power ratio. (2) 0.868589 bels; 8.68589 decibels.

Nernst effect. When heat flows across the lines of magnetic force, there is an observed electromotive force in the mutually perpendicular direction.

neutralism. The phenomenon of two different species not being affected either adversely or beneficially by a rather close association. The two organisms (plants or animals) may live in extremely confined quarters and may often come into intimate contact with one another, but their breeding behavior, nutrition, and other vital processes are generally so different that from the standpoint of relationships the other organism does not exist. Neutralism exists when abundant supplies of food and shelter are available to the inhabitants.

neutralization. The addition of an alkali to react with an acid or the addition of an acid to react with an alkali, adjusting the pH of the solution to within the desired range so the water is suitable for discharge. This basic treatment may also be necessary to establish proper conditions for completion of chemical oxidation–reduction reaction for precipitation of heavy metals as hydroxides, for proper clarification, for better adsorption, etc. This is a basic reaction for a multiplicity of waste water treatment operations. Alkalis commonly used are lime, caustic or soda ash. Sulfuric acid is most commonly used for the neutralization of alkalis.

neutrino. (1) An electrically neutral particle of very small (probably zero) rest mass and of spin quantum number $\frac{1}{2}$. When the spin is oriented parallel to the linear momentum, the particle is the antineutrino. When the spin is oriented antiparallel to the linear momentum, the particle is the neutrino. The neutrino was postulated by Pauli in explaining the beta-decay process. Whenever a beta particle (positron) is created in a radioactive decay, so is an antineutrino (neutrino). The two particles and the parent nucleus share between them the available energy and momentum. Neutrinos and antineutrinos can penetrate amounts of matter measured in light years without appreciable attenuation. (2) A particle which conserves and carries off energy released in many interactions between particles or in beta decay. Neutrinos have no charge, almost no mass, and travel with the speed of light. They are emitted during nuclear fusion on the sun and carry some of the sun's heat to the earth. As neutrinos are created in the sun's core, they move right through the mass of the sun, as if it were empty, and out into space. To a neutrino, matter practically does not exist.

neutron. A neutral elementary particle of mass number 1. It is believed to be a constituent particle of all nuclei of mass number greater than 1. It is unstable with respect to beta decay, with a half-life of about 12 minutes. It produces no detectable primary ionization in its passage through matter, but interacts with matter predominantly by collisions and, to a lesser extent, magnetically. Some properties of the neutron are: rest mass, 1.00894 amu; charge, 0; spin quantum number, $\frac{1}{2}$; magnetic moment, -1.9125 nuclear Bohr magnetons.

neutrosphere. The region of the atmosphere, extending from the earth's surface to approximately 75 km, which is electrically neutral. The neutropause is the dividing line between the neutrosphere and the ionosphere.

newton. A unit of force in the mks system of units equal to 10^5 dynes; also 0.2248 pound force, or pound weight; that force which induces in 1 kg an acceleration of 1 m/s^2.

Newton's law of cooling. The rate of cooling of a body under given conditions is proportional to the temperature difference between the body and its surroundings.

Newton's laws of motion. (1) Every body continues in its state of rest or of uniform motion in a straight line, except insofar as it may be compelled to change that state by the action of some outside force. (2) The change of motion is proportional to the force applied and takes place in the direction of the line of action of the force. (3) To every action there is always an equal and opposite reaction.

nickel (Ni). A silvery metal of atomic weight, 58.71; specific gravity, 8.9; melting point, 1452°C. Its alloys are important materials for high-temperature applications where resistance to deformation failure is required in the temperature range 800–1000°C.

NIOSH. The National Institute for Occupational Safety and Health, a division of the National Health Institutes. Among the responsibilities of NIOSH is the joint administration with the Department of Labor of the Occupational Safety and Health Act of 1970. NIOSH collects and interprets toxicological data relating to potentially hazardous substances and prepares criteria documents with recommendations by expert panels for establishing health and safety standards. OSHA promulgates and enforces the standards recommended by NIOSH, after agreement on details by the affected parties. NIOSH also publishes an annual list of toxic substances, including such information as the reported lethal dose plus monographs that detail the results of studies of the health effects of various chemicals.

nitrate. NO_3, a nitrogen compound, frequently associated with another element such as sodium and potassium, that occurs commonly in nature, but which is often a cause of water pollution and adverse health effects in humans and domestic animals. Nitrates have been added to foods since the days of ancient Rome to prevent the growth of microorganisms responsible for botulism food poisoning and because they give meat products a fresh, red color long after the meat would have otherwise acquired an unappetizing appearance. Nitrates are changed to nitrites by natural biological processes. Large doses of nitrites can cause a form of anemia called methemoglobinemia, as nitrite molecules combine with the oxygen-carrying hemoglobin molecules of the red blood corpuscles. Methemoglobinemia is also known as "blue baby disease," a form of cyanosis that is particularly dangerous to infants given water or formulas prepared with water having a high concentration of nitrates. There also is some evidence that nitrites ingested by humans can combine with amines and amides in the gastrointestinal tract to form nitrosamines, which have been identified as carcinogens. The EPA has recommended that nitrites present in concentrations greater than 1 mg/liter are hazardous and should not be used for infant feeding. Nitrate concentrations in excess of 45 mg/liter should not be permitted in domestic water supplies, according to the EPA. Groundwater supplies with high nitrate concentrations may be an indication of seepage from livestock manure deposits in the area. Effluent from secondary treatment of municipal waste water usually has a nitrate concentration of 15 mg/liter. Ordinary sewage treatment does not remove nitrates and thousands of fish are reported killed each year from the discharge of waters containing nitrates.

nitric acid. HNO_3, a colorless, somewhat toxic gas formed when high-temperature combustion takes place causing a reaction between the nitrogen and oxygen of the air.

nitric oxide. NO, a gas formed by combustion under high temperature and high pressure in an internal combustion engine. It changes into nitrogen dioxide in the ambient air and contributes to photochemical smog.

nitrification. The conversion of nitrogenous matter into nitrates by bacteria. Although some autotrophic bacteria and many heterotrophic marine bacteria can use nitrogen occurring as ammonia to synthesize their own protoplasm, it is not generally accessible in this form. The conversion of ammonia, or mostly ammonium salts, to nitrate is termed nitrification, a process which is pH dependent, occurring slowly, if at all, in acid conditions.

nitrites' cancer link. Sodium nitrite, a chemical additive used to preserve, color, and flavor cured meats, poultry, and fish, was found to produce lymph cancer in laboratory rats, as reported in a joint statement issued by the Food and Drug Administration (FDA) and the Agriculture Department. The finding came as the result of an FDA-sponsored study by the Massachusetts Institute of Technology that for the first time identified sodium nitrite, alone, as a chemical carcinogen. Until this study, it had been thought that nitrites were a threat only when combined with amines to form nitrosamines, which were known to cause cancer. The four-year research study found that 13 percent of the rats fed sodium nitrite developed lymphoma, and 8 percent of those not receiving nitrites contracted cancer. The causative factor was said to be clearly distinct from that of nitrosamines. The difference in the cancer rates were considered "significant." The danger in the continued use of nitrites presented the FDA with the need to balance risks, since nitrites inhibited growth of the bacterium that caused botulism. After weighing the hazards of botulism against the hazards of cancer, the FDA concluded that the risk of botulism was greater and stated that it would not be responsible if the government immediately banned nitrites from all the processed foods in which they were used. A phasing out of nitrites was proposed, starting with products in which nitrites were not needed to prevent botulism and progressing to a "goal" of "total" phase-out as other safe preserving methods were perfected. (Two-thirds of all beef was treated with nitrites.) It was believed that it would be impossible to remove nitrites from some products, such as canned hams and some lunch meats. Nitrites also naturally occurred in drinking water and many foods, especially leafy and root vegetables. They were present in 7 percent of the U.S. food supply, according to the FDA. The Department of Agriculture issued regulations reducing the amount of nitrites that could be used in curing bacon. Nitrosamines were formed when bacon cured with nitrites was crisply fried. A still lower nitrites requirement, for use along with other preservatives, was proposed for a year later. It was estimated that about 90 percent of the bacon marketed already met the temporary standards, but smaller companies that did not use vacuum packaging systems encountered problems in meeting the requirements. The regulation called for a reduction in the amount of sodium nitrite to 120 parts per million and required that it be used only in combination with 550 parts per million of sodium erythrobate or sodium ascorbate agents used to block nitrosamine formation. The regulation to take effect called for 40 parts per million of nitrite in combination with 0.26 parts per million of another preservative, potassium sorbate. Bacon would not be permitted to contain more than five parts per billion of nitrosamines.

4-nitrobiphenyl. $C_6H_5C_6H_4NO_2$, a regulated carcinogen used mainly in research laboratories, although it was previously used commercially. It occurs as white, needle-like crystals. The chemical is a health hazard because it has been identified as an inducer of bladder tumors in humans. It has also been found to produce tumors in dogs. The substance is regarded as particularly dangerous when industrial or laboratory workers are exposed to it at the same time as to 4-aminodiphenyl. Oral LD_{50} levels for laboratory animals have been established at around 2200 mg/kg for rats and 2000 mg/kg for rabbits. NIOSH has set exposure limits at not more than 0.1 percent by weight or volume.

nitrogen (N). An inert colorless gas of atomic weight, 14.008; density, 1.2506 g/liter; boiling point, −195.8°C. Nitrogen forms four-fifths of the earth's atmosphere. Nitrogen atoms are basic components of the protein molecules which are essential to life. Most living organisms cannot absorb nitrogen directly from the atmosphere. Certain bacteria in the soil change nitrogen from the air into compounds called nitrates, which are then absorbed by plants. When animals eat the plants, nitrogen enters their bodies.

nitrogen cycle. Approximately 78 percent of the air breathed is nitrogen. In combination with carbon, hydrogen, and oxygen, nitrogen is an important part of the proteins vital to life. Because of its role in protein synthesis, nitrogen is also one of the most important elements to plants. Absorbed as nitrate or ammonium ions, nitrogen contributes to rapid and luxuriant growth. Certain bacteria are able to "fix" nitrogen into ammonia (NH_3), using as their energy source the foods produced by the photosynthesis of a host plant. These organisms often live in swellings or nodules on the roots of plants and are capable of fixing 200 pounds of nitrogen per acre per year into ammonia. Many plants are unable to use ammonia directly; however, another group of soil bacteria is able to oxidize ammonia into nitrite (NO_2) and still another oxidizes nitrites to nitrate (NO_3). From these sources, nitrogen is made available for the synthesis of protein. The proteins are broken down by bacteria into their component amino acids, and the amino acids are in turn broken down into ammonium (NH_4) ions. Bacteria convert the ammonium into nitrite, then either degrade this to nitrogen gas or oxidize it back to nitrate, where it is absorbed by plant roots and the cycle begins again.

nitrogen dioxide. NO_2, a poisonous, yellow-brown gas used in making nitric acid. Nitric oxide is a relatively harmless gas generally emitted into the atmosphere, but varying amounts of it are converted into nitrogen dioxide, which is a considerably more poisonous gas. The oxidation of nitric oxide to nitrogen dioxide is very rapid at high concentrations in air, but is slow at low concentrations, except in the presence of hydrocarbons and sunlight. Since nitrogen dioxide is formed so readily by photochemical action, it is usually thought of as a product of the photochemical process. But actually, it may be formed whenever nitric oxide is a byproduct of a sufficiently high-temperature reaction, with or without photochemical action. It is also a product or byproduct of a number of industries, including fertilizer and explosives manufacturing. Nitrogen dioxide can significantly affect visibility, and has a pungent, sweetish odor, detectable at one to three parts per million. At sufficiently high concentrations, nitrogen dioxide can be fatal. Nitrogen dioxide reacts with raindrops or water vapor in the air to produce nitric acid (HNO_3), which even in small concentrations can corrode metal surfaces in the immediate vicinity of the source. Nitrogen dioxide contributes to the photochemical smog reaction and the formation of aerosols wherever hydrocarbons are oxidized by sunlight.

nitrogen fixation. The process of converting atmospheric nitrogen into a compound that plants and other living organisms can use. With an atmospheric concentration of 79 percent, it would appear that the nitrogen reservoir is the atmosphere, but since most organisms are unable to use atmospheric nitrogen, the crucial reservoir is the store of nitrogen occurring in both inorganic (ammonia, nitrite, and nitrate) and organic (urea, protein, nucleic acids) forms. Unlike carbon, which is readily available in reservoir quantities both in the air and water, atmospheric nitrogen must be fixed into an inorganic form, largely nitrates, before it can be tapped for biological processes.

nitrogen gas. A troublesome gas contributing to the difficulty in solving air pollution problems. The oxides of nitrogen, primarily NO and some NO_2, can basically be controlled at the source. If the source is a fuel-fired system, a combination of low excess air combustion and reduced flame temperature will reduce the nitrogen oxide levels. Catalytic oxidation to promote the reaction between a hydrocarbon (natural gas) as the fuel and nitrogen oxide as the oxidant has been successful when high nitrogen oxide levels are present.

nitrogen oxide. Oxides of nitrogen are one of the most important groups of atmospheric contaminants in many communities. They are produced during the high-temperature combustion of coal, oil, gas, or gasoline in power plants and internal combustion engines. The combustion fixes atmospheric nitrogen to produce the oxides. At these temperatures nitric oxide forms first and in the atmosphere it reacts with oxygen and is converted to nitrogen dioxide. While this oxidation is very rapid at high concentrations, the rate is much slower in low concentrations. In sunlight, especially in the presence of organic material, the conversion of nitric oxide to nitrogen dioxide is greatly accelerated. Most determinations of oxides of nitrogen combine nitric oxides and nitrogen dioxide, with a typical range of concentrations being 0.02–0.9 ppm. The hazards associated with nitrogen oxides are a direct noxious effect on the health and well-being of people and photo-chemical oxidation of organic material, which is an indirect effect. Of the oxides of nitrogen, nitrogen dioxide is considerably more toxic than nitric oxide, acting as an acutely irritating substance. In equal concentrations, it is more injurious than carbon monoxide. Nitrogen dioxide has received considerable attention as an air pollutant because it is a hazard in numerous industries. The threshold limit for an eight-hour working day has been tentatively set at 5 ppm. Since nitrogen dioxide can react with the water vapor in the air or with raindrops to produce nitric acid, small concentrations in the atmosphere can cause considerable corrosion to metal surfaces in the immediate vicinity of the source. Nitrogen dioxide is unique among the common pollutants in that it absorbs light in the visible region of the spectrum, mostly in the blue region. It is a yellow-brown gas. Because it is visible, substantial concentrations of it reduce visibility, even without the presence of aerosol particles.

nitrogenous wastes. Animal or plant residues that contain large amounts of nitrogen.

2-nitropropane. $C_3H_7NO_2$, a widely used solvent in industrial coatings and printing inks and in the manufacture of furniture, food packaging, and plastic products. An estimated 30 million pounds of 2-nitropropane are produced each year. Occupational exposure to the substance may occur in a wide variety of industries, including construction and maintenance, printing (rotogravure and flexographic inks), highway maintenance, and shipbuilding and maintenance. It also is used in electrostatic spraying. NIOSH estimates that 100,000 workers are exposed to 2-nitropropane in the various industries. Although its carcinogenic potential in humans has not been clarified, 2-nitropropane has been found to cause liver cancer in rats. No histologic changes were found in tests involving rabbits, guinea pigs, or monkeys. Cats died within 17 days after exposure to concentrations of 328 ppm and had severe liver damage and slight to moderate damage of the heart and kidneys. NIOSH has cited clinical reports of workers exposed to concentrations ranging from 20 to 445 ppm who experienced nausea, vomiting, diarrhea, dizziness, anorexia, and headaches. One worker reportedly died of exposure from painting the inside of a tank, while a fellow painter experienced liver damage. A report in the *New England Journal of Medicine* cited an excess of hepatitis among workers applying epoxy resins to the walls of a nuclear power plant. OSHA has recommended a standard for occupational exposure to 2-nitropropane of 25 ppm and advised that protective, full-body clothing be required for employees entering areas used for manufacture, filling operations, use, release, handling, or storage of 2-nitropropane.

NO$_x$. A notation meaning oxides of nitrogen. See **nitric oxide.**

noise. Any undesired sound. By extension, noise is any unwanted disturbance within a useful frequency band, such as electric waves in a transmission channel or device.

noise, acoustic. Noise from sound waves.

noise effects on hearing. Noise may affect hearing in ways that are broadly divisible into three categories: temporary threshold shift, permanent threshold shift, and acoustic trauma. In that order they indicate in a general way the degree of severity of the noise exposure which caused them. Temporary threshold shift is a short-term effect which may follow an exposure to noise, the elevation of the hearing level is reversible. The effects of a particular noise exposure in terms of temporary threshold shift are dependent on individual susceptibilities. The term "persistent threshold shift" is used to denote the threshold shift remaining after at least 40 hours. The word "permanent" is reserved for conditions which may reasonably be supposed to have no possibility of further recovery, since some recovery of hearing may be found after a weekend away from noise and probably thereafter. These effects, each detectable in man by an elevation of the hearing level, are attributable to movements of the various vibrating parts of the ear. These moving parts are seldom, if ever, at rest. Mechanical activity is their normal condition. There is another effect, the singing or ringing in the ears, known as tinnitus, which is an aftermath of noise exposure. This is due to the discharge of nerve impulses in the fibers of the auditory nerve, a sort of irritating aftereffect of the intense stimulation.

noise impulse. Impulse noise, such as that produced by a drop hammer, shows a rapid change in the level of sound. It cannot be measured with the sound-level meter because of the inertia of the meter. The peak sound-pressure level during the impulse may be as much as 20 to 30 decibels greater than the highest reading obtained on a sound-level meter. An impact-noise analyzer may be used for field measurements of peak sound pressures and approximate impulse delay times.

noise, intermittent. Intermittent noise may result from variations in the operation of equipment. It may also result from a steady noise source passing by the observer, such as traffic or aircraft noise. While the sound-level meter and the octave-band analyzer usually measure intermittent noises suitably, they may require a large number of readings to provide a statistical distribution for all the levels expected.

noise level. For airborne sound, unless specified to the contrary, the weighted sound-pressure level called sound level (the weighting must be indicated).

noise, radio. Unwanted electrical signals, including interference to radio transmission from atmospherics, cosmic noise, solar noise, and man-made noises from electrical and radio equipment.

noise rating curves. An agreed set of empirical curves relating octave-band sound-pressure levels to the center frequency of the octave bands, each of which is characterized by a "noise rating" (NR), which is numerically equal to the sound-pressure level at the intersection with the ordinate at 1000 Hz. The noise rating of a given noise is found by plotting the octave-band spectrum on the same diagram and selecting the highest noise curve to which the spectrum is tangent.

noise ratio. The ratio of the noise power at the output of a transducer to the noise power at the input.

noise reduction. A decrease of the sound-pressure level at a specified observation point, which is attributable to a designated structure. Noise reduction is also used to designate the differences in sound-pressure levels existing at two different locations at the same time when the designated structures are in position. The term "noise control" is meaningful only when noise-control components and the points of observation are fully specified.

noise, repeated-impulse.   Repeated-impulse noise, produced by such familiar industrial equipment as riveting guns, pneumatic hammers, multislide machines, headers, and automatic punch presses, constitute a situation intermediate to single impulses and continuous noise. Less than 50 impulses per minute may cause incorrect readings of a sound level meter and octave-band analyzer. More than about 200 impulses per minute usually yield meaningful readings in these instruments. In the range of 50 to 200 impulses per minute, the interpretations of measurements must be tempered with judgment gathered through experience.

noise, steady.   Steady noise has no rapid, sudden changes in level. Noises of this category include those typical of a jet engine or a weave room in a textile mill. Air moving through a duct or orifice, and other such industrial noises, will also fall into this class. In general, little or no variation for such sounds will be shown when switching from fast to slow movements in the sound-level meter.

noise wide-band and narrow-band.   Wide-band noise, such as that produced by a textile mill weave room, shows a fairly wide frequency spectrum and does not contain any outstanding pure-tone components. In most instances, octave-band measurements suffice for measuring wide-band noise. Narrow-band noise concentrates most of its energy in a narrow frequency range or ranges, and produces the sensation of pitch. Typically produced by circular saws and turbine-operated air tools, noises in this category can be investigated with an octave-band analyzer, but precise information about the frequency distribution of the noises requires analysis with a narrow-band analyzer.

noncondensable gas.   A gas whose temperature is above its critical temperature so that it cannot be liquefied by an increase of pressure alone.

nonpoint source.   A contributing factor to water pollution that can't be traced to a specific spot, like agricultural fertilizer runoff and sediment from construction.

noxious gas.   A gas which has been proved to cause ill effects in human beings. In the true sense of the word, CO_2, which is not poisonous, can be noxious in such large concentrations as to dangerously reduce the amount of available oxygen. Usually, however, noxious gases are considered to be only those which cause ill effects at low concentrations. Foul-smelling gases, if not harmful, are usually not categorized as noxious.

NTA.   Nitrilotriacetic acid, a compound proposed for use to replace phosphates in detergents.

nuclear atom.   The atom of each element consists of a small, dense nucleus which includes most of the mass of the atom. The nucleus is made up of roughly equal numbers of neutrons and protons. The positive charge of the protons enables the nucleus to surround itself with a set of negatively charged electrons which move around the nucleus in complicated orbits with well-defined energies. The outermost electrons, which are those least tightly bound to the nucleus, play the dominant part in determining the physical and chemical properties of the atom. There are as many electrons in orbit as there are protons in the nucleus.

nuclear energy.   In a nuclear power plant, fission is initiated within the reactor and the power level is regulated by the manipulation of control rods. The fission process releases heat energy; the coolant absorbs the heat and carries it off to the heat exchanger, where water is heated (without direct contact with the radioactive coolant) and becomes steam. The steam turns the turbine, which rotates the generator to produce the electrical energy that is the final product of the power plant. The radioactive coolant is recirculated in a closed loop for further use. The steam is condensed to water and the water, too, is

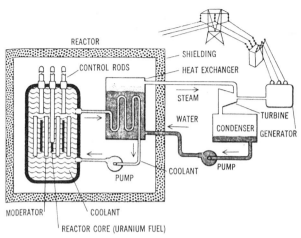

The process of energy production in a nuclear power plant.

recirculated. The moderators slow down the neutrons that bombard the fuel cores to produce more efficient absorption by the fuel.

nuclear fusion. The process whereby the nuclei of two light atoms are combined to form an atom of a new element with a heavier nucleus. This occurs when two hydrogen nuclei, each with one proton, fuse to form helium containing two protons in its nucleus. After fusion the mass of the new element is slightly less than the total mass of the particles that fused; this "loss" of mass is converted into the vast energy which binds the particles together, as predicted in Einstein's formula $E = mc^2$, where E is the energy, m is the mass of the particle, and c is the speed of light. Successful fusion requires very high temperatures (to about 2 million degrees centigrade) and tremendous pressures before the nuclei will pack together and combine. Under proper conditions, once nuclear fusion begins, the heat from the thermonuclear reaction is enough to ensure a self-sustaining reaction.

nuclear isomer. An isotope of an element having the same mass and atomic numbers, but differing in radioactive properties such as the half-life period.

nuclear neurosis. A pattern of mental health effects associated with individuals working in or living near nuclear power plants as a result of the Three Mile Island accident of 1979. A U.S. Presidential Task Force was formed on June 18, 1979, to study behavioral changes due to the accident. The study found that within the first weeks after the Three Mile Island accident, the general population distress level in the area was very nearly equivalent to that of patients in mental hospitals. However, the distress level was very transient and dropped considerably within a month. Distrust of authority and loss of faith in government and industry, which was also very high during the first weeks after the accident, was not transient and remained high long after the Three Mile Island accident. Children were found to be affected and girls experienced greater behavioral effects than boys. Symptoms included headaches, reported by 26 percent of teenagers living near the Three Mile Island nuclear plant; loss of appetite, reported by 18 percent; and increased appetite, reported by 13 percent. Among workers at nuclear power plants, the Task Force found that concern about exposure to radiation increased along with such job-related attitudes as increased tension, loss of occupational self-esteem, and worry about the future of the occupation. Demoralization was found to be higher among the rank and file nuclear workers than among the supervisors at nuclear power plants.

nuclear radiation. Corpuscular emissions, such as alpha and beta particles, or electromagnetic radiation, such as gamma rays, originating in the nucleus of an atom.

nuclear reaction. The reshuffling of the nucleons of two colliding nuclei to form different nuclei. In order to bring the two reacting nuclei close together, one of them must be accelerated to a high velocity so that it will not be turned back too soon by the strong electrostatic repulsion between the two positively charged nuclei.

nuclear reactor. A device in which controlled fission of radioactive material produces new radioactive substances and energy; a device in which fission can be maintained and controlled; an "atomic furnace" in which uranium or other "fuels" are burned to produced heat, neutrons, and radioactive isotopes. A nuclear reactor slows down and controls fission reactions so that the huge quantities of energy produced can be usefully employed. A nuclear reactor consists of the following. (a) *Fuel.* Either natural uranium or uranium enriched in ^{235}U. (b) *A moderator.* Graphite, ordinary water (H_2O), and deuterium oxide (D_2O) are the most common materials used to slow the fission neutrons down to the extremely low energies at which they are most effective in causing further fissioning. (c)

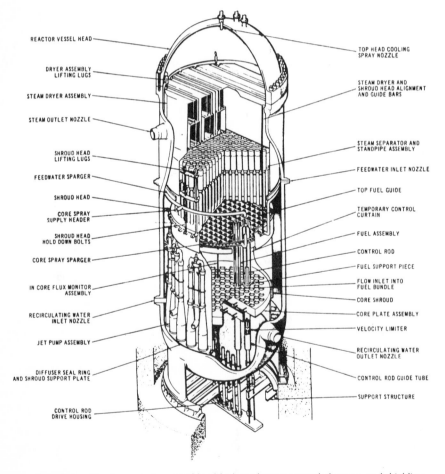

REACTOR VESSEL HEAD

DRYER ASSEMBLY LIFTING LUGS

STEAM DRYER ASSEMBLY

STEAM OUTLET NOZZLE

SHROUD HEAD LIFTING LUGS

FEEDWATER SPARGER

SHROUD HEAD

CORE SPRAY SUPPLY HEADER

SHROUD HEAD HOLD DOWN BOLTS

CORE SPRAY SPARGER

IN CORE FLUX MONITOR ASSEMBLY

RECIRCULATING WATER INLET NOZZLE

JET PUMP ASSEMBLY

DIFFUSER SEAL RING AND SHROUD SUPPORT PLATE

CONTROL ROD DRIVE HOUSING

TOP HEAD COOLING SPRAY NOZZLE

STEAM DRYER AND SHROUD HEAD ALIGNMENT AND GUIDE BARS

STEAM SEPARATOR AND STANDPIPE ASSEMBLY

FEEDWATER INLET NOZZLE

TOP FUEL GUIDE

TEMPORARY CONTROL CURTAIN

FUEL ASSEMBLY

CONTROL ROD

FUEL SUPPORT PIECE

FLOW INLET INTO FUEL BUNDLE

CORE SHROUD

CORE PLATE ASSEMBLY

VELOCITY LIMITER

RECIRCULATING WATER OUTLET NOZZLE

CONTROL ROD GUIDE TUBE

SUPPORT STRUCTURE

A nuclear reactor is a complex assembly of fuel, moderator, control elements, and shielding.

Control rods. Cadmium and boron are used to control the concentration of neutrons present in the reactor because they have a great capacity for absorbing neutrons. (d) *Coolant.* The coolant conducts heat away in order to keep the temperature at reasonable levels. (e) *Shielding.* The intense radioactivity of fission products requires the presence of thick layers of absorbing shielding. Usually water, concrete, or both are used as shields.

nuclear reactor, control rods.   Control rods are interspersed between the fuel rods in order to regulate the rate of fission, because the production of too much heat energy would burn up the reactor. Cadmium and boron are commonly used as control rods, since they capture some of the free neutrons and slow down the fission reactions. These control rods are inserted or withdrawn from the reactor by electric motors. When the control rods are completely lowered into the reactor, the number of neutrons they absorb is greatest and the least amount of fission reaction takes place. As the control rods are withdrawn, the number of fission reactions is correspondingly increased.

nuclear reactor, fuel.   The most commonly used fuel in a nuclear reactor is ^{235}U, made into solid slugs several inches long and approximately one inch in diameter, and sealed into closely fitting aluminum cans. The slugs are called fuel rods. Plutonium is also used as an atomic fuel.

nuclear reactor, moderator.   A material that surrounds the fuel rods and reduces the speed and energy of the free neutrons. Slow-moving neutrons have a better chance of hitting the nucleus of a fuel atom than fast ones and are therefore more efficient in starting and sustaining a chain reaction. Heavy hydrogen and graphite are good moderators. When plutonium is used as fuel, fast neutrons are necessary and no moderator is used.

nuclear reactor, reflector.   A reflector of ordinary water, heavy water, graphite, or beryllium surrounds the core of the atomic furnace. It reflects escaping fuel neutrons back into the core of the reactor.

nuclear-waste pollution.   When a spent nuclear-fuel rod is replaced in a nuclear reactor, it is still extremely radioactive and must be handled and stored carefully; one of the isotopes in a spent rod is ^{239}Pu, with a half-life of 24,360 years. At the moment there are no procedures to deal with high-level radioactive wastes from the civilian nuclear power program. (Low-level wastes, such as pieces of protective clothing or tools that become contaminated, from various nuclear facilities, including hospitals, are now deposited in three sites in Washington state, South Carolina, and Nevada; high-level wastes from the military nuclear weapons program are stored in special sites on remote federal reservations.) When the nuclear power industry first began, it was assumed that fuel reprocessing—spent fuel recycled into new fuel rods and wastes to be disposed of permanently—would be pursued; but, since it proved to be uneconomical, it was never developed. Also, the problems of transporting highly radioactive material long distances to a reprocessing plant were considerable, and the transportation's feasibility and safety were subject to much debate. Until a permanent solution can be found, these high-level wastes are being stored on site, usually in storage pools with a supply of cooling water. As more fuel rods are replaced, these pools are being repacked to higher density, with the Nuclear Regulatory Commission's approval. The main fear is that the cooling water supply will be disrupted, and the residual heat in the rods will rupture their cladding—the outer sheath that gives the rods structural stability—thereby releasing extremely dangerous radioactive materials to the environment. (See also **radioactive contamination.**)

nucleon.   In the classification of subatomic particles according to mass, the second heaviest type of particle; its mass is intermediate to those of the meson and hyperon. Examples of nucleons are the proton and neutron. Protons and neutrons are alike in several ways. They are both building blocks of the nucleus and have approximately the same mass. They often behave in similar ways, even though they differ in electrical charge. It is believed that protons and neutrons are particles of the same kind, with differing charges, and for this reason protons and neutrons are both called nucleons. The nucleon with no charge is the neutron, while the nucleon with a positive charge is the proton.

nucleus. (1) The dense central core of the atom, in which most of the mass and all of the positive charge is concentrated. The charge is the essential factor which distinguishes one element from another. The atomic number Z is the number of protons in the nucleus, which includes a roughly equal number of neutrons. The mass number A gives the total number of neutrons and protons. A proton has a positive electric charge which is equal to the magnitude of the negative charge on an electron. Its rest mass is 1836.1 times the rest mass of the electron. The neutron has no charge and its rest mass is 1838.6 times the rest mass of the electron. Both the proton and neutron have intrinsic angular momentums of $\frac{1}{2}h/2\pi$ and they also have small magnetic moments. It is common practice to refer to the protons and neutrons in the nucleus collectively as nucleons. (2) Within the cells of most organisms, the largest and one of the most conspicuous structural areas is the nucleus, the control center of the cell. The nucleus plays the central role in cellular reproduction, the process whereby a single cell undergoes cell division and forms two new cells. It also plays a central part in conjunction with the environment in determining what sort of differentiation the cell will undergo and what form it will exhibit at maturity, and it directs the metabolic activities of the living cell. It is from the nucleus that the "instructions" emanate that guide the life processes of the cell as long as it lives.

nuclide. Any species of atom whose nuclear structure is distinct from that of any other species. The number of neutrons in the atomic nucleus of a given element may vary; all atoms of an element will have the same chemical properties, but their masses may be different. Such atoms are called isotopes. Isotopes of an element contain the same number of protons, but different numbers of neutrons, thus each isotope of an element is a separate nuclide. There are three nuclides of hydrogen: protium (ordinary hydrogen), with one proton and no neutrons; deuterium, with one proton and one neutron; and tritium, with one proton and two neutrons.

nuisance threshold. A standard for the concentration of an air pollutant that is considered objectionable. In the case of an odor, that concentration which can be detected by a human being.

nutrient. An element or a dissolved salt essential to life may be termed a biogenic salt or a nutrient. Macronutrients include elements and their compounds that have key roles in the protoplasm and that are needed in relatively large quantities, as, for example, carbon, hydrogen, oxygen, nitrogen, potassium, calcium, magnesium, sulfur, and phosphorus. Micronutrients include those elements and their compounds also necessary for the operation of living systems, but required only in very minute quantities. At least 10 micronutrients are known to be required for primary production: iron, manganese, copper, zinc, boron, sodium, molybdenum, chlorine, vanadium, and cobalt. Nonessential elements may be of great ecological importance if they occur in quantities that are chemically toxic, if they react to bind or make unavailable essential elements, or especially if they are radioactive.

nutrient trap. At the boundary formed by a mass of freshwater coming into contact with an underlying mass of saltwater, there is a great shearing force that creates a considerable amount of horizontal turbulence resulting in eddy formation. The eddies that form contain nutrients that have been carried into the estuary and are circulated around in the boundary region; this system is referred to as a nutrient trap.

O

occlude. To take in and retain in pores or other openings; to absorb (used particularly with respect to the absorption of gases by certain substances which do not lose their characteristic properties).

occluded front. A front in which cold air has pushed into a warm air mass, lifting the warm air out of contact with the ground.

ocean thermal conversion. The use of temperature differentials between warm and cold parts of the ocean to drive a low-pressure turbine connected to a generator to produce electricity.

octave. (1) A pitch interval of 2:1. (2) The tone whose frequency is twice that of the given tone. (3) The interval of an octave together with the tones included in that interval.

odor. The perception of smell, referring to the experience, or that which is smelled, referring to the stimulus. The experience of smell can be taken to mean any perception that results from nasal inspiration, including common chemical irritations such as those produced by acid fumes, or the true odor sensations perceived via the receptors of the olfactory epithelium, which has an area of about 2.5 cm^2 within the upper nasal cavity. In industrial hygiene and community air pollution applications, irritants and stenches are usually grouped together as objectionable atmospheric contaminants that should be removed even though they may not be specifically toxic. Odor sources may be confined to a specific emission, such as a vent, stack, or exhaust duct. Generally, an odor may be said to be confined when its rate of discharge to the atmosphere can be measured and when the atmospheric discharge is amenable to representative sampling and to physical or chemical processing for purposes of odor abatement. Generally, odor in water is due to the presence of dissolved gases, such as hydrogen sulfide, and volatile organic compounds. When the threshold odor value exceeds three units, based on n-butyl alcohol calibration tests, it is generally considered objectionable. The criterion for water quantity with respect to odor is simply that it not be objectionable. Taste and odor in water supplies may result from natural phenomena such as decaying vegetation, algae, or bacterial slime. As in the case of color removal, taste and odor may also be removed through processes for the treatment of, or the control of, other waste constituents. Processes employed for taste and odor removal are coagulation, carbon adsorption aeration, and oxidation with chlorine or other oxidizing agents.

odor control. There are two distinctly different approaches to odor control: The odorant may be reduced in concentration so that its smell is less intense, with the result that any objectionable effects are diminished, or the odorant may be changed in quality so that it becomes more pleasant or acceptable to people. The first approach includes reducing the source of the odor, diluting the odor by ventilation or dispersal, or removing odorant from the atmosphere by adsorption or scrubbing. The second approach attempts to make an odor more pleasant by counteraction or masking. Ventilation is the most common method of removing odorous air from enclosed spaces, and dispersal, especially with the aid of stacks, is the most common method for odor abatement outdoors. The control of atmospheric odors by adsorption is for all practical purposes limited to the use of activated carbon as the sorbent. Oxidation systems can be fully effective methods of odor control if the oxidation is completed. Equipment used for odor control by air oxidation comprises direct-flame incinerators and catalytic combustion systems. The chemical conversion of odorants is practically always accomplished by oxidizing agents. Ozone converts organic matter by oxidative degradation usually to aldehydes, ketones, and acids. Ozone treatment has been used to deodorize exhaust gases in stacks where no human exposure is involved. Permanganate oxidants are used for odor control by direct treatment of odor sources by scrubbing with aqueous permanganate solutions or by the use of granular beds of absorbent impregnated with permanganates. Air scrubbers or washers may also effectively remove odorants.

odor threshold. The concentration of an odor in air that can be detected by a human being; i.e., in the case of sulfides emitted from a Kraft pulping process, an observer can smell these compounds at and above the odor threshold, a concentration of about one part in 10^9 parts of air.

oersted. The cgs electromagnetic unit of magnetic intensity existing at a point where a force of one dyne acts upon a unit magnetic pole at that point, i.e., the intensity 1 cm away from a unit magnetic pole.

off-road vehicle. A form of motorized transportation that does not require prepared surfaces—it can be used to reach remote areas.

oil combustion emission factors (in lb/1000 gal of oil burned).

Pollutant	Large sources (1000 hp or more)	Small sources (1000 hp or less)
Aldehydes	0.6	2
Benzo[a]pyrene	5000 µg/1000 gal	40,000 µg/1000 gal
Carbon monoxide	0.04	2
Hydrocarbons	3.2	2
Oxides of nitrogen	104	72
Sulfur dioxide	157 S[a]	157 S[a]
Sulfur trioxide	2.4 S[a]	2 S[a]
Particulate	8	12

[a]S equals the percentage of sulfur in oil.

oil desulfurization. Heavy fuel oil (residual fuel oil) contains quantities of sulfur usually varying between 0.5 and 5 percent, depending on the source and subsequent refinery treatment. Effective processes of oil desulfurization which have been developed in recent years essentially involve the treatment of residual oil with hydrogen at high pressures in the presence of a suitable catalyst and these are called "hydrodesulfurizations" or HDS processes. The sulfur in fuel oil is present as either straight-chain or cyclic sulfur compounds, the general trend being an increase in the proportion of aromatic sulfur compounds in the high-boiling fraction, and a predominance of thiols, sulfides, and disulfides in the lower-boiling distillates. Sulfur can be removed from fuel oil by several different processes of varying effectiveness, removing different proportions of sulfur. A minor degree of desulfurization may be achieved by a simple distillation process, while delayed coking, a solvent deasphalting, and residual HDS processes will give increasing degrees of sulfur removal.

oil "fingerprinting". A method that identifies oil spills so they can be traced back to their sources.

oil gasification. The oil gasification process essentially involves the thermal or catalytic cracking of oil molecules to produce gas molecules and carbon. Hydrogen can be added during the process, either directly or in the form of steam. Carbon is produced in the process; in some processes it is collected as carbon black, whereas in others it is deposited in the catalyst, which is subsequently removed either by using a cyclic process with a fixed catalyst, burning off carbon with air during a "blow," or by recycling continuously in a fluid-bed-type plant, removing part of the catalyst and burning off the carbon in a separate vessel. Simple thermal cracking processes such as the "Jones" process produce town gas (550–750 Btu/ft^3) and in some modifications, carbon black, in a cyclic system, and use chambers with a nonactive cracking. The "Hall" process produces gas with much higher calorific values (above 1000 Btu/ft^3). Catalytic steam cracking processes are most likely applicable to the gasification of high-sulfur residual fuels, and can be either continuous or cyclic. The continuous processes are able to use light oil or gaseous feeds, low in sulfur for cracking and reforming operations, while the cyclic processes can use a wide range of fuels, including residuals, with no limit to their sulfur content.

oil spill. An accidental discharge of oil into bodies of water, controlled by chemical dispersion, combustion, mechanical containment, and absorption.

olefin. A member of a group of unsaturated hydrocarbons. Most olefins appear to have no direct effect on animal life, although some cause a general reduction in plant growth and a drying out of orchid sepals. In addition, olefins take part in the photochemical reaction along with nitrogen oxides and several other classes of compounds.

oligotrophic lake. A deep, clear lake with a low nutrient supply. It contains little organic matter and has a high dissolved oxygen level.

oncogenic. A substance that causes tumors, whether benign or malignant.

opacity. The amount of light obscured by an object or substance. A window has zero opacity; a wall, 100 percent opacity.

opacity rating. A measurement of the opacity of emissions, defined as the apparent obscuration of an observer's vision to a degree equal to the apparent obscuration of smoke or a given rating on the Ringelmann chart.

open burning. The occasional burning of outdoor stores of lumber, sawdust, scrapped cars, tires, textiles, and open dumps.

open burning emission factors (in lb/ton of refuse burned).

Pollutants	Burning dump[a]	Backyard dump
Aldehydes	4	3.6
Benzo[a]pyrene	250,000 g/ton	350,000 g/ton
Carbon monoxide	not available	not available
Hydrocarbons	280	280
Oxides of nitrogen	0.6	0.5
Oxides of sulfur	1.2	0.8
Ammonia	2.3	1.6
Organic acids	1.5	1.5
Particulates	47	150

[a]Three pounds per capita per day of refuse burned is assumed.

open dump. See **dump.**

open-hearth furnace. A reverberatory furnace containing a basin-shaped hearth, for melting and refining suitable types of pig iron, iron ore, and scrap for steel production. In all open-hearth furnaces, a large amount of dust from ore and other materials and splashings from slag are carried away by the waste gases. This is especially serious in the basic process; to prevent choking of the regenerators, and the fluxing and glazing of the checkerbricks, a supplementary chamber is used for collecting slag and dust (slag pocket or dust catcher). In open-hearth furnaces the substance is in contact with the gaseous products of combustion. They are used for melting steel, purifying it, and alloying other substances with it. The fuel burnt is usually oil, but it sometimes uses producer gas from coal, a long luminous flame being required above the bath of molten metal. Atmospheric pollution is similar to that from coke ovens, with the addition of smoke if insufficient air is used in proportion to the producer gas.

open space. A relatively undeveloped green or wooded area provided usually within an urban development to minimize feelings of congested living.

ORD. An abbreviation for the Office of Research and Development for the U.S. Environmental Protection Agency. The main function of ORD is to provide the EPA with scientific data in support of proposal, review, and enforcement of EPA standards and regulations. ORD research efforts are designed to define pollution hazards and their sources and to develop, test, and evaluate methods of controlling pollution threats. The studies may involve the use of *in vitro*, *in vivo*, and whole animal experiments. ORD employs nearly 2000 scientific and support personnel at 20 laboratories and other facilities from Alaska to Florida with a budget of approximately $300 million per year. About half the research personnel are involved in air- and water-quality studies, although the greatest proportion of funding is directed toward research in energy extraction and processing and the related health efforts of energy production.

organic. Of, relating to, or derived from living organisms; in chemistry, a carbon-containing compound.

organic and inorganic compounds. A large number of organic and inorganic substances have been identified in the air. If classified according to physiological effects, a division of four groups might be made: general irritants, systemic toxicants, carcinogens, and allergens. The group of general irritants might include sulfur dioxide, sulfuric acid, acid sulfates, nitrogen oxides, ozone, hydrogen sulfides, the halogens, fluorides and chlorides, aldehydes, and some of the products of photochemical and oxidative reactions of hydrocarbons. Sulfur oxides are widely prevalent and of unquestioned physiological significance at higher concentrations. Concentrations as high as one part per million are occasionally reported, more usual are concentrations on the order of a few parts per hundred million. Of the nitrogen oxides, nitrogen dioxide appears to be the only one of probable health importance as an air pollutant. It is widely found in concentrations on the order of a few parts per hundred million. Ozone is widely prevalent, appearing to be formed to a considerable extent by photochemical reactions in the air. Concentrations up to 0.5 parts per million have been occasionally reported, but the concentrations are usually much lower. Besides showing the usual properties of an irritant gas, ozone has been stated to produce pulmonary fibrosis with continuous exposure. There are a large number of organic substances, especially hydrocarbons in the air, such as aldehydes, acids, hydroperoxide, peroxides, nitroolefins, and perhaps free radicals. These compounds may well include a number which are of health significance. Systemic toxicants are materials which produce some form of systemic disease on absorption. Many metals capable of producing poisoning have been found in the air. Their concentrations, however, are for the most part far below levels of concern. Carbon monoxide concentrations above a few parts per million are rare, and such levels are quite harmless. However, concentrations approaching the threshold limit of 100 parts per million have been recorded near heavy traffic arteries.

organic contaminant. Fuel combustion is a major source of organic contamination, the amount emitted being a function of the completeness of the combustion process. Among the process categories of major importance for organic emission are those involving the transportation and production of heat and power. Waste disposal by incineration and the emission of organic gases and vapors by chemical industries are additional significant sources of organic atmosphere contaminants. Interest in the organic fraction of air contaminants has been stimulated by the fact that some have been shown to be carcinogenic.

organic sediment. A sediment formed from constituents that were once dissolved in water and later extracted through the activity of plants and animals. Many organisms are inorganic substances in the development of their protective and supporting structures, such as bones, shells, and tests. These structures contain in varying amounts such constituents as phosphates, sulfides, iron oxides, calcium and magnesium carbonates, and silica, which accumulate as sediments when the organisms perish. Other organisms bring about chemical reactions that lead to the precipitation of sediments. Peat and coal are composed of the altered remains of plants.

organism. Any living thing.

organochlorine (biocides). Chlorinated hydrocarbons, organochlorines, include the best known of all the synthetic poisons: endrin, heptachlor, aldrin, toxaphene, dieldrin, lindane, DDT, chlordane, and methoxychlor. There is no clear understanding of just how organochlorines work. Apparently, the central nervous system is affected, for typical symptoms of acute poisoning are tremors and convulsions. Chronic levels have various effects. In aquatic organisms, which are especially sensitive to this class of compounds, the uptake of oxygen through the gills is disrupted and death is associated with suffocation, rather than with a nervous disorder.

organophosphate (biocides). Parathion and malathion are organophosphates. Nerve impulses are conducted across the gap between adjoining nerve fibers by a compound called acetylcholine. As soon as the impulse has bridged the gap, an enzyme, cholinesterase, destroys the acetylcholine present, preventing further impulse transmission. Organophosphates deactivate the cholinesterase, allowing a stream of impulses to flow uninterrruptedly along the nervous system, resulting in spastic uncoordination, convulsions, paralysis, and death in short order.

oscillation. The variation, usually with time, of the magnitude of a quanitity with respect to a specified reference when the magnitude is alternately greater and smaller than the reference.

oscillator. Any system, such as a pendulum or a piano string, capable of carrying out regular oscillations. If the frequency (the number of oscillations per second) is independent of the amplitude (the amount by which the system happens to be oscillating), we speak of a harmonic oscillator.

OSHA. The Occupational Safety and Health Administration. The agency was created by the Occupational Safety and Health Act of 1970 to develop guidelines for the control of environmental hazards in the workplace. It was estimated in a government report at the time of the congressional action that working environment hazards accounted for 14,000 deaths on the job, 100,000 additional deaths attributable to occupational diseases associated with exposure to chemicals and other physical hazards, and 2,000,000 injuries each year. A reorganization of OSHA in 1977 redirected its objectives to emphasize the prevention of occupational diseases, to concentrate its efforts on high-risk industries, and to eliminate or modify nearly 900 regulations not considered beneficial to workers or employers. A 1979 change in the charter of OSHA requires employers to open their medical records to the company's workers.

osmosis. The passage of solvent molecules from the lesser to the greater concentration of solute when two solutions are separated by a membrane which selectively prevents the passage of a solute molecule, but is permeable to the solvent.

outfall. The point or location where sewage or drainage discharges from a sewer, drain, or conduit.

outgassing. The emission of gases by materials and components, usually during exposure to elevated temperatures or reduced pressures.

overall reduction. The percentage reduction in the final effluent, as compared with the raw sewage.

overfire. Overfire air jets are streams of high-velocity air issuing from nozzles in a furnace enclosure, providing turbulence and oxygen to aid combustion, or to provide cooling air.

overturn. The period of mixing (turnover), by top to bottom circulation, of previously stratified water masses. This phenomenon may occur in spring and/or fall; the result is a uniformity of physical and chemical properties of the water at all depths.

oxidant. An oxidizing agent; the capacity of certain oxygen containing substances to react chemically in polluted air to form new products; the chemical substance that makes oxygen available for this reaction. Any oxygen-bearing compounds, such as nitrogen dioxide, that takes part in a photochemical reaction can be called an oxidant. This is because ozone is an early and continuing product of the photochemical smog reaction, and the presence of ozone in the air assures the continuation of the oxidizing process. For these reasons ozone is the chemical whose presence is used to measure the oxidant level of the atmosphere at any time. Severe oxidants (smog) produce reduced visibility and lachrymation, and cause characteristic injuries to the leaves of certain plants. Leaf lesions due to smog are fairly characteristic, but plants vary in their responses. Oxidant (smog) injury to leafy vegetable and field crops has been described as a silvering, glazing, and bronzing of the lower surface, sometimes with concomitant necrosis. Injury to the upper leaf surface typically occurs only after damage to the lower surface is extensive and excessive amounts of water are lost. Similar symptoms are produced on weeds, which may be used as indicator plants for the presence of smog. Likewise, chlorotic blotched and streaked areas develop on the leaves of cereals and forage grasses. If the exposure involves a high concentration of the toxicant, or lasts for a long period of time at low levels, light tan to brown areas develop in the leaves. Alfalfa, together with several other plants and trees, is sensitive to smog. It has been observed that the sensitivity of vegetation to smog injury is influenced by several factors. Plants are more susceptible to injury when the light, temperature, and soil moisture conditions are adequate. Mature leaves are more susceptible than either the immature or older leaves. In general, factors which tend to increase succulence also increase the plant's susceptibility to injury to smog.

oxidation. Any chemical reaction involving the loss of one or more electrons by an atom or ion. The particles which lose the electron(s) are said to be oxidized. The oxidation state of an element is represented by a signal number called an oxidation number, which somewhat arbitrarily indicates the number of electrons which may be assumed to be lost, gained, or shared by an atom in compound formation.

oxidation, direct. The oxidation of substances in sewage without the benefit of living organisms by the direct application of air or oxidizing agents such as chlorine.

oxidation number. The charge which an atom appears to have when electrons are counted according to some rather arbitrary rules: (a) Electrons shared between two unlike atoms are counted with the more electronegative atom; (b) electrons shared between two like atoms are divided equally between the sharing atoms. In free elements, each atom has an oxidation number of zero, no matter how complicated the molecule is. In simple ions (which contain one atom), the oxidation number is equal to the charge on the ion. In compounds containing oxygen, the oxidation number of each oxygen is generally -2. In compounds containing hydrogen, the oxidation number of hydrogen is generally $+1$. All oxidation numbers must be consistent with the conservation of charge.

oxidation pond. A holding area where organic wastes are broken down by aerobic bacteria.

oxidation rate. The rate at which the organic matter in sewage is stablized.

oxidation–reduction. The term oxidation refers to any chemical change in which there is an increase in oxidation number. When hydrogen, H_2, reacts to form water, H_2O, the hydrogen atom's oxidation number changes from 0 to $+1$. The H_2 is said to undergo oxidation. The term reduction applies to any decrease in oxidation number. When oxygen, O_2, reacts to form H_2O, the oxygen atom's oxidation number changes from 0 to -2. Thus there is a decrease in oxidation number; hence O_2 is said to undergo reduction. In oxidation and reduction the increase and decrease of oxidation numbers result from a shift of electrons. The oxidizing agent is the substance that does the reducing; it is the substance containing the atom which shows an increase in oxidation number.

oxidation–reduction reaction. One substance cannot gain electrons unless another substance loses electrons. If oxidation occurs during a chemical action, then reduction must occur simultaneously. If one kind of particle is oxidized, another kind of particle must be reduced to a comparable degree. Any chemical process in which there is a transfer of electrons, either partial or complete, is an oxidation–reduction reaction. This time is often shortened to "redox" reaction.

oxidation, sewage. The process whereby, through the agency of living organisms in the presence of oxygen, the organic matter that is contained in sewage is converted into a more stable mineral form.

oxides of nitrogen. Compounds made up of oxygen and nitrogen. These include nitric anhydride or nitrogen pentoxide (N_2O_5), nitrogen tetroxide (N_2O_4), nitrous anhydride or nitrogen sesquioxide (N_2O_3), nitrogen trioxide (NO_3), nitrogen dioxide (NO_2), and nitrous oxide (NO). Oxides of nitrogen are produced in combustion reactions such as occur in diesel engines and, in smaller amounts in gasoline engines. Air contains approximately 21 percent O_2 and 79 percent N_2, by volume. When fuel oil is oxidized with air at high temperatures, the composition of the combustion products is about 12 percent CO_2, 5 percent O_2, and 83 percent N_2. Other compounds called NO_x are formed with these at lower concentrations.

oxides of sulfur. Compounds made up of sulfur and oxygen. A major source of SO_2 in regard to atmospheric pollution is the combustion of coal and oil. This gas is a principal air pollutant and has been extensively studied in urban and industrial areas. The studies revealed distinct seasonal trends of SO_2 emission that coincide with the increased consumption of coal and oil in the winter months. The major sulfurous emissions from coal combustion are sulfur oxides.

oxygen (O). A colorless, odorless gas with atomic weight 16.00; boiling point, -183°C; and density, 1.429 g/liter. The most abundant and widely distributed of the earth's elements, oxygen composes about 20 percent of the atmosphere at sea level. Oxygen is chemically very active and supports combustion, the combination of oxygen and other substances generally producing heat. It is the basic factor in combustion, and in combining with fuel to produce heat, it forms several byproducts, among which is water vapor. Oxygen is taken from the Greek words meaning "acid producing."

oxygen, available. The quantity of atmospheric oxygen dissolved in the water of a stream; the quantity of dissolved oxygen available for the oxidation of organic matter in sewage.

oxygen, consumed. The quantity of oxygen taken up from potassium permanganate in solution by a liquid containing organic matter; commonly regarded as an index of the carbonaceous matter present. The time and temperature must be specified.

oxygen, dissolved. Usually designated as DO, the oxygen dissolved in sewage, water, and other liquids, usually expressed in parts per million or percent of saturation.

oxygen in water. One of the most critical chemical factors in an aquatic environment is the amount of oxygen in water, because most living organisms require oxygen for respiration. There are three recognizable zones with regard to oxygen concentrations: a surface stratum, where the oxygen tends to be in equilibrium with the atmosphere above, and an intermediate stratum below this surface stratum of variable depth, where oxygen values fluctuate in accordance with existing factors. Respiration, decomposition of organic materials (stagnant ponds), and stream pollution all tend to reduce the amount of available oxygen, while photosynthetic activity will often balance or more than balance the oxygen loss. The deepest layers of water in the deeper lakes and oceanic areas will usually have a very low oxygen concentration because the continual decomposition of organic debris, the respiration of organisms inhabiting these deeper waters, and the complete absence of photosynthetic activity in the lower strata tends to deplete the oxygen concentration.

oxygen pulse. The oxygen concentration which fluctuates in shallow marine and fresh-water areas within a 24-hour period; particularly noticeable in ponds, shallow lakes, and intertidal areas. Photosynthetic activity during the day results in a maximal oxygen concentration before dusk, but in the course of the nocturnal interval, constant respiratory activity reduces the available oxygen to a minimum before dawn. The following day, oxygen gradually increases, becoming supersaturated in regions where plant life is abundant.

oxygen sag curve. A curve that represents the profile of dissolved oxygen content along the course of the stream, resulting from deoxygenation associated with the biochemical oxidation of organic matter, and the reoxygenation through the absorption of atmospheric oxygen and through biological photosynthesis.

ozone. A colorless gas (or dark-blue liquid), O_3. It has a molecular weight of 48; a melting point of -251°C; and a boiling point of -112°C. Ozone is a powerful oxidizer which absorbs ultraviolet radiation from the sun, and, thus, protects life on earth from excessive doses; it is also a major air pollutant. Ozone is oxygen in molecular form, with three atoms of oxygen forming each molecule. Atmospheric oxygen is in molecular form, but each molecule contains two atoms of oxygen. Ozone is formed by passing high-voltage electric charges through dry air. The third atom of oxygen in each molecule of ozone is loosely bound and easily released. Ozone is sometimes used for the disinfection of water, but more frequently it is used for the oxidation of taste-producing substances such as phenol in water and the neutralization of odors in gases of air. Ozone is a poisonous form of oxygen that irritates the mucous membranes of the breathing system, causing coughing, choking, and impaired lung function. It aggravates chronic respiratory diseases like asthma and bronchitis. Ozone causes structural and chemical changes in the lungs and some alterations of blood components. Peroxyacetyl nitrate (PAN) and other photo-chemical oxidants that accompany ozone are powerful eye irritants. All oxidants are formed in the air by the chemical combination of nitrogen oxides and hydrocarbons, using the energy of sunlight. The threshold for direct action is around 1 ppm. While the natural concentration of ozone is about 0.02 ppm, it may range as high as 0.5 ppm or more in severe photochemical smog. Ozone attacks the palisade layer in plant leaves, causing brown flecks to appear. Natural vegetation is affected by ozone as well.

ozone plant pollution. *Sources*: photochemical reactions of products from petroleum burning stratospheric advection. *General effects*: bleached stippling on the upper surface of leaves; acute exposure possibly causing necrosis; the leaves of some plants dropping prematurely. *Examples*: Eastern white pine needles damaged by exposure to 60 ppm ozone for six hours; several hours of exposure to 0.4 ppm ozone producing visible leaf injury in grapevines.

ozonosphere. The general stratum of the upper atmosphere in which there is an appreciable ozone concentration and in which ozone plays an important part in the radiation balance of the atmosphere. The region lies between 10 and 50 km, with a maximum ozone concentration at about 20 to 25 km (also called ozone layer).

P

packed bed.   In gas chromatography, a device used for separating the components of a gaseous mixture, consisting of a solid mesh support. The components of the mixture are separated by volatizing a sample into a carrier gas stream passing through and over the bed of packing. The surface of the bed is usually coated with a relatively nonvolatile liquid, designated as the stationary phase, hence the term gas–liquid chromatography; if the liquid is not present, the process is dry, a gas–solid chromatography. The different components move through the packing at different rates, appearing one after another at the effluent end, where they can be detected and measured separately.

packed column.   A vertical column used for distillation, absorption, and extraction, containing packing, e.g., Raschig rings, Berl saddles, or crushed rock, which provides a large contacting surface area between phases. Normally, gas flow is countercurrent to liquid flow in this type of equipment. The separating mechanism is believed to be primarily the impingement of the particle on the packing itself, with the liquid medium acting to clean the surface of the packing material (also called packed scrubber or packed tower).

packed tower.   A pollution control device that forces dirty air through a tower packed with crushed rock or wood chips while liquid is sprayed over the packing material. The pollutants in the air stream either dissolve or chemically react with the liquid.

packed tower gas scrubber.   A prime method of scrubbing a true gas, used for the removal of gaseous fumes, noxious gases, and entrained droplets in gas-cleaning installations for various chemical processes. Basically, the packed tower is a vertical vessel in which various fill materials are wetted. The surface area provided by the various packings offers a basis for inducing interaction of the liquid and gas phases. The air or gas enters the bottom of the tower and receives a preliminary washing, as the scrubbing liquid drains in an opposing flow from the packed, irrigated bed. The liquid which is pumped into the top of the tower flows down over the packing bed. En route, it covers the surface areas of the packing with a liquid film that accomplishes most of the collection. Finally, the air stream passes through a mist eliminator section before it is permitted to exhaust. Diagrams of the weir box and spray assemblies used for liquid distribution in counterflow packed towers are shown on p. 233.

PAN.   Peroxyacetyl nitrate, a pollutant created by the action of sunlight on hydrocarbons and nitrogen oxides in the air; an ingredient of smog.

pandemic.   Widespread throughout an area.

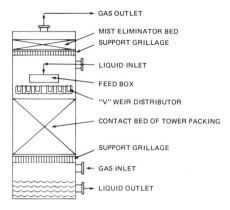

Counterflow packed tower, conventional "V" weir-type distributor.

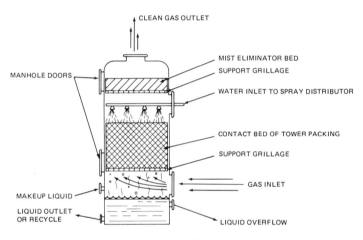

Counterflow packed tower using retractable spray assembly-type distributor.

paper. A thin flexible material made usually in sheets from a pulp prepared from rags, wood, or other fibrous material. Most paper is made from wood pulp. The cellulose of wood is separated from a noncellulosic substance called lignin. The latter is converted into an alkali-soluble substance by treatment with calcium hydrogen sulfite, $Ca(HSO_3)_2$, and removed from the insoluble cellulose fibers. The cellulose fibers, after removal of lignin, are washed and removed from the mixture and placed as a matting on a large flat filter. Compression of the matting, followed by drying, produces paper.

paper mill wood residue conversion. The Bowater Southern mill in Calhoun, Tennessee consumes pulpwood at the rate of 1600 cords per day to produce market pulp and large quantities of newsprint. Energy for mill operations is derived from a variety of sources including oil, natural gas, and purchased electricity, but nearly 20 percent of the energy requirements are satisfied by burning wood residues from the pulp mill and nearby sawmills. The basic fuels and energy sources are purchased bark and wood residues, bark and wood fines from the manufacturing process, natural gas, bunker C fuel oil, and purchased electricity. In addition to these, substantial amounts of energy are obtained from chemical recovery associated with the Kraft pulping process. Based on purchase, handling, and equipment maintenance expenses, the cost of wood residues at the time of the study amounted to $8.50 per ton. When compared to fuel oil, then at $10.00 per barrel, the annual gross savings added up to roughly $750,000, assuming 1976 consumption rates of 150,000 tons per year. The benefits of the residue burning operation included reduction of wood residue disposal problems for local operators of small sawmills, displacement of nonrenewable fuels (oil), and a positive monetary return on investment.

parasitism. A disoperation affecting a large segment of a community in which the adverse activity requires a highly specialized organism possessing the structures and physiology that enable it to successfully extract essential materials from a host plant or animal. In many cases parasitism has modified population growth because the invasion of a host organism eventually shortens its life cycle, weakening the plant or animal, or drastically inhibiting reproductive activities.

parent. The radioactive nucleus that disintegrates to make a different kind of nucleus, which is called the "daughter."

partial node. A point, line, or surface in a standing-wave system where some character-istic of the wave field has a minimum nonzero amplitude.

particle. A small discrete mass of solid or liquid matter such as in aerosols, dusts, fumes, mists, smokes, and sprays. (1) *Aerosol.* A dispersion of solid or liquid particles of microscopic size in a gaseous media, such as smoke, fog, or mist. (2) *Dust.* A dispersion of solid particles, formed by disintegration processes such as crushing, grinding, and demolition, normally greater than 2 μm in diameter, but sometimes as small as 0.1 μm in diameter. The most noticeable dusts and grits are above 75 μm in diameter. (3) *Fume.* A dispersion of solids generated by condensation of vapors, sublimation, condensation, foundry processes, and chemical reactions. These solids are often metals or metal oxides and they may differ materially from the substances from which they originated. They are less than 1 μm in diameter. (4) *Mist.* A dispersion of liquid droplets less than 10 μm in diameter, generated by condensation. (5) *Smoke.* A product of incomplete combustion, consisting of minute carbonaceous particles which remain suspended in the air until removed by precipitation or gravitation. The vast majority of smoke particles are less than 1 μm in diameter and require an electron microscope for detailed examination. (6) *Spray.* A dispersion of liquid droplets, greater than 9 μm in diameter, created by some mechanical disintegration process.

particle concentration. A concentration expressed in terms of the number of particles per unit volume of air or other gas.

particle detector. A device used to detect subatomic particles. There are two main categories—counters and track detectors. The counters usually contain a gas or a liquid. When a charged particle passes through a counter, a signal is produced which can be recorded electronically. A track detector shows the track of a charged particle by indicating its ionized train in a special medium. Usually this train is photographed for study. Among track detectors are the cloud chamber, the bubble chamber, and the spark chamber.

particle fall. A measurement of air contamination consisting of the mass rate at which solid particles deposit from the atmosphere; a term used in the same sense as the older term "dust fall" and "soot fall," but without any implication as to the nature or source of the particles.

particle size. The size of liquid or solid particles expressed as the average or equivalent diameter.

particle size distribution. The relative percentage by weight or number of each of the different fractions of particulate matter.

particles and their antiparticles.

Particle	Symbol	Antiparticle	Symbol
Proton	p	Antiproton	$\bar{p}$
Neutron	n	Antineutron	$\bar{n}$
Sigma	Σ^+	Antisigma	Σ^-
	Σ^0		$\bar{\Sigma}^0$
	Σ^-		Σ^+
Pion	π^+	Pion	π^-
	π^0		π^0
	π^-		π^+
Kaon	K^+	Kaon	K^-
	K^0	Antikaon	$\bar{K}^0$
	K^-	Kaon	K^+
Muon	μ^+	Muon	μ^-
	μ^-		μ^+
Positron	e^+	Electron	e^-
Electron	e^-	Positron	e^+
e-Neutrino	ν_e	Anti-e-neutrino	$\bar{\nu}_e$
μ-Neutrino	ν_μ	Anti-μ-neutrino	$\bar{\nu}_\mu$

particulate. A fine liquid or solid particle, such as in dust, smoke, mist, fumes, or smog, found in the air or emissions.

particulate collector. The simplest form of particulate collector is the dry inertial collector. The most prevalent type of collector in air pollution control is the cyclone, where the entire mass of a gas stream with entrained particulates is forced into a constrained vortex in the cylindrical portion of a cyclone. By virtue of their rotation with the carrier gas around the axis of the tube and their higher density with respect to the gas, the entrained particulates are forced toward the wall by centrifugal force. Here they are carried by gravity and/or secondary eddies toward the dust outlet at the bottom of the tube. The flow vortex is reversed in the lower portion of the tube, leaving most of the entrained particulates behind. The cleansed gases then pass through the central or exit tube and out of the collector.

particulate collector, fabric filters. In essence, a very large vacuum cleaner with filters of various configurations made of porous fabrics which can withstand the thermal, chemical, and mechanical rigors of individual applications. The most common form of fabric filter comprises a number of cylinder bags, usually inflated by the gases to be cleaned, with the gas passing through the fabric filter medium from the inside. Particles suspended in the gas stream impinge on and adhere to the filter medium and are thus removed from the gas stream.

particulate effects. Very small aerosol particles (from 0.001 to 0.1 μm in diameter) can act as nuclei on which vapor condenses relatively easily. Fog, ground mists, and rain may be thus increased and prolonged. Particles less than 2 or 3 μm in size (about half of the particles suspended in urban air are estimated to be that small) can reach deep into the part of the lung that is unprotected by mucus, and can attract and carry such harmful chemicals as sulfur dioxide with them. Sulfur dioxide alone would be dissolved in the mucus before it reached the vulnerable tissue. Particulates can act as catalysts; an example of this is the conversion of sulfur dioxide into sulfuric acid, helped on by catalytic iron oxides.

particulate emission factors for coal combustion without control equipment.

Type of unit	Particulate emission (lb/ton of coal burned)	Benzo[a]pyrene emission (μg/ton of coal burned)
Pulverized — general		
Dry bottom	16A[a]	600
Wet bottom	17A[a]	
Without reinjection	13A[a]	
With reinjection	24A[a]	
Cyclone	2A[a]	6000
Spread stoker		
Without reinjection	13A[a]	700
With reinjection	20A[a]	
All other stokers	5A[a]	100,000
Hand-fired equipment	20	12 × 10[6]

[a]A equals the percentage of ash in coal. For example, if 10 percent of the coal is ash, the ash emission for a cyclone unit would be 2 × 10 or 20 lb/ton of coal burned.

particulate gas sampling (bubble tube meter). Gas is drawn from a stack through a sampling train and a cylindrical tube of known volume. A soap solution at the bottom of the tube is caused to form bubbles by the passage of gas through the liquid. The rate of rise per unit time provides a direct measurement of the volumetric flow rate. Bubble tubes are usually small-flow devices with rates of less than one liter per minute, because high flows and large-diameter bubbles cause surface tension effects which tend to promote bubble breakage. Bubble tubes are extensively used wih gas chromatographs and coulometric titrators.

particulate gas sampling (dry gas meters). The dry gas meter is a total volume displacement device which uses a gas stream moving a diaphragm as a means of flow measurement, where the rate of flow is the volume displaced per unit time. Dry gas meters are high flow rate devices (0.5–2.0 cubic tons/min) extensively used in particulate sampling. They are usually inaccurate for low flow rates except where small meters are used.

particulate gas sampling (orifice flow meter). Gas is drawn from a stack through a sampling train followed by a restricted opening in the line. Here the rate of flow is proportional to the pressure drop across the restriction, which limits the flow to a certain maximum value. Corrections may be made to stack conditions, after a calibration curve is set up for a given orifice, with air at ambient conditions, by determining the variation in flow rate with pressure drop, with the sampling train in place.

particulate gas sampling (rotameter). The rate of flow through a rotameter is proportional to the height of rise for a metering float in a transparent vertical tube. Rotameters are low, essentially constant-pressure-drop devices which can be used in both gas and particulate sampling. Rotameters suffer from float oscillations with various vacuum sources and should be operated at or near atmospheric pressure, or else calibrated at conditions substantially different from atmospheric. They are subject to condensation, which will render flow measurement useless. This problem may be eliminated by placing a drying agent upstream of the rotameter or heating the rotameter.

particulate gas sampling (spirometer). An absolute calibration device which consists of two vertical cylinders, one closed at the top and fitting inside the other, which is closed at the bottom and filled with water. The inner tube is attached to pulleys and counter-weighted and caused to slide up and down a calibrated measuring bar. Gas is added to the floating inner tube at a point just above the fluid level. The volume is the product of the cross-sectional area of the inner tube and the distance of displacement, while the rate of flow is the volume displaced per unit time. Spirometers provide an absolute measure-ment of volume, but they are cumbersome and generally used only for laboratory calibration.

particulate gas sampling (wet test meter). The wet test meter is composed of a case containing rotor vanes divided into compartments of equal volume, where gas is passed consecutively into each compartment, thereby causing the rotors to turn. Wet test meters are total volume devices, where the rate of flow is the volume displaced per unit time. They are specifically intended for laboratory calibration of flow measuring devices only and should not be used for field sampling.

particulate loading. The introduction of particulates into ambient air.

particulate matter. Airborne matter consisting of solid and liquid particles (over 20 µm in diameter) and aerosols or suspensions of fine particles (down to less than 0.05 µm). The larger particles are usually solids, fly ash, coarse dirt, and dirt particles larger than 10 µm in diameter, and these normally settle out of the air rapidly. They are composed of a variety of materials, some soluble in water, some not. In heavily polluted areas, the soluble part is usually high in sulfate and is relatively acidic. Fume or smoke particles, ranging in size from 0.1 to 5 µm or smaller may be liquid or solid. They may form mechanically stable suspensions in the air. The smaller particles, generally smaller than 1 µm, are readily transported by wind and currents and have motions similar to the surrounding gases and may be treated as such. These very fine particles may often be found in metal oxide fumes such as those of lead, zinc, arsenic, cadmium, and beryllium. Particles may be complex in their chemical composition. The inorganic fraction of collected samples of airborne particles usually contains a few dozen metallic elements in addition to carbon or soot and tarry organic material. The most common metallic elements are silicon, calcium, aluminum, and iron; relatively high quantities of magnesium, lead, copper, zinc, sodium, and manganese may also be found. The organic fraction of collected samples of particles is usually even more complex and may contain a large number of aliphatic and aromatic hydrocarbons, acids, bases, phenols, and many other compounds.

particulate matter sources. Both organic and inorganic matter emanate from a number of sources such as industrial operations, modern transportation facilities, and domestic combustion processes. Major sources of dust include coal- and oil-burning power plants, iron and steel mills, cement mills, and oil refineries. In addition, small sources, such as automobiles and incinerators, contribute significantly to the dust load of the atmosphere because they are so numerous. Smoke (dust and droplets) is produced during combustion or destructive distillation, and fume (dust) is formed by high-temperature volatilization or by chemical reaction. A large number of extremely fine particles are emitted from automobile exhaust systems, with approximately 70 percent in the size range of 0.02–0.06 µm. These particles consist of both inorganic and organic compounds of high molecular weight. The quantity of solid and droplet material produced in the exhaust amounts to a few milligrams per gram of gasoline burned. Most gasolines contain additives. The most common additive contains tetraethyl lead or tetramethyl lead together with organic chlorides and bromides, and the resulting lead compounds comprise some of the exhaust particles. Another source of lead pollutant is the melting of scrap metals in foundaries. Highly refined liquid fuels do not ordinarily contain appreciable amounts of dust-forming ash, except when additives are blended to improve quality. Nevertheless, poor combustion of any fuel produces soot and tars, and heavy heating oils also contain some metals, as well as substantial amounts of sulfur.

particulate plant pollution. *Source:* carbon from combustion processes and dusts from various sources such as cement manufacturers. *Effects:* possible leaf spotting resulting from the wetting of acidic and alkaline dusts, possible damage due to interference with light and gas transmissions. *Example:* Dusting with certain cement dusts at a loading of 1.8–3.8 g/m³/day caused wilting of bean leaves six hours after wetting, and necrosis and leaf roll within two days.

particulate removal of air pollutants. The removal of particulate air pollutants from their gaseous media, accomplished using a number of mechanisms or forces, including gravitational, centrifugal, electrostatic, and magnetic forces, thermal diffusion, and Brownian diffusion. The magnitudes of these forces are generally dependent upon the particle size.

particulate removal of air pollutants (centrifugal collectors). Cyclones are probably the most common particulate collection devices currently used. The conventional cyclone is relatively simple to construct and has no moving parts. In a cyclone, the gas stream moves in a continuous spiral, as compared to the single angular diversion in inertial separators. The rotational motion of the gas stream can be applied in several ways, including passing the gas through curved vanes in a duct, rotating it in a turbine, or having it flowing tangentially into a cylindrical duct. The tangential inlet cyclone is the most common type of cyclone.

particulate removal of air pollutants (electrostatic precipitator). Electrostatic precipitation involves the use of an electric field to remove electrically charged particles from their gaseous media. An electrostatic precipitator has a discharge electrode (positive) of large surface area, such as a plate or tube. There are essentially four steps in the electrostatic precipitation of particles: (a) placing an electrical charge on the particles, (b) migration of the charge in the electrical field to the collecting surface, (c) neutralization of the electrical charge on the particle after impacting on the collection electrode, and (d) removal of the precipitated particles from the collection electrode. The particles are charged by collision with air molecules ionized at the discharge electrode. The neutralized particles are removed from the collecting surface by rapping, scraping, brushing, or washing. There are two general types of electrostatic precipitators, low voltage (two stages) and high voltage (single stage). The low-voltage precipitator consists of an ionizing stage followed by a collecting stage. Two-stage electrostatic precipitators are generally used for the removal of oil smoke or in air conditioning. High-voltage or single-stage electrostatic precipitators combine ionization and collection and are used in removing particulates from highly concentrated emissions.

particulate removal of air pollutants (filter). The mechanisms of collection of airborne particles upon the obstacles in filters are similar to those in wet collectors and include interception, inertial impaction, Brownian diffusion, electrostatic migration, and thermal diffusion. A filter is essentially a porous structure which removes the particulate matter from the fluid passing through it. Generally, there are two types of filters: fibrous or deep-bed filters (low efficiency) and cloth or paper filters (high efficiency). Fibrous filters are commonly used in air conditioning and have low pressure drops (less than 0.5 in. of water). Cloth and paper filters are capable of high efficiencies and have pressure drops in the range of 0.5 to 6 in. of water. The cloth or fabric filters used in air pollution control are usually arranged as bags. The filter bags are cleaned periodically using systems such as shaking, air jets, or reverse air flow. The fabric filters are composed of various materials including cotton, wood, asbestos, glass, and synthetics. The design of filter systems involves the gas to cloth ratio, which is defined as the gas filtered, in cubic feet per minute, per square foot of filter area and is expressed in feet per minute. The gas to cloth ratios range from 1 to 25 ft/min. A typical gas to cloth ratio is 3 ft/min for intermittently cleaned bags and 12 ft/min for reverse-jet cleaning.

particulate removal of air pollutants (settling chamber). Settling chambers are simple in construction, have low pressure drops, and are relatively low in their efficiency for collecting small particles (less than 50 μm in diameter). The design parameters of a settling chamber include the gas velocity, the chamber length, and the chamber width. The gas velocity should be less than 10 ft/s to prevent particle reentrainment.

particulate removal of air pollutants (wet collector). A wet collector is essentially a gas–liquid contacting device in which airborne particulates are transferred from suspension in the gaseous phase to a liquid phase or solution. The liquid is usually water. To maximize the gas–liquid contact area (and hence the mass transfer rate), the liquid is usually either in the form of droplets (sprays) or bubbles (foam). The advantage of wet collectors include the absence of particulate reentrainment, the ability to handle high-humidity gases, the space requirements are usually small, the ability to simultaneously collect particulate and gaseous pollutants, and the ability to cool high-temperature gases. The disadvantages are that one has to remove the particulates from the polluted scrubbing liquid, the washed air has a high relative humidity, causing steam plume problems, and one has to dispose of the waste scrubbing liquid.

particulate size. Coarse dust particles larger than 10 μm in diameter and fly ash composed of the impurities remaining after coal is burned settle out of the air quickly. Fume, dust, and smoke particles, ranging in size from under 1 up to 10 μm, travel farther, the distance covered depending mainly on the size. Particles less than 1 μm in diameter (generally referred to as aerosols) move as easily and as far on wind or air currents as gases do. Smoke describes both solid and liquid particles under 1 μm in diameter. It can be produced during all forms of combustion and in such processes as distillation, the removal of impurities from liquids by heating them to their boiling point and then condensing the vapors. The designation "smoke" may include fume. Fume indicates solid particles under 1 μm in diameter that are formed as vapors condense or as chemical reactions take place. Fumes are emitted by many industrial processes, including smelting and refining. Dust is a more general term than fume. When solid particulates are more than one micron in size, they are generally referred to as dust. Mist is made up of liquid particles up to 100 μm in diameter. These may be released industrially in such operations as spraying and impregnating or formed by the condensation of vapor in the atmosphere, or by the effect of sunlight on automobile exhaust. As mist evaporates, more concentrated liquid aerosols are formed.

parts per million (ppm). The number of milligrams per liter divided by the specific gravity. It should be noted that in water analysis ppm is always understood to imply weight/weight ratio, even though in practice a volume may be measured instead of a weight. By contrast, "percent" may be either a volume/volume or a weight/weight ratio.

Pascal's law. Pressure exerted at any point upon a confined liquid is transmitted undiminished in all directions.

passive solar system. An assembly of natural and architectural components including collectors, thermal storage devices, and transfer fluids which converts solar energy into thermal energy in a controlled manner and in which no pumps are used to accomplish the transfer of thermal energy.

pathogenic. Capable of causing disease.

PCB's. (See also **polychlorinated biphenyls**). A group of toxic, persistent chemicals used in transformers and capacitors. Further sale or new use was banned in 1979 by law.

peak concentration. During a period of monitoring an air pollutant or pollutants, the highest concentration of the pollutant or pollutants during that period.

peak to mean ratio. The ratio of the peak concentration to the mean concentration of an air pollutant, or air pollutants, which have been monitored over an established period of time.

pelletized industrial fuel made from biomass residues. Using surplus forest residues, Woodex Inc. of Brownsville, Oregon is producing a dense pelletized fuel which can be utilized in coal-fired equipment. The fuel itself has a specific gravity of 1.3 and a heat content of 18 million Btu/ton, which compares favorably with coal. The current production of 100 tons per day is sold to nearby industrial customers as a replacement for coal. Fuel pellets are fabricated from pulverized biomass material which has been dried to a prescribed moisture content in rotating-drum driers. The pelletizing process involves continuous extrusion under conditions of controlled temperature and pressure to produce a dense pellet which is resistant to moisture absorption. Energy is recovered in conventional coal-burning furnaces, without sulfur emissions and with only a 2.8 percent loss.

Peltier effect. When a current flows across the junction of two unlike metals, it gives rise to an absorption or liberation of heat. If the current flows in the same direction as the current at the hot junction in a thermoelectric circuit of two metals, heat is absorbed; if it flows the same direction as the current at the cold junction of a thermoelectric circuit, heat is liberated.

pentachloronitrobenzene. $C_6Cl_5NO_2$, a nitrogen compound of the aromatic hydrocarbon category. It is a colorless crystal at room temperatures, but when heated it decomposes to emit highly toxic fumes of chlorides and nitrogen oxides. The threshold limit value of pentachloronitrobenzene has been established as 1 mg/m³. The LD_{50} has been reported to range from 135 mg/kg to as much as 1650 mg/kg, depending upon the species of animal and the route of administration. It has been found to be a teratogen, a carcinogen, and a neoplastic agent when administered via oral and dermal routes. Other clinical effects observed include damage to red blood cells, hyperthermia, central nervous system depression, kidney and liver damage, and bladder and skin irritation. Pentachloronitrobenzene has been listed by the EPA Carcinogen Assessment Group and accepted by OSHA for review as a suspected occupational carcinogen.

pentachlorophenol. PCP, C_6HCl_5O, a compound that has been used as an insecticide, mainly in an effort to eradicate termites. A health hazard has been associated with the substance as a result of applications of PCP to termite-infested soil and the resultant seepage of PCP into local groundwater supplies. PCP has also been employed as a herbicide, a fungicide, a molluscicide, and a wood preservative ingredient. The chemical is highly toxic, with liver and kidney damage resulting from chronic exposure. Acute poisoning can produce weakness and adverse changes in respiratory, circulatory, and renal functions, as well as skin damage, and, in severe cases, convulsions and prostration. LD_{50}'s in laboratory animals range from 50 mg/kg in rats administered PCP orally to 105 mg/kg in dermal-exposure studies of rats. Pentachlorophenol decomposes to produce highly toxic chloride fumes when heated to temperatures of 310°C. OSHA recommends for time-weighted average exposure a maximum concentration of 0.5 mg/m³ and 1.5 mg/

m^3 for short-term exposure. In 1980, the Food and Drug Administration established an action level of 3.0 ppm PCP in the edible portions of fish or shellfish, as a result of spills of commercial PCP's in the Mississippi River–Gulf of Mexico marshes, bayous, and lakes.

perched groundwater. Groundwater present in a position above the main zone of saturation. The upper surface of this layer is called the perched water table. A perched mass of groundwater is the result of an impervious layer of material, usually calcium carbonate, located some distance from or above the main water table, which prevents the normal passage of gravitational water down through this zone.

percolation. The downward flow or filtering of water through pores or spaces in rock or soil.

perfect gas. A gas which has the following characteristics: (a) it obeys the Boyle–Mariott law and the Charles–Gay-Lussac law, thus satisfying the equation of state for perfect gases; (b) its internal energy is a function of temperature alone; and (c) it has specific heats with values independent of temperature.

perfectly diffuse radiator. A body that emits radiant energy in accordance with Lambert's law. The radiant intensity emitted in any direction from a unit area of such a radiation varies as the cosine of the angle between the normal to the surface and the direction of radiation. When viewed from a distance, an incandescent, perfectly diffuse radiator appears as a uniformily illuminated flat surface, regardless of its actual shape or orientation.

period, aeration. (1) The theoretical time, usually expressed in hours, that a mixed liquor is subjected to aeration in an aeration tank undergoing activated sludge treatment; equal to the volume of the tank divided by the volumetric rate of flow of the sewage and return sludge. (2) The theoretical time that water is subjected to aeration.

period, detention. The theoretical time required to displace the contents of a tank or unit at a given rate of discharge (volume divided by rate of discharge).

periodic law. Elements, when arranged in the order of their atomic weights or numbers, show regular variations in most of their physical and chemical properties.

periodic table. A listing of the chemical elements by the sizes of their atoms in such a way that elements having similar properties fall together and properties repeat according to a recognizable pattern as one proceeds from lighter to heavier atoms. There are 92 elements found in nature. Under ordinary temperatures and pressures, roughly from 65 to 90°F, three elements are liquids (mercury, gallium, and bromine), eleven are gases, and the rest are solids. When elements are grouped by their atomic numbers, they fall into a regular pattern of families with similar properties. This pattern forms the periodic table of elements.

periodic variations. Hourly, daily, seasonal, or annual variations in the concentration of air pollutants due to meteorological, technological, and/or sociological factors. Thus nitric oxide concentrations in the atmosphere have been shown to reflect the same three factors, seasonally and diurnally, namely, the rate of conversion to NO_2, the dilution capacity of the atmosphere, and the strength of combustion sources, the three factors, in turn, being influenced by sunlight patterns.

permanganate oxidation. The oxidation of odors by potassium permanganate solutions. Rendering plant odors, asphalt plant fumes, fish processing odors, sewage plant odors, and odors from many chemicals can be substantially reduced or eliminated using this method.

permeability. (1) The ability to permit penetration or passage. In this sense the term is applied to substances which permit the penetration or passage of fluids. (2) For a magnetic material, the ratio of the magnetic induction to the magnetic field intensity in the same region.

permeability coefficient. The steady-state rate of flow of gas through a unit area and thickness of a solid barrier per unit pressure differential at a given temperature. The permeability coefficient is usually expressed in volume or mass per unit time, per unit area of cross section, per unit thickness, per unit pressure differential across the barrier.

permeance. The reciprocal of magnetic reluctance. A unit permeance is the permeance of a cylinder one square centimeter in cross section and one centimeter in length taken in a vacuum.

permeation. As applied to gas flow through solids, the passage of gas into, through, and out of a solid barrier having no holes large enough to permit more than a small fraction of the gas to pass through any one hole. The process always involves diffusion through the solid and may involve various surface phenomena such as sorption, dissociation, migration, and desorption of the gas molecules.

permissable dose. The amount of radiation which may be received by an individual within a specified period with no harmful result to himself.

permissible immission concentration. The concentration of foreign matter (or the deposition of dust) in the layers of the atmosphere close to the ground which is tolerable for man, animal, or plant, according to present experience, for a given duration and frequency of effect.

Permutit. A synthetic zeolite used in softening water.

peroxyacetyl nitrates (PAN's). A homologous series of organic nitrogen compounds that are formed when sunlight acts on air which is polluted with trace concentrations of organic compounds and nitrogen oxides. They are of interest because of their biological activity. Exposures to concentrations measured in parts per hundred million (pphm by volume) can cause visible damage to agricultural crops. Concentrations in the parts per million to parts per billion range can cause eye irritation. Concentrations of 100–200 ppm are lethal for mice for a two-hour exposure. Peroxyacetyl nitrate is the family name given these compounds. Gasoline vapors, automobile exhaust, pure hydrocarbons, and aldehydes all produce this compound. Other investigations have shown that olefins (with the exception of ethylene), aromatics (with the exception of benzene), and aldehydes (with the exception of formaldehyde) all produce these compounds.

persistent pesticide. A pesticide that does not break down chemically and remains in the environment after a growing season.

pesticide. Any substance used to control pests ranging from rats, weeds, and insects to algae and fungi. Pesticides can accumulate in the food chain and can contaminate the environment if misused.

pesticide tolerance. The amount of pesticide residue allowed by law to remain in or on a harvested crop. By using various safety factors, EPA sets these levels well below the point where the chemicals might be harmful to consumers.

petroleum. A naturally occurring, complex mixture of liquid hydrocarbon compounds, consisting mostly of carbon and hydrogen, with minor quantities of nitrogen, oxygen, and sulfur. Most petroleum contains 82 to 87 percent carbon and 11 to 15 percent hydrogen. These elements are combined in several types of molecules, many of which have a very complex structure. Distilling off the light fractions of crude petroleum leaves a residue of light-colored paraffin wax, a dark tarry asphalt, or a mixture of the two.

petroleum refinery emission factors.

Processes	Units for emission factors	Emission factor
Boilers and process heaters	lb hydrocarbon/1000 bbl oil burned	140
	lb hydrocarbon/1000 ft³ gas burned	0.026
	lb particulate/1000 bbl oil burned	800
	lb particulate/1000 ft³ gas burned	0.02
	lb NO_2/1000 bbl oil burned	2900
	lb NO_2/1000 ft³ gas burned	0.23
	lb CO/1000 bbl oil burned	Negligible
	lb CO/1000 ft³ gas burned	Negligible
	lb HCHO/1000 bbl oil burned	25
	lb HCHO/1000 ft³ gas burned	0.0031
Fluid catalytic units	lb hydrocarbon/1000 bbl of fresh feed	200
	lb particulate/ton of catalyst circulation	1.8[a]
	lb NO_2/1000 bbl of fresh feed	63
	lb CO/1000 bbl of fresh feed	13,700
	lb HCHO/1000 bbl of fresh feed	19
	lb NH_3/1000 bbl of fresh feed	54
Moving-bed catalytic cracking units	lb hydrocarbon/1000 bbl of fresh feed	87
	lb particulate/ton of catalyst circulation	4[b]
	lb NO_2/1000 bbl of fresh feed	5
	lb CO/1000 bbl of fresh feed	3800
	lb HCHO/1000 bbl of fresh feed	12
	lb NH_3/1000 bbl of fresh feed	5
Compressor internal combustion engines	lb hydrocarbon/1000 ft³ of fuel gas burned	1.2
	lb NO_2/1000 ft³ of fuel gas burned	0.86
	lb CO/1000 ft³ of fuel gas burned	Negligible
	lb HCHO/1000 ft³ of fuel gas burned	0.11
	lb NH_3/1000 ft³ of fuel gas burned	0.2
Miscellaneous process equipment		
1. Blowdown system	lb hydrocarbon/1000 bbl refinery capacity	
a. With control		5
b. Without control		300
2. Process drains	lb hydrocarbon/1000 bbl waste water	
a. With control		8
b. Without control		210

Processes	Units for emission factors	Emission factor
3. Vacuum jets a. With control b. Without control	lb hydrocarbon/1000 bbl vacuum distillation capacity	Negligible 130
4. Cooling towers	lb hydrocarbon/1,000,000 gal cooling water capacity	6
5. Pipeline valves and flanges	lb hydrocarbon/1000 bbl refinery capacity	28
6. Vessel relief valves	lb hydrocarbon/1000 bbl refinery capacity	11
7. Pump seals	lb hydrocarbon/1000 bbl refinery capacity	17
8. Compressor seals	lb hydrocarbon/1000 bbl refinery capacity	5
9. Others (air blowing, blend changing, and sampling)	lb hydrocarbon/1000 bbl refinery capacity	10

[a]With electrostatic precipitator.
[b]With high-efficiency centrifugal separator.

pH. A symbol denoting the negative logarithm of the hydrogen ion concentration in a solution. pH values run from 0 to 14. The number 7 indicates neutrality, while numbers less than 7 indicate increasing acidity and numbers greater than 7 indicate increasing alkalinity.

phenacetin. $C_{10}H_{13}O_2N$, a glistening, crystalline powder commonly used in analgesic mixtures with caffeine and aspirin. The chemical name of the compound is *para*-acetphenetidin. Adverse health effects result mainly from oral ingestion of the chemical which may produce central nervous system symptoms and cyanosis, which is due to the conversion of the hemoglobin molecule in red blood cells to methemoglobin. The red blood cell change is the result of oxidation of the ferrous ion of the hemoglobin to the ferric form. Cyanosis signs indicate that at least 15 percent of the hemoglobin has been converted to methemoglobin. However, headache, dizziness, weakness, and breathing difficulty usually do not appear until a greater proportion of red blood cell changes have occurred. Chronic exposure to phenacetin can cause insomnia, loss of weight, and aplastic anemia. The lowest recorded toxic dose in humans is on the order of 1000 mg/kg; the oral LD_{50} in the rat is approximately 1650 mg/kg. Human fatalities resulting from phenacetin are rare, but the chemical has been associated with tumor formation and the substance has been recommended for review by the U.S. Occupational Safety and Health Administration as a possible occupational carcinogen.

phenols. Organic compounds that are byproducts of petroleum refining, tanning, textile, dye, and resin manufacture. Low concentrations of phenols can cause taste and odor problems in water, higher concentrations can kill aquatic life.

phenylenediamene. $C_6H_4(NH_2)_2$, a compound that occurs in several forms employed in hair dyes, eyelash dyes, fur dyes, and photographic chemicals. Of the three phenylenediamine forms—*m, o,* and *p*—*p*-phenylenediamine is regarded as the most toxic isomer, with at least one fatality associated with exposure to it. Asthma and other respiratory disorders and liver damage have been reported among workers exposed to the chemical, although skin irritation and conjunctivitis probably are more commonly reported effects. The dermatitis signs are characterized by weeping, crusting, and itching. Skin sensitization may also result from continued exposure. The lowest reported oral LD_{50} in rats exposed to phenylenediamene is 100 mg/kg; a threshold limit value of 0.1 mg/m³ has also been reported in toxicology studies. The chemical has been recommended for review by OSHA as a potential occupational carcinogen.

phon. The unit of measure of the loudness level when a standard pure tone of frequency 1000 Hz is produced by a sensible plane, sinusoidal, progressive sound wave coming from directly in front of the observer. The sound-pressure level in the free progressive wave is expressed in decibels above 2×10^{-5} N/m².

phonometer. An instrument for measuring the intensity or frequency of sounds.

phosphate. A chemical compound containing phosphorus.

phosphate rock. An earthy substance, varying in appearance from a hard rock to a granular, loosely consolidated mass, consisting of more or less impure noncrystalline calcium fluorophosphate. Phosphate rock is used as a fertilizer.

phosphor. A fluorescent substance, such as zinc sulfide and calcium tungstate, that emits light when excited by radiation or ionizing particles, or converts ultraviolet radiation into visible light.

phosphorescence. The emission of radiation by a substance as a result of previous absorption of radiation of a shorter wavelength. In contrast to fluorescence, the emission may continue for a considerable time after cessation of the exciting irradiation.

phosphorus (P). A nonmetallic element, a white, phosphorescent, waxy solid, becoming yellow when exposed to light. Unnaturally high phosphorus concentrations may be the result of animal and plant processing or fertilizer and chemical manufacturing operations. Generally, the total phosphorus concentrations above 50 mg/liter contribute to the overgrowth of objectionable plant forms. When the concentration of complex phosphates is greater than 100 mg/liter, coagulation processes may be adversely affected. Another effect of high phosphorus concentrations is that it may bring about oxygen depletion in a body of water. This results from the requirement of 160 mg of O_2 to completely oxidize 1 mg of phosphorus from an organic source.

phosphorus cycle. As water runs over rocks, it gradually wears away the rock surface and carries off a variety of minerals, some in solution and some in suspension. Some of these minerals, such as phosphates, sulfates, calcium, magnesium, and others, are necessary for the growth of plants and animals. Phosphorus is taken in by plants as inorganic phosphate and converted to a great variety of organic phosphate compounds. Animals get their phosphorus as inorganic phosphates in the water they drink, or as inorganic plus organic phosphates in the food they eat. The phosphorus cycle is not completely balanced, for phosphates are being carried into the sediments at the bottom of the sea faster than they are being returned by the action of marine birds and fish.

photochemical air pollution. Among the substances responsible for photochemical air pollution are unsaturated hydrocarbons, saturated hydrocarbons, aromatics, and aldehydes. These are emitted during the incomplete combustion of all fuels (including rubbish and agricultural field wastes), but automobile exhaust is the major source. Hydrocarbons and other organic gases are also expelled during the production, refining, and handling of gasoline and from such manufacturing apparatus as industrial dryers and ovens, and furnaces used for baking paints, enamels, and printing ink. The oxides of nitrogen, the other partners in the formation of photochemical smog, are formed during high-temperature combustion, chiefly in automobiles and large power plants. The temperature produced by the explosion of gasoline in an automobile engine causes a few thousand parts per million of nitric oxide to form from the mixture of nitrogen and oxygen in the air. The naturally occurring ozone that exists in the upper atmosphere limits the sunlight radiation at the earth's surface to wavelengths longer than the near ultraviolet, at about 2900 angstroms. In order to produce a chemical reaction, light must be absorbed; and among the more common pollutants the most important light absorber is nitrogen dioxide, a yellow-brown gas that absorbs light in the blue and near ultraviolet part of the spectrum, and is thereby dissociated into nitric oxide and atomic oxygen. At a concentration of 0.10 ppm, the rate of photolysis of nitrogen dioxide by average sunlight amounts to about 2 ppm per hour. This reaction is the fastest of all known primary photochemical processes in polluted air. Other contributing reactions are the photochemical decomposition of aldehydes and the formation of excited oxygen molecules.

photochemical air pollution (eye irritation). Studies by eye specialists have not indicated that repeated irritation results in any permanent injury to the eye, but severe photochemical smog can be highly distressing and can temporarily hinder normal functions. The precise composition of all the substances responsible for eye irritation has not definitely been determined. Ozone at observed levels in photochemical air pollution does not irritate the eyes or respiratory membranes. Peroxyacetyl nitrate is a powerful eye irritant at concentrations in the 1-ppm range, and there is some belief that, along with acrolein and formaldehyde, it is probably responsible for photochemical eye irritation.

photochemical oxidant. An air pollutant formed by the action of sunlight on oxides of nitrogen and hydrocarbons.

photochemical smog. A pollution product of the photochemical process, chemical changes due to the radiant energy of the sun. The sun plays a major part because its energy can be absorbed by nitrogen dioxide in the presence of some hydrocarbons. In the photochemical process, the nitrogen dioxide separates into nitric oxide and atomic oxygen. The atomic oxygen reacts with the oxygen molecules and other constituents of automobile exhaust to form a variety of products, including ozone. Ozone is harmful in itself and is also a participant in a highly complex series of continuing reactions. Hundreds of chemical transformations thus begun take place at different rates and continue as long as there is ozone or nitrogen dioxide and sunlight. New, equally undesirable chemicals result, including peroxyacetyl nitrate (PAN) and formaldehyde, and throughout, nitrogen dioxide is reformed and continues to function as the primary light-energy absorber. When a sensitive plant confronts this chemical mixture, its leaves react: The stomata, the leaf openings through which a plant must draw life-giving carbon dioxide, close up. Even when concentrations of these chemicals do not cause irreversible outward damage, chronic exposure to smog appears to slow down the growth of many plants. Seedlings are especially susceptible. Two different injury syndromes have been identified as smog damage: One, caused by ozone, affects the upper surfaces of plant leaves; the other, probably caused by peroxyacetyl nitrate (PAN), damages the lower part of plant leaves.

photochemistry. The branch of chemistry that deals with the effect of light in causing or modifying chemical changes. Important examples are: natural photosynthesis, the production of a photographic image, reactions of chlorine with hydrocarbons and other organic compounds, polymerization and crosslinking reactions, and various degradation processes.

photoelectric cell. A transducer which converts electromagnetic radiation in the infrared, visible, and ultraviolet regions into electrical quantities such as voltage, current, or resistance.

photoelectric effect. A process by which a photon ejects an electron from an atom. All the energy of the photon is absorbed in ejecting the electron and in imparting kinetic energy to it.

photoelectric particle counter. An instrument such as a scintillation counter which consists of several transparent phosphors together with a photomultiplier tube that detects ionizing particles or radiation by means of the light flash emitted when the radiation is absorbed in the phosphors.

photoelectron. An electron which has been ejected from its parent atom by an interaction between that atom and a high-energy photon. Photoelectrons are produced when electromagnetic radiation of sufficiently short wavelength is incident upon metallic or other solid surfaces (photoelectric effect) or when radiation passes through a gas.

photoemission. The emission of electrons from the surface of various substances, including sodium, potassium, lithium, rubidium, and cesium, when irradiated by light. Each substance has a threshold frequency for incident light, below which no emission takes place. Thin films of alkali metals adsorbed on suitable surfaces exhibit a strong selective photoeffect.

photoionization. The removal of an electron from an atom (or molecule) by a photon, whose entire energy is absorbed by the ejected electron, leaving behind a positively charged atom (or molecule).

photolysis. The splitting of a molecule under the action of light; e.g., in photosynthesis, the cleavage of water by the radiant energy absorbed by chlorophyll.

photometer. An instrument for measuring the intensity of light or the relative intensity of a pair of lights. If the instrument is designed to measure the intensity of light as a function of wavelength, it is called a spectrophotometer. Photometers may be divided into two classes: photoelectric photometers, in which a photoelectric cell is used to compare electrically the intensity of an unknown light and that of a standard light, and visual photometers, in which the human eye is the sensor.

photon. A basic particle of electromagnetic radiation, a quantum of electromagnetic energy. The higher the frequency of a radiation, the more energy its photons possess (their energy is equal to Planck's constant times the frequency of radiation). Photons have no rest mass; they exist only when moving at the speed of light. In motion, their mass can collide with and affect other subatomic particles. Photons are generated in collisions between nuclei or electrons and in any other process in which an electrically charged particle changes its momentum; conversely, photons can be absorbed (i.e., annihilated) by any charged particle.

photoperiodism. The length of daylight period in proportion to the period of darkness, which has an effect on all living organisms. The photoperiod varies continually in the natural environment according to the season of the year. A longer photoperiod will ensure a greater amount of photosynthesis by plants, for there will generally be a greater net receipt of solar energy; it also means an increase in temperature, which often increases the metabolic activities of plants. Many animals are also directly affected by changes in the photoperiod. The breeding cycles of many mammals and birds are governed by the differential rate of hormonal production, which in turn is controlled by the photoperiod. Migrations of birds, mammals, and insects are thought to be triggered by the photoperiod. The general condition of the pelt or plummage has been correlated with changes in the photoperiod.

photoreversal. The partial recovery of body tissues or cells damaged by ultraviolet radiation upon exposure to visible light. Full recovery is never accomplished; the damaged cells or tissues react as though they had received a smaller dose of radiation than that actually administered. This reaction is referred to as the dose-reduction principle.

photosynthesis. The process by which plants manufacture carbohydrates with the aid of sunlight, using carbon dioxide and water as the raw materials, and chlorophyll as the catalyst.

photovoltaic cell. A photocell which sets up a potential difference between its terminals when exposed to radiant energy. It is a self-contained current and voltage generator, a transducer which converts electromagnetic radiation into electric current. The basic principle causing photovoltaic cells to produce electricity are basically as follows. Sunlight contains tiny energy particles called photons. When these photons strike the surface of materials called semiconductors, some of these photons cause electrons in the semiconductors to move around. Since the flow of electrons is, by definition, an electric current, the action of photons from sunlight generates electricity. The basic operation of a solar cell is shown in the diagram.

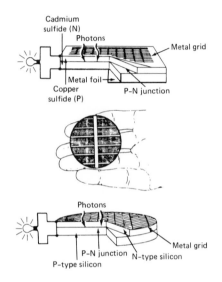

Operation of a solar cell.

phytotoxicant. An agent which produces a toxic effect on vegetation.

piezoelectric effect. Exhibited by certain crystals, the phenomenon of expansion along one axis and contraction along another when subjected to an electric field. Conversely, the compression of certain crystals generates an electrostatic voltage across the crystal. Piezoelectricity is only possible in crystal classes which do not possess a center of symmetry.

pig. A container, usually lead, used to ship or store radioactive materials.

pile. A nuclear reactor.

pilot balloon. A small balloon filled with hydrogen or helium that is released by an observer who, with an instrument called a theodolite, is able to determine wind speed and direction aloft by means of the balloon's movement as it ascends.

pit incinerator. An incinerator that eliminates the grate of a conventional incinerator and uses only overfire air for combustion. An open pit incinerator's construction consists of a refractory-lined box with appropriate cleanout doors at either end and an open top. An air manifold with a number of nozzles set at 30° to the top plane of the incinerator is fed by a forced-draft blower which forces air down across the refractory-lined box.

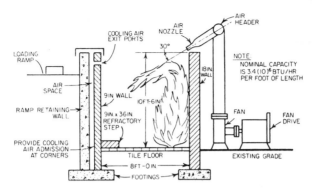

Pit incinerator. (*E. I. du Pont de Nemours & Co.*)

Waste material is dumped into the bottom of the pit to approximately one-third of its total depth, ignited, and allowed to burn with air circulating over its surface. This type of incinerator has been successful on wood and plastic materials.

pitch.   The attribute of auditory sensation in terms of which sounds may be ordered on a low–high scale. Pitch depends primarily upon the frequency of the sound stimulus, but it also depends upon the sound pressure and waveform of the stimulus.

Planck's constant.   A universal constant of nature which relates the energy of a quantum of radiation to its frequency. Planck's constant has the dimensions of action (energy × time) and is expressed by $E = h\nu$, where E is the energy of the quantum and ν is its frequency; its numerical value is $(6.62517 \pm 0.00023) \times 10^{-27}$ erg/s.

Planck's law.   An expression for the variation of monochromatic radiant flux per unit area of source as a function of the wavelength of black body radiation at a given temperature; it is the most fundamental of the radiation laws.

plankton.   Tiny plants and animals that live in water.

plant.   The greatest interest in plants are those which contain organs for seed production, the spermatophytes. These plants contain roots, and root hairs, stems, and leaves. Because of their exposure to air, their large surface area relative to the rest of the aerial parts of a plant, their high metabolic activity, and the fact that gas exchange takes place through the surfaces, the leaves of a plant are usually the most obvious and most important sites of air pollution damage. Leaves constitute the principal site for the synthesis of food in a plant. This is accomplished through the capture of light by the chlorophylls and the subsequent use of this energy to "fix" carbon dioxide. Geochemists have expressed concern that man is now liberating stored CO_2 at a rate faster than the vegetation cover of the earth can "fix" it. The result of such an imbalance would be an increased greenhouse effect (the trapping of heat in the earth's atmosphere through the capture of long wavelength radiation by carbon dioxide) with a consequent increase in surface temperatures. Other environmental parameters affected by the presence of plants include air humidity, geostrophic winds, soil and air temperatures in the vicinity of vegetation, and soil composition, moisture content, and chemical characteristics. Plants may also contribute particulate material in the form of pollens into the air. The fact that many common air pollutants react with vegetation indicates that plants are removing these substances from air and hence act as natural air-cleaning devices. Atmospheric contaminants associated with plant damage are most commonly gases, but examples of both particulate and solubilized materials exist. The nature of the injuries to plants by various air pollutants can be divided into two basic categories: visible effects and suppression of growth.

plasma. An electrically conductive gas comprised of neutral particles, ionized particles, and free electrons, but which, taken as a whole, is electrically neutral.

plastcrete. A material used in concrete that can replace sand; it is made from recycled plastic.

plastic. A nonmetallic compound that results from a chemical reaction, and is molded or formed into rigid or pliable structural material.

platinum (Pt). A dense, white metal which looks like silver and tin, a precious metal costing more than gold. Metals of the platinum group include platinum, palladium, iridium, osmium, rhodium, and ruthenium. Platinum and platinum alloys are employed in many delicate electrical and laboratory instruments, in electronic tubes, voltage regulators, relays, and high-tension magnetos. Palladium is used widely in the contacts of telephone relays. Ruthenium and osmium are employed in many hard alloys for the tips of fountain pens and phonograph needles. Chemical laboratories use platinum utensils and equipment. Platinum–gold and platinum–rhodium alloys are used to make spinnerettes (nozzles with tiny holes) for making rayon fibers from viscose. Platinum in spongy or colloidal form is used as a catalyst in the synthesis of sulfuric acid and the oxidation of ammonia to nitric acid. Platinum and palladium alloys are used extensively in dentistry for anchorage, pins, and dentures. Platinum is never found in nature as a pure metal, but is always associated with gold, copper, nickel, iron, chromite, and the metals of the platinum group.

plenum. A condition in which the air pressure in a closed space or duct is greater than that of the outside atmosphere; the opposite of a vacuum.

plume. The path taken by a continuous discharge of products from a chimney. The shape of the path and the concentration distribution of gas plumes is dependent on turbulence. Looping, coning, fanning, fumigating, and lofting are among the designations of various plume shapes. The geometrical configuration of plumes diffusing in the atmosphere is a function of vertical air stability. (a) *Looping plume.* Looping plumes occur when there is a superadiabatic lapse rate and solar heating. The large thermal eddies in the unstable air may bring the plume to the ground periodically; however, the dilution of the plume with the surrounding air generally occurs rather rapidly. (b) *Coning plume.* Coning plumes result when the vertical air-temperature gradient is between dry adiabatic and isothermal, the air being slightly unstable with some horizontal and vertical mixing occurring. Coning is most likely to occur during cloudy or windy periods. (c) *Fanning plume.* Fanning plumes spread out horizontally, but do not mix vertically. Fanning plumes occur when the air temperature increases with altitude (inversion). The plume rarely reaches the ground level, unless the inversion is broken by surface heating or the plume encounters a hill. (d) *Lofting plume.* Lofting plumes diffuse upwards but not downwards and occur when there is a superadiabatic layer above a surface inversion. A lofting plume will generally not reach the ground surface. (d) *Fumigation.* Fumigation results in the high pollutant concentration plume reaching the ground level along the full length of the plume and is caused by a superadiabatic lapse rate beneath an inversion. The superadiabatic lapse rate at the ground level occurs because of solar heating. This condition is favored by clear skies and light winds. (See diagram p. 251.)

plutonium (Pu). A heavy, radioactive, man-made, metallic element with atomic number 94. Fissionable ^{239}Pu, produced by neutron irradiation of ^{238}U, is used as a reactor fuel and in weapons.

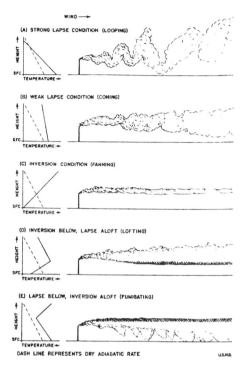

Plume behavior and related weather.

plywood residue conversion. Producing more than 100 million square feet of plywood annually (⅜-in. basis), the Crown Zellerbach plant at Omak, Washington is equipped with large veneer dryers which consume substantial quantities of heat. Originally, the dryers were heated with propane gas burners, but rising fuel costs coupled with a surplus of hog fuel resulted in a decision to convert to wood-burning heating equipment. Scraps from plywood trimming operations plus materials from an onsite furniture factory are collected by blowpipes which convey them to a large storage bin. Fuel is prepared by passing the scraps through one of two hammer mills, followed by screening and storage. Prepared fuel is then burned in two double-vortex, suspension-type burners supplied by Energex Ltd. of San Marcos, California. Combustion gases at 2000°F are blended with recycled air from the veneer dryers before being distributed in stainless-steel ducts. Any wood residues not used for veneer drying are fed to boilers in the steam plant, which produces steam for turboelectric generators.

PNdB. The unit of perceived noise level, equal to the sound level of the reference sound expressed in dB above 2×10^{-5} N/m^2.

podzol soil. A light-colored, relatively infertile soil. Podzol soils develop in relatively cool, humid climates, where percolation of water is sufficient to cause the leaching of chemicals from surface layers (*A* horizon). These soils are quite acid as a result of acid organic detritus in the upper layers. This organic debris may accumulate to a depth of 12 inches or more on the surface. Mineral soil in the *A* horizon is characteristically gray (and acid); the *B* horizon is brown or dark brown in color. Iron and aluminum compounds are leached in large quantities from the *A* horizon and accumulate in the *B* horizon. Podzol soils extend from central Alaska and Canada across the northern lake states and northern New England.

podzolic soil, red. Characteristic of the drier areas of southeastern United States and West Indies, red podzolic soils have a thin layer of unincorporated organic matter above a gray or brownish-gray *A* stratum. A yellowish or pinkish-gray, more-or-less-sand A_2 layer lies above a red or brownish-red clay soil of the *B* horizon, which may be three or more feet in thickness.

podzolic soil, yellow. Yellow podzolic soils develop under more humid conditions than do red podzol soils. The A_1 layer is often dark in color, owing to the inclusion of organic debris, but this grades into a pale-yellow A_2 stratum. The *B* horizon is yellow or light reddish-yellow in color.

point source. A stationary location where pollutants are discharged, usually from an industry.

poise. A unit of viscosity, defined as the tangential force per unit area (dyn/cm^2) required to maintain a unit difference in velocity (1 cm/s) between two parallel planes separated by 1 cm of fluid; 1 poise = 1 dyn s/cm^2 = 1 g/cm s.

polar air. Cold, dry air having its source in the polar regions.

polar aurora. A luminous phenomenon which appears in the high atmosphere in the form of arcs, bands, draperies, or curtains.

polar front. The frontal zone between air masses of polar origin and those of tropical origin.

polarized light. Light which exhibits different properties in different directions at right angles to the line of propagation. Specific rotation is the power of liquids to rotate the plane of polarization. It is stated in terms of specific rotation or the rotation in degrees per decimeter per unit density.

pollen. A fine dust produced by plants; a natural or background air pollutant.

pollutant. Any introduced substance that adversely affects the usefulness of a resource.

pollutants, industrial.

Aeroallergens (pollens)
Aldehydes (includes acrolein and for-
  maldehyde)
Ammonia
Arsenic and its compounds
Asbestos
Barium and its compounds
Beryllium and its compounds
Biological aerosols (microorganisms)
Boron and its compounds
Cadmium and its compounds
Chromium and its compounds (in-
  cludes chromic acid)
Chrlorine gas
Ethylene

Hydrochloric acid
Hydrogen sulfide
Iron and its compounds
Manganese and its compounds
Mercury and its compounds
Nickel and its compounds
Odorous compounds
Organic carcinogens
Pesticides
Phosphorus and its compounds
Radioactive substances
Selenium and its compounds
Vanadium and its compounds
Zinc and its compounds

pollution. A term rather loosely used to indicate substances introduced into the environment that are potentially harmful or that interfere with man's use of his environment.

Water pollution by industrial and domestic wastes, air pollution (smog), and soil pollution (including excessive erosion) are three types of special concern. A fundamental distinction between two basic types of wastes can be made: those that involve an increase in the volume or rate of introduction of materials already present in natural ecosystems, and those involving poisons and chemicals that are not normally present in nature.

pollution load. A measure of the strength of a wastewater in terms of its solids or oxygen-demanding characteristics and/or other objectionable physical and chemical characteristics, or in terms of harm to receiving waters.

pollution, water. The introduction into a body of water or substances of such a character and of such quantity that the water's natural quality is altered so as to impair its usefulness or render it offensive to the senses of sight, taste, or smell.

polybrominated biphenyls (PBB's). A mixture of more than 200 possible chemical entities with the general formula $C_{12}X_{10}$, where X may represent either hydrogen or bromine in the two-ring structure. The chemical is manufactured primarily for use as a fire-retardant material, thus constituting a potential health hazard in the event of a fire, which could cause the dispersal of polybrominated biphenyls into the surrounding environment. PBB's were the source of a pollution problem in Michigan in 1973 when PBB fire-retardant material was accidentally mixed with a food supplement for dairy cows. The cows fed the PBB mixture lost their appetites, grew thin and weak, developed open sores, and either became sterile or produced dead calves. Before the error was discovered, the cows also produced PBB-contaminated milk that in turn contaminated much of the food supply of the upper Midwest. Before the chemical pollution could be brought under control, nearly 30,000 dairy cattle had been killed or destroyed, in addition to 1.5 million chickens, and about 7500 hogs and sheep lost through contamination of other animal feeds. A study by a team of medical scientists from Mt. Sinai Hospital in New York found that virtually every person living in Michigan at that time had significant amounts of PBB's in their body tissues and were in generally poorer health than individuals living in surrounding states. Symptoms of the Michigan residents exposed to PBB's included headaches, fatigue, joint pains, and numbness in the hands and feet. Subsequent studies of persons as far away as New Jersey and Arkansas showed the presence of PBB's in their tissues. The original PBB error was extended by the disposal of an estimated 100 tons of PBB's in a landfill site from which the chemical leached into surrounding groundwater supplies.

polychlorinated biphenyls (PCB's). A series of isomers and compounds used mainly as plasticizers, flame retardants, and insulating materials. More than 30 substances are included in the category and variations in composition, texture, and other qualities exist. Some are oily liquids, others are white crystals, and still others occur as hard resins. All are potentially toxic and carcinogenic. Toxic effects generally involve damage to the skin and liver. A type of skin lesion associated with exposure to PCB's is called chloracne and is characterized by the appearance of dark pigmentation and small pustules on dermal surfaces that have been in contact with one of the substances. This form of contact dermatitis may also be marked by the development of acne-like comedones. The minimum lethal dose for rabbits exposed to PCB's on the skin has been reported as approximately 3200 mg/kg. Liver effects appear to vary in severity directly with the amount of chlorine in the PCB, as well as the dose. Other factors affecting the degree of liver damage include an increased toxicity with the concurrent exposure to carbon tetrachloride and the amount of oxidized material in the substance. The hepatic damage is marked by the appearance of a yellowed atrophy of the liver cells. The patient usually experiences nausea and vomiting, weight loss, jaundice, abdominal pain, and fluid accumulation in the body. The adverse effect may progress to loss of consciousness and death. The oral LD_{50} for rats has been reported to be from 1315 mg/kg to as much as 11,300 mg/kg, depending upon the PCB mixture. The Food and Drug Administration in 1980 allowed a temporary tolerance of 2 ppm of PCB's in food.

polychlorinated biphenyls contamination. The Food and Drug Administration notified health officials in Minnesota, Montana, Idaho, Washington, North Dakota, and Utah on September 17, 1979, that poultry and hog feed contaminated with PCB's had been sent to farms in their states since June. PCB's, or polychlorinated biphenyls, were oily synthetic chemicals used as a coolant in electrical equipment. They have been found to cause reproductive problems in human beings and cancer in laboratory animals. Use of the chemicals had been banned on October 12, 1976. Agriculture Department officials had found samplings of contaminated chickens and eggs in Utah, July 7, 1979. The source of contamination was tracked to the Pierce Packing Company in Billings, Montana, an animal feed processing concern. It was discovered that sometime in June 1979 a transformer in a storage shed had leaked the PCB's into a drain line. The firm recovered oil and fat from the drainage system and reused it in the animal feed manufacturing process. It was determined that six states had received shipments totaling 1.9 million pounds from Pierce Packing Co. with about 90 percent of the feed going to Montana. Three supermarket chains removed all eggs produced in Montana from their stores. The eggs were taken off the shelves at 61 stores operated by major supermarkets in Montana, and five supermarkets in Wyoming. Farmers in Idaho and Utah had destroyed more than 400,000 contaminated chickens and more than one million poisoned eggs by September 17, 1979. Some bakeries and meat processing plants had also been forced to destroy contaminated food. A Pepperidge Farms factory in Richmond, Utah found traces of PCB's in 75,000 frozen cakes, and at a Swift and Co. plant in Clinton, Iowa, meat products made from 28,000 contaminated chickens were destroyed. Officials said the contaminated food contained more than 50 times the allowed amounts of PCB's. A U.S. Environmental Protection Agency official said that some of the contaminated chickens had up to 167 ppm of the chemical.

polyclimax theory. The polyclimax theory recognizes a number of different climax communities in any geographic region: climatic climax, preclimax, postclimax, disclimax, adaphic climax, pyric climax, and topographic climax. Climax communities are all self-reproducing and can be maintained for an indefinite period of time. In this sense, the term climax merely refers to the final or terminal community that will appear in that localized situation.

polyelectrolytes. Used as a coagulant or a coagulant aid in water and wastewater treatment (activated carbon is another coagulant aid), it is a synthetic polymer having a high molecular weight. Anionic: negatively charged. Nonionic: carrying both a negative and a positive charge. Cationic: positively charged.

polyethylene. A plastic, chemically inert substance made by polymerizing ethylene under pressure; molecular weight, 1000–3000; melting point, 115–125°C.

polymer. A compound, usually of high molecular weight, formed by the linking of simpler molecules, monomers; a substance in which the original molecules have been linked together to form giant molecules. Natural rubber is a polymer of isoprene.

polymerization. The union of two or more molecules of a compound to form a more complex molecule.

polyvinyl chloride. A plastic that releases hydrochloric acid when burned.

pond, sewage oxidation. A natural or artificial pond into which partly treated sewage is discharged and where natural purification processes take place under the influence of sunlight and air.

pooling, filter. The formation of pools of sewage on the surface of filters, caused by surface clogging.

pore volume. The difference in the volumetric displacements by granular activated carbon in mercury and helium at standard conditions.

positron. A particle of the same mass as an ordinary electron. It has a positive electrical charge of exactly the same magnitude as that of an ordinary electron (which is sometimes called negatron). Positrons are created either by the radioactive decay of certain unstable nuclei or, together with a negatron, in a collision between an energetic (more than one MeV) photon and an electrically charged particle (or another photon). A positron does not decay spontaneously, but on passing through matter it sooner or later collides with an ordinary electron and in this collision the positron–negatron pair is annihilated. The rest energy of the two particles, which is given by Einstein's relation $E = mc^2$ and amounts to 1.0216 MeV altogether, is converted into electromagnetic radiation in the form of one or more photons.

potable water. Water free from amounts of impurities sufficient to cause disease or harmful effects to health.

potash salt. A commercial term for any of the compounds of potassium used in the fertilizer and munition industries. The naturally occurring compounds are mainly chlorides, sulfates, and carbonates. Large deposits of potash of inorganic origin are products of the evaporation of saline waters, such as seawater or the waters of some highly saline lake or inland seas.

potassium–argon. One of the isotopes of potassium, ^{40}K, is radioactive. It changes at a steady rate to argon, ^{40}Ar, and to calcium, ^{40}Ca, by electron capture, without much change in mass. This calcium cannot be distinguished from calcium of another origin. Since argon is an inert gas, most of it remains within the crystal lattice of the potassium-bearing minerals where it was formed. The minerals biotite, muscovite, orthoclase, and glauconite are possible sources of argon.

potential difference. The work required to carry a unit of positive charge from one point to another.

potential energy. The energy inherent in a mass because of its position with reference to other masses; e.g., a rock at the edge of a precipice has potential energy owing to its height with respect to the ground below.

potential ionization. The potential necessary to separate one electron from an atom resulting in the formation of an ion pair.

potential temperature. The temperature a parcel of dry air would have if brought adiabatically from its initial state to the (arbitrarily selected) standard pressure of 1000 millibars.

power level. In decibels, 10 times the base 10 logarithm of the ratio of a given power to a reference power. The reference power must be indicated.

ppm. Parts per million; the number of parts of a given pollutant in a million parts of air.

prairie soil. A soil that develops in cool, moderately humid climates, with rather weak podzolization. It supports tall grass vegetation and is extensive in the Midwest, the east coast of South America, and appears in small, isolated, thin strips in Eastern Europe, Russia, Africa, and Asia. Although a small amount of leaching occurs in prairie soils, the grasses return sufficient calcium to surface layers. The surface strata are usually dark brown or gray in color.

preaeration. A preparatory treatment for sewage, consisting of aeration to remove gases and oxygen, or to promote the flotation of grease, and to aid coagulation.

prechlorination. (1) The chlorination of water prior to filtration. (2) The chlorination of sewage prior to treatment.

precipitate. A solid that separates from a solution because of some chemical or physical change.

precipitation. (1) The condensation of water vapor in the atmosphere so that it falls to the earth as rain, snow, sleet, or hail. (2) The formation of solid particles in a solution; also, the settling out of small particles in either a liquid or gaseous medium. (3) The process of softening water by the addition of lime and soda ash as the precipitants.

precipitation classification. The classification of forms of precipitation by droplet size, rate of fall, and physical state. (1) *Drizzle.* Drizzle consists of small, liquid water droplets with diameters between 0.1 and 0.5 mm. Drizzle usually falls from low stratus clouds at a rate rarely exceeding 1 mm/hr (0.04 in./hr). (2) *Rain.* Rain consists of liquid water drops larger than 0.5 mm. In the United States, rainfall rates are reported at three intensities; light, for rates up to 2.5 mm/hr (0.10 in./hr); moderate, from 2.8 to 7.6 mm/hr (0.11 to 0.30 in./hr); and heavy, over 7.6 mm/hr (0.30 in./hr). (3) *Glaze.* A generally clear and smooth ice coating, formed by the freezing of water droplets on surfaces. (4) *Rime.* A white, opaque, granular deposit of ice, formed by the rapid freezing of droplets on an exposed surface. (5) *Snow.* Snow is composed of white or translucent ice crystals, usually in a complex aggregate form. The specific gravity of fresh snow varies from 0.05 to 0.20 and is often assumed to be 0.1. (6) *Snow pellet.* Snow pellets consist of white, opaque, rounded or conical ice particles of snowlike structures, having a diameter of 2 to 5 mm (0.1 to 0.2 in.). (7) *Hail.* Balls of ice produced in convective clouds. Hailstones may be spherical, conical, or irregular in shape and range in size from 5 mm to over 125 mm in diameter. (8) *Ice pellet.* Ice pellets are composed of transparent or translucent ice. They resemble hail but are less than 5 mm in diameter. Sleet and hail are both forms of ice pellets.

precipitation, electrical. The ionization of particles suspended in transmitted air for collection at an electrode. Potential differences of 12,000 to 30,000 V dc (can be as high as 100,000 V) are required between spaced electrodes in the electrical precipitation of particles.

precipitation, electrostatic. The separation of dust or droplets suspended in a gas or in air by electrical means. In this process, the material to be precipitated is already present in a phase distinct from that of the medium, and it is only necessary to provide the electrical charge by means of ions from the gas to bring about precipitation.

precipitation, polluting. The precipitation of air-polluting substances from the atmosphere in grams per square meter per time.

precipitation, thermal. A process of separation which operates on the principle that a thermal force (a force greater than that caused by convection) acts on a body suspended in a gas migrate from a zone of high temperature to one of low temperature, because of fluid creep. In general, the thermal force is negligible if the gradient is less than about 750°C/cm. In a "thermal precipitator" (a sampling instrument), the air or gas is drawn slowly through a narrow chamber across which extends a heated wire, particulate matter being deposited upon the adjacent collecting surface.

precipitation, ultrasonic. A process consisting of the separation of particulate matter from air and other gases, following agglomeration induced by an ultrasonic field.

precipitator. Any of a number of devices using mechanical, electrical, or chemical means to collect particulates. Precipitators are used in measurement, analysis, or control. Several aerosol-collecting devices operate on the principle that particles drawn through a strong electrical or thermal gradient will be deposited on a collecting surface. In electrostatic precipitators, particles entering the tube with the high-voltage electrodes are electrically charged and tend to collect on the inside walls of the cylinder which carries the opposite charge.

precursor. (1) A chemical compound which is a gaseous air pollutant and reacts with other substances in the atmosphere to produce different pollutants, such as the case where photochemical reactions producing ozone involve preliminary reactions of NO and NO_2 with the oxygen of the air. (2) A substance which precedes another substance in a metabolic pathway; a substance from which another substance is synthesized.

pressure spectrum level. For a sound at a specified frequency, the effective sound-pressure level for the sound-energy contained within a band 1 cps wide, centered at a specified frequency. Ordinarily this has significance only for sound having a continuous distribution of energy within the frequency range under consideration. The reference pressure should be explicitly stated.

pretreatment. Any waste water treatment process used to partially reduce the pollution load before the waste water is introduced into a main sewer system or delivered to a treatment plant for substantial reduction of the pollution load.

primary air. Air admitted to a furnace during the first part of the firing cycle, i.e., for firing fresh fuel.

primary cosmic ray. A stream of high-energy particles originating from outside the earth's atmosphere. Primary cosmic rays appear to come from all directions in space. Their energy appears to range from 10^9 to more than 10^{17} eV.

primary emission. Pollutants emitted directly into the air from identifiable sources. Primary emissions may be characterized as follows: fine solids (less than 100 μm in diameter), coarse particles (greater than 100 μm in diameter), sulfur compounds, organic compounds, nitrogen compounds, oxygen compounds, halogen compounds, and radio-active compounds.

primary treatment. A process used to substantially remove all floating and settleable solids in waste water and to partially reduce the concentration of suspended solids. This process can include screening, grit removal, sedimentation, sludge digestion, and sludge disposal.

principal quantum number. The first quantum number for an electron, describing the radius of its orbit. The principal quantum number is designated by the letter n. The principal quantum number has the integral values 1, 2, 3, 4, etc., which correspond to the similarly numbered energy levels of the Bohr scale.

probe. A tube used for sampling or measuring pressures at a distance from the actual collection or measuring apparatus. It is commonly used for reaching inside stacks and ducts.

process, activated sludge. A biological sewage treatment process in which a mixture of sewage and activated sludge is agitated and aerated. The activated sludge is subsequently separated from the treated sewage (mixed liquor) by sedimentation and wasted or returned to the process as needed. The treated sewage overflows the weir of the settling tank in which the separation from the sludge takes place.

process, biological. The process by which the vital activities of bacteria and other microorganisms in search for food break down complex organic materials into simple, more stable substances. Self-purification of sewage polluted streams, sludge digestion, and all the so-called secondary sewage treatments result from this process. Also called biochemical process.

process, oxidation. Any method of sewage treatment for the oxidation of putrescible organic matter; the usual methods are biological filtration and the activated sludge process.

process weight. The total weight of all materials, including fuel, used in a manufacturing process. It is used to calculate the allowable rate of emission of pollutant matter from the process.

producer gas. The effluent gas consisting chiefly of carbon monoxide and nitrogen resulting from an appropriate amount of air blown through red-hot coke at about 1000°C. In all large coke-burning producers a proportion of steam is introduced with the air blown through the red-hot coke. The products then contain a little water gas. The thermal efficiency of the process is raised, the calorific value of the gases is increased by the presence of hydrogen, and troubles due to the formation of clinkers are reduced because of the heat absorbed by the water-gas reaction. Though its calorific value is relatively low (about 150 Btu/ft³), producer gas has the advantage over water gas in that it is made in a continuous, rather than intermittent, process. It can be made from all solid fuels and from waste such as cotton seed. Industrial producer gas is often made from coal, in which case it is enriched by coal distillates. A major use of producer gas is for heating retorts, chambers, or coke ovens in gas works.

promethium (Pm). A metallic element with an atomic weight of 147 and a high source of radiotoxicity. The isotope decays to samarium, producing beta emissions. It is used in the manufacture of gunsights that permit guards, soldiers, and other marksmen to see their targets in darkness. Persons using weapons treated with promethium may become exposed to the radioactive material. Radioactive promethium has been found on the hands and shoes of guards using the gunsights in law enforcement training exercises at Argonne National Laboratory near Chicago. Traces of promethium have also been found in patrol cars and security headquarters of guards who have undergone training with the gunsights.

propane. $CH_3CH_2CH_3$, a colorless gas with a molecular weight of 44.06; melting point, $-187.7°C$; boiling point, $-42.1°C$. It is the third member of the gas family having three carbon and eight hydrogen atoms; it is found in most natural gases and is the first product found naturally in crude petroleum. Generally it is mixed with butane in varying proportions to make LP gas, bottled gas, etc.

proportional counter. An instrument used in radiation monitoring, employing a gas-filled chamber and two electrodes, with the gas being slightly pressurized. The chamber may be incorporated in the instrument, or be in the form of a probe connected to the electronic circuit by a cable. Ionization of the gas by radiation produces a pulse amplified by secondary ionization, which is proportional to the original ionization. This pulse is fed into a count-rate meter and indicated in counts per minute on the meter. Particles may be shielded out by means of screens.

β-proprionolactone. $C_3H_4O_2$, a colorless liquid with a pungent odor, used in the production of acrylic acids, plastics, and esters. It is also employed in sterilizing blood plasma, vaccines, tissue grafts, and surgical instruments, and is used as a vapor-phase disinfectant. It is a strong irritant and is considered the most toxic of all lactones. The oral LD_{50} for rats is 50 mg/kg; the lowest recorded lethal dose for β-proprionolactone administered to mice via the intravenous route is 3 mg/kg. It is readily absorbed through the skin and produces tumors in rats, mice, guinea pigs, and hamsters. NIOSH has determined that it also may be toxic via the inhalation route and has classified β-proprionolactone as a known carcinogen with exposure limits of not more than one percent by weight or volume to solid or liquid mixtures of the substance.

protocooperation. A type of heterogeneous association between two organisms where each benefits from the other, but neither individual must remain associated with the other in order to survive, i.e., several marine crabs use living coelenterates, sponges, and other invertebrates as a form of camouflage to cover the dorsal surface of the exoskeleton. The crab avoids detection, while the invertebrate benefits by being transported to areas containing greater quantities of food and oxygen.

proton. An elementary particle having a positive charge equivalent to the negative charge of the electron, but possessing a mass approximately 1837 times as great. The proton is in effect the positive nucleus of the hydrogen atom. Protons are constituents of all atomic nuclei and occur in numbers characteristic of each element.

proton–proton reaction. A thermonuclear reaction in which two protons collide at very high velocities and combine to form a deuteron. The resultant deuteron may capture another proton to form tritium and the latter may undergo proton capture to form helium.

PSI. An abbreviation for Pollutant Standard Index adopted by the EPA in 1978 to implement a 1977 Clean Air Act amendment requiring a uniform air quality index system. The PSI system is based primarily on the National Ambient Air Quality Standards. A PSI value of more than 100 indicates the air quality is "unhealthful." A PSI value of more than 300 is interpreted as an air pollution situation that is hazardous to the general population. A PSI of 400 would constitute an emergency in which all persons should remain indoors, keeping doors and windows closed, and avoid physical exertion. Premature death would occur among ill and elderly persons and normally healthy individuals would experience adverse health effects. At a PSI of 400, pollutant levels would be on the order of: SO_2, 2000 μg/m³ for 24 hr; CO, 46 mg/m³ for 8 hr; O_3, 1000 μg/m³ for 1 hr; and NO_2, 3000 μg/m³ for 1 hr.

psia (pounds per square inch absolute). A unit for measuring pressure, whether hydraulic (liquids) or pneumatic (gases), without including the effect of atmospheric pressure, which is 15 psi at sea level.

psychrometer. An instrument consisting of two thermometers, a wet-bulb thermometer (with a muslin wick over its bulb) and a dry-bulb thermometer. Psychrometers are used in the calculation of dew points and relative humidity.

pulmonary. Pertaining to or affecting the lungs or any components of the lungs.

pulmonary emphysema. An anatomical change in the lungs that shows itself in the shortness of breath. It is characterized by a breakdown of the walls of the alveoli, the tiny air spaces beyond the terminal bronchioles. As the disease progresses, the alveoli become enlarged, lose their resilience, and their walls disintegrate, with irreversible tissue changes. No single factor can be said to be the original cause of emphysema. Asthma and chronic bronchitis are often found in individuals with emphysema; both can increase the severity of the disease. The symptoms of all these diseases are aggravated by air pollution.

pulmonary function. In medicine, a term used in describing the adequacy of the lung's performance.

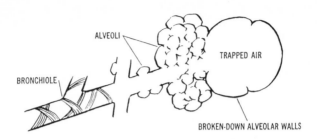

ALVEOLI

TRAPPED AIR

BRONCHIOLE

BROKEN-DOWN ALVEOLAR WALLS

Pulmonary emphysema. The loss of elasticity and the deterioration of the alveoli walls deters the exhalation of carbon dioxide.

pulp. A material used in papermaking processes. Pulp can be obtained from wood fibers, or from waste paper that has been cleaned and treated in a hydrapulper.

pulp and paper mill wood residue conversion. The Abitibi Paper Company mill at Pine Falls, Manitoba produces 178,000 tons of newsprint annually, using large quantities of electricity and steam in a sulfite pulping process. For more than 50 years, bark and wood refuse have been recovered and burned to yield useful energy. Recent improvements have resulted in the processing of all wastes to yield 10 percent of total energy requirements for process steam. Energy requirements for steam are met by burning Saskatchewan lignite plus 100 percent of the woodwaste products derived from sawing and debarking operations. Each cord of pulp wood yields roughly 200 lb of soaking wet bark and 40 lb of sawdust, which are processed to yield shredded wood residues (hog fuel) suitable for combustion in a Dutch oven. Water removed from the bark is recycled through the "drum barker."

pulp mill dry barking. The Boise Cascade pulp and paper mill in Rumford, Maine produces pulp and high-quality specialty papers using both hardwoods and softwoods in order to avoid serious pollution problems associated with bark removal. Special equipment was installed to remove bark from fresh logs without employing water. This dry barking procedure not only eliminated a pollution hazard, but enhanced the recovery of energy from bark by reducing moisture content and eliminating the need for drying equipment.

pulp mill residue conversion. The Smooth Rock Falls Division of the Abitibi Paper Company operates a small stud mill plus a bleached draft pulp mill with a rated capacity of 120,000 tons per year. Approximately 90 percent of the energy for a generating process steam is recovered from mill residues. Additional studies have demonstrated the feasibility of generating all the steam from mill residues or even generating electric power, a portion of which would be surplus and available for sale. The Kraft pulping process involves the chemical treatment of wood chips in "white liquor" to remove interfiberous lignin, thus facilitating fiber separation with minimal mechanical treatment. Spent pulping liquor (black liquor) contains substantial quantities of valuable inorganic chemicals, which are separated in a recovery process involving the burning of accompanying organic matter in a recovery furnace. Energy thus liberated is used to generate process steam.

pulverization. The crushing or grinding of materials into small pieces.

pumping station. A machine installed on sewers to pull the sewage uphill. In most sewer systems waste water flows by gravity to the treatment plant.

pure air. Air free of nongaseous suspensoids (dust, hydrometeors) and of such gaseous contaminants as industrial effluents. Since the composition of the atmosphere is slightly variable with respect to certain components, the term "pure air" has no precise meaning, but it is commonly used as defined above. By far the most important gas in air from the meteorological standpoint is water vapor. "Dry air" denotes air from which all water vapor has been removed. Pure air has a density of 1.2923 g/cm³ at a pressure of 1013.25 mbar and a temperature of 0°C, a specific heat at constant volume of 0.1707 cal/g °C, a specific heat at constant pressure of 0.2396 cal/g °C, and its gas constant (per gram) is 2,870,400 ergs/g °C. The percent by volume of those gases found in relatively constant amounts in dry air is very nearly as follows: N_2, 78.084; O_2, 20.946; Ar, 0.934; CO_2, 0.033; Ne, 0.0018; He, 0.000 524; CH_4, 0.0002; Kr, 0.000 114; H_2, 0.000 05; N_2O, 0.000 05; Xe, 0.000 008 7.

purification, degree of. (1) A measure of the completeness of destruction or removal of objectionable impurities such as bacteria and hardness from water, by natural means (self-purification) or by treatment. (2) A measure of the removal, oxidation, or destruction of solids, organic matter, bacteria, or other specified substances effected by sewage treatment processes.

Purkinje effect. A phenomenon associated with the human eye, making it more sensitive to blue light when the illumination is poor (less than about 0.1 lm/ft²) and to yellow light when the illumination is good.

putrefaction. The biological decomposition of organic matter accompanied by the production of foul-smelling products associated with anaerobic conditions.

putrescibility. (1) The relative tendency of organic matter to undergo decomposition in the absence of oxygen. (2) The susceptibility of waste waters, sewage, effluent, or sludge to putrefaction. (3) A term used in water sewage analysis to define the stability of polluted water or raw or partially treated sewage.

putrescible. A substance that can rot quickly enough to cause odors and attract flies.

pyramid of numbers. The pyramid of numbers and biomass indicate that the producer element is ingested in large numbers by a smaller number of primary consumers; these primary consumers, in turn, are eaten by a still larger number of secondary consumers, and so on, yielding a pyramid of numbers. In some cases certain levels of the pyramid may be inverted because at times a large number of smaller organisms may feed on a relatively large producer, herbivore or carnivore, depending upon their position in the food series; i.e., a large number of aphids (herbivores) may feed on a single plant (producer). The pyramid of energy is always conventional in shape because the transfer of energy from one trophic level to the next is rather inefficient, as some energy will always be lost; i.e., of the total solar energy available, only about 4/100 of 1 percent is utilized by plants in the production of food.

pyranometer. An actinometer which measures the combined intensity of incoming direct solar radiation and diffuse sky radiation. The pyranometer consists of a recorder and a radiation-sensing element which is mounted so that it views the entire sky.

pyrheliometer. A meteorological tool used to indirectly measure the intensity of solar radiation. The pyrheliometer's operation is based upon the temperature effects of radiation falling on an absorbing element. The sun's radiation is soaked up by an absorbent material inside the instrument, and a rise in temperature results. The temperature reading is then converted into calories.

pyrite. A pale, brass-yellow, hard (6 to 6.5 on the Mohs' scale), brittle iron sulfide, FeS_2, called fool's gold because of its slight resemblance to gold. Pyrite's streak is greenish black, and scattered cubes and other crystal forms occur in many different rocks.

pyrolysis. (1) In destructive distillation, a process in which organic material is decomposed at elevated temperatures in either oxygen-free or low-oxygen atmospheres. Unlike incineration, which is inherently a highly exothermic combustion reaction with air, pyrolysis requires the application of heat, either indirectly or by partial oxidation or other reactions, in the pyrolysis reactor. Unlike incineration, which produces mainly carbon dioxide and water, the products of pyrolysis are normally a complex mixture of primarily combustible gases, liquids, and solid residues. (2) A recycling process that breaks down burnable waste by combustion in the absence of air. High heat is usually applied to the wastes in a closed chamber, evaporating all moisture and breaking down materials into various gases and solid residues. The gases are collected, used, and sold. The residue can be further processed into useful materials, or it can be used in a landfill.

pyrolytic material. A material deposited on a heated surface by the thermal decomposition of a gas.

pyrometer. An instrument used for measuring temperatures; generally temperatures above 600°C.

Q

quantization. The restriction of certain physical quantities such as energy or angular momentum to only certain discrete values, under certain circumstances. Thus a light wave of frequency hv, or a harmonic oscillator of that frequency, can gain or lose energy only in multiples of hv, where h is called the quantum of action, or Planck's constant, $h = 6.62 \times 10^{-32}$ J s. Similarly, the angular momentum of a freely rotating object is quantized in steps of $h/2\pi$.

quantum. The smallest indivisible quantity of radiant energy; a photon; a unit of radiant energy absorbed or emitted by an atom. Energy is radiated only in quanta and the quantum is the smallest amount of energy which can be transmitted at any given wavelength. Its value is hv, where v is the frequency of the radiation and h is Planck's constant. A light quantum is a photon.

quantum mechanics. The branch of physics that deals with the behavior of particles whose specific properties are given by quantum numbers.

quantum number. A number that is used to describe a specific property of a subatomic particle in mathematical terms. Quantum numbers are used in special equations to compute the energy of a particle, such as the electron.

quantum theory. The concept that energy is radiated intermittently in units of definite magnitude called quanta, and absorbed in a like manner.

quartz. Silicon dioxide, SiO_2, colorless or white when pure, although commonly tinted. Quartz grains are abundant in granite and most sands. Crystal quartz has a piezoelectric property; that is, it has an electric polarity resulting from the pressure used to form the crystal.

quench tank. A water-filled tank used to cool incinerator residues or hot materials during industrial processes.

R

rad. The unit of absorbed radiation equivalent to an energy deposition of 100 erg s/g, i.e., a measure of the energy which ionizing radiation imparts to matter per unit mass of irradiated material.

radiance. In radiometry, a measure of the intrinsic radiant intensity emitted by a radiator in a given direction. It is an irradiance (radiant flux density) produced by radiation from a source upon a unit surface, oriented normal to the line between source and receiver, divided by the solid angle subtended by the source and receiving surface. It is assumed that the medium between the radiator and receiver is perfectly transparent; therefore, radiance is independent of attenuation between source and receiver. If the radiant source is a perfectly diffuse radiator (that is, emits exactly according to Lambert's law), then its radiance is equal to its emittance per unit solid angle. The radiance of a light source is termed luminance.

radiant energy. The transmission of energy through space or a material medium in the form of electromagnetic waves. Radiant energy or radiated heat cannot be easily distinguished from light. It is known that radiant energy acts as if it were transmitted in the form of waves similar to but much shorter than radio waves. In meteorology, radiation is usually of such a form and the problems are of such a nature that treatment from the wave point of view is preferable. Some of the radiation is in what is called the visible range, i.e., it can be detected by the human eye. Radiation with a wavelength too short to be observed by the human eye may be classed as ultraviolet radiation and the long-wavelength type, as infrared radiation. Wavelengths of radiation are usually given in microns (μm) or in angstrom units. A micron is 10^{-4} cm (0.0001 cm) and an angstrom (Å) is 10^{-8} cm (0.000 000 01 cm). The wavelengths of the visible range are those lying between 0.4 and 0.7 μm, or 4000 and 7000 Å. The sun radiates its maximum energy at wavelengths within the visible range.

radiation. The emission and propagation of energy through space or through a material medium in the form of waves. The term may be extended to include streams of subatomic particles such as alpha rays, or beta rays, and cosmic rays, as well as electromagnetic radiation. The term "radiation" is often used to designate the energy alone, without reference to its character. In the case of light this energy is transmitted in bundles (photons).

radiation, annihilation. The photons produced when an electron and a positron unite and cease to exist. The annihilation of a positron–electron pair results in the production of two photons, each of 0.51-MeV energy.

radiation, background. The radiation arising from radioactive material other than the one directly under consideration. Background radiation due to cosmic rays and natural radioactivity is always present. There may also be background radiation due to the presence of radioactive substances in other parts of the building, in the building material itself, etc.

radiation, characteristic. The radiation originating from an atom following removal of an electron or excitation of the nucleus. The wavelength of the emitted radiation is specific, depending only on the element concerned and the particular energy levels involved.

radiation counter. An instrument used for detecting or measuring moving subatomic particles by a counting process.

radiation counter, Geiger–Mueller. A radiation counter consisting of a tube filled with gas in which the ionization takes place. A single ionization in the gas causes an "avalanche" of secondary ionizations and this produces a pulse of the same size, regardless of the energy of the ionizing particle. Beta particles may be shielded out by means of a shield on the tube cover.

radiation counter, ionization chamber. A gas-filled chamber where one of the walls acts as a charged electrode and the other acts as a central wire. Ions produced in the gas by radiation are collected at the electrodes generating a current flow. The ionization chamber is designed in such a manner as to measure the magnitude of this current flow, which is proportional to the radiation intensity. The current flow is then read out on an appropriate meter scale in terms of the dose or dose rate. Discrimination between alpha and beta particles and gamma rays of different energies is accomplished by means of a shielding built into the chamber walls.

radiation counter, neutron. A proportional counter employing a gas-filled tube as a detector. Since neutrons are electrically neutral and cannot produce ionization of the gas directly, the inner walls of the tubes are lined with a material which will absorb the neutron and emit a charged particle. A hydrogenous substance (polyethene, etc.) is used for fast neutron adsorption, emitting a proton to cause ionization. Boron is used to absorb slow or thermal neutrons and emit an alpha particle to cause ionization in the gas.

radiation counter, proportional. A radiation counter employing a gas-filled chamber and two electrodes, with the gas slightly pressurized. The chamber may be incorporated in the instrument or be in the form of a probe connected to the electronic circuit by a cable. Ionization of the gas by radiation produces a pulse amplified by secondary ionization, which is proportional to the original ionization. This pulse is fed into a count rate meter and indicated in counts per minute. Particles may be shielded out by means of screens.

radiation counter, scintillator. A radiation counter employing the principle that ionization caused by radiation produces small flashes of light in certain materials. These flashes are "sensed" by a photomultiplier tube and transformed into current, which is then relayed to a meter for readout. Different materials are or can be made to be more sensitive to a particular type of radiation, and this fact is employed to discriminate between the different kinds of radiation.

radiation dosage. The amount of harmful radiation received in a given time measured in roentgens. The International Commission on Radiological Protection (July 1953) established the absorbed dose of any ionizing radiation as the amount of energy imparted to matter by ionizing particles per unit mass of irradiated material at the place of interest.

This is expressed in rads. The term "rem" is, however, in common use.

rem = rad × RBE

rad = 100 ergs of energy being absorbed in 1 g of irradiated material and for soft tissue is roughly equal to the roentgen unit.

RBE = relative biological effectiveness.

RBE of x radiation is unity; of slow x particles, 10–20; and of slow neutrons, 2–10.

Median lethal dose (for man), the LD_{50}, is about 500 rems; the maximum permissible dose is five rems per year.

radiation, infrared. Invisible radiation of a thermal nature whose wavelength is longer than those of the red segment of the visible spectrum.

radiation, ionizing. Any electromagnetic or particulate radiation capable of producing ions, directly or indirectly, in its passage through matter.

radiation laws. The four physical laws which together fundamentally describe the behavior of black body radiation: (a) The Kirchhoff law is essentially a thermodynamic relationship between emission and absorption of any given wavelength at a given temperature. (b) The Planck law describes the variation of intensity as a function of wavelength of black body radiation at a given temperature. (c) The Stefan–Boltzmann law relates the time rate of radiant energy emission from a black body to its absolute temperature. (d) The Wien law relates the wavelength of maximum intensity emitted by a black body to its absolute temperature.

radiation, monochromatic. Electromagnetic radiation of a single wavelength, or one in which all the photons have the same energy.

radiation, monoenergetic. Radiation of a given type (alpha, beta, neutron, gamma, etc.) in which all particles or photons originate with and keep the same energy.

radiation particle. A particle of matter, such as an atomic nucleus, that has been radiated at high velocities from a given source.

radiation pressure. The pressure exerted upon a surface exposed to radiation. The pressure of the sun's radiation upon the earth's surface is about 2 lb/mile^2. For bodies below a certain critical diameter the radiation pressure can be greater than the gravitational force.

radiation, secondary. Radiation originating as the result of absorption of other radiation in matter; it may be either electromagnetic or particulate in nature.

radiation shielding. The principal method of radiation shielding is the use of multiple cover plates. Between a hot surface and cool surface there is an effective heat transfer resistance which depends on the emissivities and temperatures of the two surfaces. Multiple covers essentially stack these resistances in series, thus reducing radiation loss from the absorber to the sky.

radiation sickness. A self-limited syndrome characterized by nausea, vomiting, diarrhea, and psychic depression, following exposure to appreciable doses of ionizing radiation, particularly to the abdominal region. Its mechanism is unknown and there is no satisfactory remedy. It usually appears a few hours after a treatment and may subside within a day.

radiation standards. Regulations that govern exposure to permissible concentrations of radioactive materials and their transportation.

radiator. (1) Any source of radiant energy, especially electromagnetic radiation. (2) A device that dissipates the heat from some source, as from water or oil, not necessarily by radiation only.

radical. A group of atoms of different elements acting as a single unit (normally incapable of separate existence) in a chemical reaction. A radical may be negatively charged, positively charged, or without charge.

radioactive. Emitting radiation. The chemical activity of an atom is generated by the electrons orbiting around its nucleus. The nucleus is the heart of the atom, and it generally remains stable. Only when it is unbalanced does the nucleus break down naturally. When it does, radiation is emitted and the atom is said to be radioactive. Radioactivity is the unstable atom's attempt to regain stability.

radioactive contamination. The first part of the reactor-fuel cycle, the production of nuclear fuel, involves the mining, milling, and refining of uranium or thorium ore, the chemical separation of uranium and thorium from their daughter products and later from each other, a possible isotopic separation, and the manufacture of fuel elements. The air pollution problems of the uranium industry up to the point of reactor operation are minimal, and would be of concern primarily from the standpoint of occupational exposure. Once the fuels obtained have been introduced into nuclear reactors, the second part of the reactor-fuel cycle takes place. Radioactive wastes that are formed during the operation of reactors include fission products, which normally remain incorporated in the fuel, and extraneous activation products that are found mainly in the coolants. Both the fuel elements and coolants are thus potential sources of radioactive atmospheric pollution. Such pollution can be caused by the release of the radioactive gases produced by fission, such as iodine, xenon, and krypton, into the atmosphere; by active aerosols that contain primary or secondary fuels (uranium, thorium, and plutonium); or by the release of fission products, such as ^{90}Sr, ^{44}Ce, ^{140}Ba, and ^{95}Zr. During the normal operation of all reactors a small amount of gaseous radioactive pollutants is released into the atmosphere. Of all the operations that are involved in the reactor-fuel cycle, chemical reprocessing plants, to which the irradiated, spent reactor fuel is taken, present the most important potential source of air pollution. The fuel must be dissolved and the solutions processed in order to separate the unfissioned uranium and plutonium from the radioactive waste products. Because of the large amount of radioactivity, the processes of chemical separation and dissolution present substantial potential hazards which have been overcome by the development of a highly specialized technology enabling the plants to operate safely. Nevertheless, during processing, fission-product gases are released into the atmosphere, including ^{85}Kr, ^{133}Xe, and ^{131}I. Although most of the iodine is removed by various processes, some of it does escape.

radioactive effects. ^{90}Sr is abundantly formed in the fission process, it has a long life (a half-life of about 28 years), and is chemically similar to calcium, therefore readily absorbed by living things. Thus it may easily pass through the food chain to man by depositing on soil, where it is absorbed by plant roots, and by depositing directly on vegetation. When man eats the vegetation or drinks the milk from animals that have eaten it, ^{90}Sr is deposited in the skeleton. The potential hazards of ^{90}Sr and ^{89}Sr are primarily to the bone and bone marrow. Experiments have shown that, when deposited in sufficient amounts in the bones of animals, these isotopes will produce leukemia, bone cancer, and other skeletal effects. The food-chain process operates in the same way for other nuclides, with the exception of those short-lived fission products that produce only external radiation. ^{137}Cs, which irradiates the body externally and internally, is formed in slightly greater amounts than ^{90}Sr and has a half-life of about 27 years. It differs, however, from ^{90}Sr in that it does not become fixed in the body. Its principal significance to man is that it is a gamma emitter and as such it is a major contributor of long-lived gamma activity. The potential hazard of ^{137}Cs is primarily genetic, resulting from irradiation of gonadal tissue. ^{131}I is a very short-lived nuclide, with a half-life of eight days. It irradiates the body as an internal and external emitter. Within two to four days after it is deposited on pastures and consumed by dairy cattle, levels of ^{131}I in milk reach a maximum. When this milk is consumed, the iodine finds its way into the thyroidal glands. The total radiation dose to the gland is proportional to the concentration of ^{131}I per gram of thyroid.

radioactive effects on man. Radionuclides that are present in radioactive debris from fallout having the greatest significance in terms of their effects on man are: (1) ^{90}Sr and ^{89}Sr, which are beta emitters and principally irradiate the skeleton; (2) ^{137}Cs, which is a beta−gamma emitter and concentrates in soft tissues, resulting in internal, whole-body irradiation; (3) ^{14}C, a beta-gamma emitter which accumulates in the body and delivers whole-body radiation; (4) ^{131}I, a beta−gamma emitter, which concentrates in the thyroid gland; and (5) a number of short-lived fission products that produce external, whole-body irradiation when deposited on the ground.

radioactive element characteristics. All radioactive elements have certain common characteristics. (a) They affect the light-sensitive emulsion on a photographic film. (b) They produce an electric charge in the surrounding air. The radiations from radioactive elements knock electrons from the gas molecules in the air surrounding the radioactive material, leaving the gas molecules with a positive charge. (c) They produce fluorescence with certain other compounds. Radiations from radioactive elements produce a series of bright flashes when they strike certain compounds. The combined effect of these flashes is a glow of fluorescence given off by the affected material. (d) Their radiations have special physiological effects, the ability to destroy the germinating power of plant seeds and kill bacteria. (e) The atoms of all radioactive elements continually decay into simpler atoms and simultaneously emit radiations.

radioactive gas. In atmospheric electricity, any one of three radioactive inert gases— radon, thoron, and actinon—which contribute to atmospheric ionization by virtue of the ionizing effects on the alpha particles which each gas emits upon atomic disintegration. The three gases are isotopic to each other, all having atomic number 86.

radioactive materials (maximum permissible concentrations in air).

Element	Radiation emitted	Maximum permissible concentration (Ci/ml)[a]
Uranium, natural	α	3×10^{-11}
^{239}Pu	α	6×10^{-13}
^{210}Po	α	7×10^{-11}
^{226}Ra	α	10^{-11}
^{222}Rn	α	10^{-8}
Tritium (as H_2O)	β	5×10^{-6}
^{90}Sr	β	10^{-10}
^{14}C	β	10^{-6}
^{35}S	β	9×10^{-8}
^{131}I	β, γ	3×10^{-9}
^{135}Xe	β, γ	10^{-6}

[a]The values given above are for a 24-hour per day exposure.

radioactive nuclides. An atom that disintegrates by emission of corpuscular or electro-magnetic radiation. The rays most commonly emitted are alpha, beta, or gamma rays. The three classes are: (a) primary atoms, which have lifetimes exceeding 10^8 years (these may be alpha- or beta-emitters); (b) secondary atoms, which are formed in radioactive transformations starting with ^{238}U, ^{235}U, or ^{232}Th; (c) induced atoms, having geologically short lifetimes and formed by induced nuclear reactions occurring in nature. All these reactions result in transmutations.

radioactivity. The process whereby certain nuclides undergo spontaneous disintegrations in which energy is liberated, generally resulting in the formation of new nuclides. The process is accompanied by the emission of one or more types of radiation, such as alpha particles and gamma photons. Artificial radioactivity is the property induced in certain elements under controlled conditions. Induced radiation is that radioactivity produced in a substance after bombardment with neutrons or other particles. Natural radioactivity is the property of radioactivity exhibited by more than 50 naturally occurring isotopes (natural radionuclides).

radiobiology. The study of the principles, mechanisms, and effects of radiation on living things.

radiocarbon (^{14}C). A carbon isotope formed in the atmosphere by the collision of cosmic rays (neutrons) with atoms of nitrogen, ^{14}N. Radiocarbon is then taken up by plants and passed on to animals. By continuous exchange during the life of organisms, it remains in equilibrium in a fixed ratio with ordinary carbon, ^{12}C. When the organisms die, exchange ceases, and the organic radiocarbon begins to operate as a clock as it reverts to nitrogen. Its half-life is only about 5570 years, so its usable range in measuring time is limited to about the last 40,000 years.

radioecology. The study of the effects of radiation on plants and animals in natural communities.

radioisotope. A radioactive isotope of an element. Some radioisotopes are the products of natural radioactive decay, but a far greater number have been artificially prepared. Reactors produce radioisotopes during the fission process. In addition, radioisotopes can be prepared by inserting elements into a reactor while it is operating. The many uses of radioisotopes can be divided into three categories: effects of radioisotope radiation on materials; effects of materials on radioisotope radiation; and tracing materials with radioisotope radiation. In the first category, a radioisotope is used as a source of radiation, just as x rays and radium are used. In the second category of radioisotope utilization, the effect of the material on radiation yields information about the material. In the third category, a carefully selected radioisotope acts as a tracer that makes it possible to follow the course of some complicated chemical or biological process.

radiology. The branch of medicine which deals with the diagnostic and therapeutic applications of radiant energy, including x rays and radioisotopes.

radiosonde. A meteorological tool used to gather data about the upper atmosphere. Its major elements, carried to high levels by a gas-filled balloon, are numerous electronic recording instruments and a transmitter. During the balloon's ascent, the weather-sensing devices gather the desired data on pressure, humidity, and temperature and convert it into electrical impulses. These are sent out by the radio transmitter in the form of precise tone signals, which are picked up and recorded by electronic equipment at a ground station. When the radiosonde balloon reaches the limit of its ascent, it bursts. A parachute attached to the package of weather instruments lowers it to earth.

radon daughters. The radioactive decay product "daughter" concentrations of radon gas. In 1976, NIOSH reported that measurement of radon daughters in a number of

National Park Service caves frequented by the public approached the limits established by OSHA for the occupational safety of uranium miners. The OSHA standards for uranium miners are based on a working level (WL) of alpha-particle radiation. At alpha radiation 0.1 WL or above, all underground smoking should be prohibited. From 0.1 to 0.2 WL, the workplace should be monitored at least once a year. Between 0.2 and 0.3 WL, the workspace should be monitored quarterly. Above 0.3 WL, the workspace must be monitored weekly and exposure records of employees maintained. Above 1.0 WL, OSHA recommends that immediate corrective action be taken to lower the concentrations below the 1.0 WL. EPA has calculated that a cumulative individual exposure of 4.0 WL will increase the risk of lung cancer by 100 percent after 10 to 20 years of work in the areas of radon daughters. National Park Service caves in which air samples showed 0.3 WL or higher included Carlsbad Caverns National Park (New Mexico), Lehman Caves National Monument (Nevada), Mammoth Cave National Park (Kentucky), Oregon Caves National Monument (Oregon), and Round Spring Cave in Ozark National Scenic Waterways (Missouri). In addition, NIOSH reported that buildings above the ground at Mammoth Cave, cooled with air from the cave, had an atmosphere of 0.6 WL alpha radiation. Studies of uranium miners have shown that alpha radiation emitted by radon daughters causes an increase in lung cancers which become evident 10 years or more after the initial exposure.

rain. The precipitation of liquid water particles, either in the form of drops more than 0.5 mm in diameter or in the form of smaller, widely scattered drops. Freezing rain is recorded when the drops freeze on impact with the ground, with objects on the earth's surface, or with aircraft in flight.

rainbow. A group of concentric areas, with colors ranging from violet to red, produced on a "screen" of water drops (raindrops, droplets of drizzle, or fog) in the atmosphere by light from the sun or moon.

random noise. A sound or electrical wave whose instantaneous amplitude occurs as a function of time according to a normal (Gaussian) distribution curve. Random noise is an oscillation whose instantaneous magnitude is not specified for any given instant of time. The instantaneous magnitudes of random noise are specified only by probability functions giving the fraction of the total time that the magnitude, or some sequence of the magnitude, lies within a specific range.

Rankine cycle. The complete expansion cycle of a gas brought into a structure like a turbine engine, where the gas is compressed and then rapidly expands to give off energy.

Rankine cycle–inverse Brayton cycle system. Refrigeration in such a system is obtained by air compression, heat removal from the compressed air, expansion of the air through a turbine to extract work, and discharge of the cool air into the space. The power to operate this system has to be obtained from a heat engine such as a Rankine cycle heat engine.

Rankine cycle–vapor compression system. A system consisting of a Rankine cycle heat engine which produces a mechanical output to drive a conventional air conditioning compressor. In the vapor compression portion of the system, refrigerant vapor is compressed and subsequently cooled until it condenses. This liquid, high-pressure refrigerant can then be expanded to produce the cooling effect before it is recompressed and the cycle continues. The heat engine portion of the system is connected only mechanically to the refrigeration system. It uses heat energy to vaporize a working fluid, which is subsequently expanded through a turbine or piston device to produce the mechanical power output. This working fluid is then condensed and pumped back into the vaporizer to close the heat engine cycle.

Raoult's law. When dissolved in a definite weight of a given solvent, molar weights of nonvolatile nonelectrolytes under the same conditions lower the solvent's freezing point, elevate its boiling point, and reduce its vapor pressure equally for all such solutes.

rasp. A machine that grinds waste into a manageable material and helps prevent odor.

raw sewage. Untreated waste water.

Rayleigh atmosphere. An idealized atmosphere consisting of only those particles, such as molecules, that are smaller than about one-tenth the wavelength of all radiation incident upon that atmosphere. In such an atmosphere, simple Rayleigh scattering would prevail. This model atmosphere is amenable to reasonably complete theoretical treatment, and hence has often served as a useful starting point in descriptions of the optical properties of actual atmospheres. The polarization of skylight, for example, exhibits almost none of the complexities found in the real atmosphere.

Rayleigh scattering. Any scattering process produced by spherical particles whose radii are smaller than about one-tenth the wavelength of the scattered radiation.

reaction, chain. Any chemical or nuclear process in which some of the products of the process or energy released by the process are instrumental in the continuation or magnification of the process.

reaction, endothermic. A reaction which absorbs energy specifically in the form of heat.

reaction, thermonuclear. A nuclear reaction in which the energy necessary for the reaction is provided by colliding particles that have kinetic energy by virtue of their thermal agitation. Such reactions occur at appreciable rates only for temperatures of millions of degrees and higher, the rate increasing enormously with temperature.

reactivation. The removal of adsorbate from spent, granular activated carbon, which will allow the carbon to be reused. Also called regeneration and revivification.

reactor, fast. A nuclear reactor in which there is little moderation and fission is induced primarily by fast neutrons that have lost relatively little of the energy with which they were released. The slowing down of neutrons that does occur is due largely to inelastic scattering, instead of elastic scattering. About 100,000 eV is regarded as the minimum value of mean energy of neutrons inducing fission for a reactor to be considered fast, with $\frac{1}{3}$ to $\frac{1}{2}$ MeV being more common.

reactor, homogeneous. A nuclear reactor in which the fissionable material and moderator (if used) are combined in a mixture such that an effectively homogeneous medium is presented to the neutrons. Such a mixture is represented either by solution of fuel in the moderator or by discrete particles having dimensions small in comparison with the neutron mean free path.

reactor, intermediate. A nuclear reactor in which fission is induced predominantly by neutrons whose energies are greater than thermal energies, but much less than the energy with which neutrons are released in fission. From 0.5 to 100,000 eV may be taken as roughly the energy range of neutrons inducing fission in intermediate reactors. The neutron absorption resonances of the fuel may be important in this range.

reactor, nuclear. An apparatus in which nuclear fission may be sustained in a self-supporting chain reaction. It includes fissionable material (fuel) such as uranium or plutonium, and moderating material, unless it is a fast reactor, and usually includes a reflector to conserve escaping neutrons, provision for heat removal, and measuring and control elements.

reactor, thermal. A nuclear reactor in which fission is induced primarily by neutrons of such energy that they are in substantial thermal equilibrium with the material of the core. A representative energy for thermal neutrons is often taken as 0.025 (2200 meters per second), which corresponds to the mean energy of neutrons in a Maxwellian distribution at 293°K, although most thermal reactors actually operate at higher temperatures. A moderator is an essential element of a thermal reactor.

receiving water. Any body of water where untreated wastes are dumped.

recharge. The process by which water is added to the zone of saturation, as in recharge of an aquifier.

reciprocating engine. An engine, especially an internal combustion engine in which a piston or pistons move back and forth, working upon a crankshaft or another device to create rotational movement.

recirculating baffle collector. A collector where the gas to be cleaned is introduced at a high velocity under a horizontal baffle made up of rods spaced about a half inch apart. To pass between the rods to reach the cleaned-gas outlet chamber, the dirty gas must make a sudden, high-velocity turn. The low-specific-gravity gas can easily make the sharp turn. The heavier dust, driven by inertia, is unable to make this sharp turn and is restrained below the baffle until it is captured by the dust slot. The dust is conveyed at a lowered velocity as the dust slot expands over the collector hopper into which it settles, as in a settling chamber. The circulating flow is controlled at a nominal velocity by the expanding dust slot and the circulating flow control baffle.

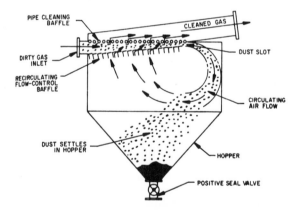

Recirculating baffle collector.

recruitment. The condition (usually characteristic of nerve deafness) where faint or moderate sounds cannot be heard, while at the same time there is little or no loss in the sense of loudness of loud sounds. It explains why old people, whose deafness is usually a gradual, high-toned nerve deafness, complain in one moment that they cannot hear a speaker and at the next moment when he raises his voice, that he is shouting too loudly at them.

recycling. Reusing waste materials to produce materials that may or may not be similar to the original.

red tide. A proliferation of ocean plankton that may kill large numbers of fish. This natural phenomenon may be stimulated by the addition of nutrients.

reduction. Any process which increases the proportion of hydrogen or base-forming elements or radicals in a compound. Reduction is also the gain of electrons by an atom, ion, or element, thereby reducing the positive valence of that which gained the electron.

reflection coefficient. The ratio of the light reflected from a surface to the total incident light; also the reflectivity. The coefficient may refer to diffuse or specular reflection. In general, it varies with the angle of incidence and with the wavelength of the light.

reflective loss. The energy which strikes a surface and is absorbed but reflected from it.

reflectivity. The fractional part of the incident radiation reflected by a surface (total or monochromatic); it is a dimensionless number between zero and unity.

reflector. A substance such as ordinary water, heavy water, graphite, or beryllium placed around a nuclear reactor core to reflect escaping fuel neutrons back into the core of the reactor.

refraction. The process in which the direction of energy propagation is changed as the result of a change in density within the propagation medium, or as the energy passes through an interface representing a density discontinuity between two media. In the first instance the rays undergo a smooth bending over a finite distance. In the second case the index of refraction changes through an interfacial layer that is then compared to the wavelength of the radiation; thus the refraction is abrupt and essentially discontinuous.

refractory. A material that can resist high temperatures and large changes of temperature without melting, crumbling, or cracking, and stay relatively unaffected by the flow of very hot materials over its surface. Alumina, magnesia, zirconia, some silicon compounds, carbon, and graphite are some of the substances used in materials required for the highest temperatures.

refuge, wildlife. An area designated for the protection of wild animals, within which hunting and fishing are either prohibited or strictly controlled.

refuse. A general word for solid waste.

refuse reclamation. The conversion of solid waste into useful products, e.g., composting organic wastes to make a soil conditioner.

regulated system. An operational procedure required by NIOSH for industries that manufacture, process, use, repackage, release, handle, or store chemicals identified as carcinogens. Regulated systems are generally classified as closed or isolated. A closed system is an operation in which the walls of the tanks, pipes, reactors, or other equipment prevent the release of a carcinogenic chemical. An isolated system is one which is fully enclosed in a structure such as a closed room, a glove box, or a laboratory hood that prevents a substance from entering the environment if the container in which it is held should leak or spill. No longer permitted under NIOSH regulations is the open-vessel system, in which a carcinogen is in an open container and not in an isolated or closed system that would prevent the carcinogen from entering the environment of the workplace or the surrounding public area.

reheating furnace. A furnace used for heating steel and other metals for cogging, rolling, forging, pressing, and other methods of working them. The fuel used may be coal, pulverized coal, gas, oil, or electricity, and the furnace may be intermittent or continuous. In the intermittent furnace a batch of material is heated to the required temperature, when its place is taken by a second batch. In the continuous furnace the material is moved from the coolest to the hottest zone at a suitable rate; cold material is continuously being put in at one end and hot material taken from the other.

reinjection. A method of disposing of fly ash where it is reinjected in a furnace for further burning.

Reinluft process. A dry scrubbing process for the adsorptive separation and recovery of gaseous oxides of sulfur, nitrogen, and phosphorus from industrial gases. This involves bringing the oxide-containing gas together with free oxygen into contact with a solid adsorption medium of carbon of low surface activity. The oxide is converted to a higher oxidation stage for which the adsorbing medium, an inexpensive char made from peat, has greater adsorbing power. (Oxidation and, consequently, adsorption can be accelerated by application of metal catalysts which favor oxidation.) The char travels downward through a large three-stage adsorber, against the rising flow of flue gas. The dioxide (SO_2) is first adsorbed at about 200°F; the trioxide (SO_3) is adsorbed in a smaller lower chamber at about 320°F. In the lowest chamber, N_2 or CO_2 liberates the adsorbed dioxide from the char at 750°F and carries it off in a 50 percent dioxide stream for further processing. The oxides, then being reduced, are separated out, and the medium is cooled and ready for reuse. The simultaneous presence of water vapor increases the efficiency of adsorption, making possible the formation of the corresponding acids of H_2SO and HNO_3.

relative humidity. The ratio of the quantity of water vapor present in the atmosphere to the quantity which would saturate the air at the existing temperature; also the ratio of the pressure of water vapor present to the pressure of saturated water vapor at the same temperature.

relative momentum. The product of the mass of a particle and its related velocity; in the case of a fluid, the product of density and relative velocity.

rem. The unit of measurement concerned with internal radiation. It equals the radiation dose in rads multiplied by the relative biological effectiveness (RBE) of the particular radiation; e.g., a given number of rads of the closely packed ions of alpha rays has an effect on the tissue 20 times greater than an equal number of rads of gamma rays.

reoxygenation. The replenishment of oxygen in a stream from dilution water entering the stream, biological oxygenation through the activities of certain oxygen-producing plants, and atmospheric reaeration.

rep. A unit of measurement of any kind of radiation absorbed by man. It has a value slightly less than the rad.

reserve. The tonnage in a specific block of coal that is known to be minable with the presently available technology. For example, a strippable reserve is the tonnage of coal which, because of its thickness, depth, quality, and other attributes, is known to be minable by strip mining methods. A recoverable reserve that can actually be extracted by a particular mining method.

reservoir. A pond, lake, basin, or other space, either natural in origin or created in whole or in part by the construction of engineering structures. It is used for the storage, regulation, and control of water.

residual fuel. The liquid or semiliquid product obtained as residue from the distillation of petroleum and used as fuel.

residual oil. When crude oil is extracted from the ground, it is a combination of many organic molecules. In the refining process, the smaller and lighter molecules are evaporated and sold as petroleum gas. The slightly heavier molecules become gasoline and diesel oil. Intermediate molecules are "cracked" under high temperatures and pressures to yield additional gasoline. Still heavier oils become lubricants. Paraffin, asphalt, etc., are also extracted from crude oil. The remaining residual oil, which is not economically usable for other industrial purposes, is sold for burning.

resistance. A property of conductors depending on their dimensions, material, and temperature which determines the current produced by a given difference of potential. The practical unit of resistance, the ohm, is that resistance through which a difference of potential of one volt will produce a current of one ampere. The international ohm is the resistance offered to an unvarying current by a column of mercury at 0°C, 14.4521 grams in mass, of constant cross-sectional area, and 106.300 centimeters in length; it is sometimes called the legal ohm.

resonance. A property, inherent in a vibrating system such as a structure or electrical circuit, of responding to one or more frequencies to a greater extent than others. A resonance curve is a curve of amplitude over a range of frequencies, showing maxima at the resonant frequencies.

resource recovery. The process of obtaining matter or energy from materials formerly discarded, e.g., solid waste and wood chips.

respiration. The taking in and giving off of carbon dioxide and all steps involved in the process.

respiratory damage from air pollutants. Four major types of respiratory damage from air pollutants are bronchitis, bronchial asthma, emphysema, and lung cancer. Chronic bronchitis is characterized by permanent damage to the bronchial tubes, resulting in the reduction or failure of ciliary action, and overproduction of mucus by gland cells. Because cilial action cannot dislodge the extra mucus, a chronic cough develops. The mucus also constricts the opening of the bronchial system, causing shortness of breath. Bronchial asthma is usually the result of the allergic reaction of the bronchial membranes to foreign protein or other materials. The membranes swell and make the expulsion of air from the lungs difficult. Emphysema follows the constriction of the finer branches of the bronchial tubes, the bronchioles. When air is exhaled, more air remains in the tiny sacs of the lungs (alveoli) than there should; when new air is inhaled, the overinflated sacs balloon larger and larger until they explode. This causes two adjacent sacs to unite. The gradual reduction in the number of air sacs destroys the capillaries through which oxygen is taken up by the red blood cells and slowly pushes out the chest. The loss of the oxygen exchange capacity of the lungs leads to slow oxygen starvation of the entire body and chronic shortness of breath.

respiratory system. The group of organs concerned with the exchange of oxygen and carbon dioxide in organisms. In higher animals this consists successively of the air passages through the mouth, nose, and throat, the trachea, the bronchi, the bronchioles, and the alveoli of the lungs. The human respiratory system is a complex mechanism with the principal function of taking in molecular oxygen (O_2) for subsequent use, a process called internal respiration, and ridding the body of carbon dioxide (CO_2) produced as a result of this biochemical process. The term respiration used to apply solely to the gross process of exchanging O_2 and CO_2 between the air and the body (external respiration), but under present usage it also applies to the process occurring inside body cells whereby O_2 acts as the oxidizing agent in the reactions with organic compounds, with a resulting liberation of energy which the body uses for its function.

Structure, Nature, and Function of Elements
Comprising the Human Respiratory System

Structural Unit	Description	Function
External nares	Nostrils	Openings through which air enters the system, and discharges from the sinuses, tear ducts, and upper respiratory system pass
Nasal cavity	Above mouth cavity and below brain, lined with ciliated, mucus-secreting tissue	Contains sense organs of smell, and warms and cleans air
Internal nares	Internal to nasal chamber and lined with similar tissue	Same as nasal chamber
Pharynx	Paths of air and food cross	Air passage
Epiglottis	Flap of tissue	Folds over opening to the lower respiratory system whenever liquids or solids are swallowed
Larynx	The "voice box" consisting of folds of tissue	Speech and air passage
Trachea	Windpipe made rigid by rings of cartilage	Air passage
Bronchi	Branches of the trachea at the level of the first rib, lined with ciliated, mucus-secreting tissue	Air cleaning
Bronchioles	Smaller branches of the bronchi	Same as bronchi
Air sacs	Small sacs located at the end of the terminal bronchioles, the ultimate cavities of the respiratory system	Same as bronchioles
Alveoli	Cup-shaped cavities off the air sacs	Gas exchange between blood and air in the lungs
Pleura	Membranes which surround the lungs	Maintain integrity of the respiratory system

respiratory system, airborne stresses. As materials enter the respiratory system they may, if in the form of droplets or large particulates, be filtered by the hairs of the nostrils. If this barrier is passed, exclusion from the lower respiratory tract may be effected. When irritant vapors, e.g., NH_3^+, enter the pharynx and larynx, impulses are sent to the brain to inhibit breathing, and hence advection of the material into the lungs occurs. Other materials, especially particulates, may elicit a more passive response. The irritation of or impingement on respiratory tract tissue containing mucus-secreting cells produces a liberation of mucus. The trapped irritant is then transported upward along the mucus in a flow induced by the beating of cilia. Irritants which pass these barriers and enter the trachea and bronchi may cause a constriction of the muscle fibers which lie between the cartilaginous rings. The difficulties in breathing experienced by persons suffering from asthma are due to such passage constriction.

response. For a device or system, the motion (or other output) resulting from an excitation (stimulus) under specified conditions.

retrofit. The fitting of solar equipment to an existing building.

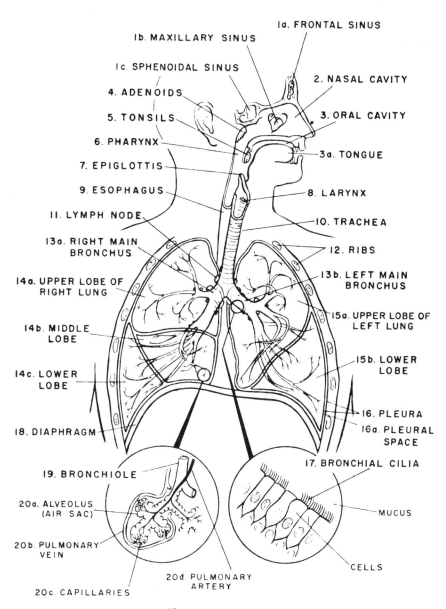

The respiratory system.

reverberant field. A region where the measured pressure levels of incident sound at some distance from the source are essentially independent of the direction and distance to the source.

reverberation. The echoes of a sound that persist in an enclosed space after the sound source has stopped.

reverberation time. For a room, the time required for the mean-square sound pressure level therein, originally in a steady state, to decrease by 60 dB after the source has stopped.

reverse osmosis. An advanced method of waste treatment that uses a semipermeable membrane to separate water from pollutants.

Reynolds number. A nondimensional parameter representing the ratio of the momentum forces to the viscous forces in a fluid flow.

ridge. An elongated area of relatively high pressure extending from the center of a high-pressure region.

rime. A deposit of ice composed of grains more or less separated by trapped air, sometimes adorned with crystalline branches.

Ringelmann chart. A chart used in air pollution evaluation for assigning an arbitrary number, referred to as "smoke density," to smoke emanating from any source. This chart is designed specifically for the subjective estimation of the density of black smoke and is not applicable to other emissions. The observer compares the grayness of the smoke with a series of shade diagrams, formed by horizontal and vertical lines on a white background.

riparian. Pertaining to anything connected with or adjacent to the banks of a stream or other body of water.

riparian rights. The entitlement of a land owner to the water on or bordering his property, including the right to prevent diversion or misuse of it upstream.

river basin. The land area drained by a river and its tributaries.

Robert's smoke chart. A disk chart having radial black lines on a white background. When the disk is revolved, solid shades are produced. The chart has an eyelet center and can be spun on a broad awl, nail, or other convenient shaft.

rock creep. The slow, downhill slipping of large, joint blocks wherever well-jointed, massive formations crop out along a slope. By this process a joint block gradually widens the gap between itself and the parent outcrop and eventually tilts to the angle of slope of the surface.

rodenticide. A chemical or agent to destroy rats or other rodent pests, or to prevent them from damaging food, crops, etc.

roentgen. A unit of radiation; that quantity of x rays or gamma rays which will produce, as a consequence of ionization, one electrostatic unit of electricity in one cubic centimeter of dry air measured at 0°C and standard atmospheric pressure; a unit used to measure exposure to radioactivity and to determine its effect on living organisms.

roentgen-equivalent-man. A unit of radiation which, when absorbed by a human being, produces the same effect as the absorption of one roentgen of high-voltage x rays.

roentgen-equivalent-physical. A unit measuring a purely physical effect of radiation by the number of ion pairs produced per unit volume of target material per unit time. One rep is equivalent to the absorption of 93 ergs per gram of tissue.

roof monitor. A cupola, skylight, or raised roof shed with windows designed to provide natural lighting and ventilation in buildings. Roof monitors can also serve as heat traps in the winter by admitting the winter sun's rays directly into the house interior. Roof monitors are excellent sources of natural light and in the summer months they are ideal for natural ventilation, because heat naturally rises to the top of the building.

room air temperature. The temperature measured in degrees centigrade at 0.75 m above the floor in the middle of the room, but not further than 2.5 m from the outside walls.

room constant. The name given to the expression $Sx/(1-x)$, where S is the total area of the bounding surfaces of the room in square feet and x is the average absorption coefficient of the absorption present in the room.

root-mean-square sound pressure. The root-mean-square volume of the instantaneous sound pressure over a time interval at the point under consideration; also the effective sound pressure. In the case of periodic sound pressures, the interval must be an integral number of periods or an interval that is long compared to the period. In the case of nonperiodic sound pressures, the interval should be long enough to make the value obtained essentially independent of small changes in the length of the interval. The term "effective sound pressure" is frequently shortened to "sound pressure."

rotary kiln. One of the oldest types of incinerators, applicable to both solid wastes and sludges. A rotary kiln consists of a long horizontal cylinder that rotates on steel "tires" which rest on trunnions. Waste material is fed at one end, tumbled by the kiln to provide attrition and exposure to combustion air and drying, and gradually burned as it moves toward the exit, where ash is removed. Afterburners are usually required for this type of system because the volatized gases from the refuse cannot be adequately mixed with the air in the kiln. This type of incinerator is used in central plant disposal facilities where there will be a wide variety of waste materials.

rotometer. A device for measuring the rate of fluid flow, based on Stoke's law. It consists of a tapered, vertical tube having a circular cross section. A float inside moves in a vertical path to a height dependent upon the rate of fluid flow upward through the tube.

rough fish. Those species not prized for game purposes or for eating: gar, suckers, etc. Most rough fish are more tolerant of changing envionmental conditions than game species.

rubbish. Solid waste, excluding food waste and ashes, from homes, institutions, and workplaces.

rubidium (Rb). A soft, white metallic element inflammable in air; it has an atomic weight of 85.48, a specific gravity of 1.5, a melting point of 38.8°C, and a boiling point of 710°C.

rud. A reflectance unit of dirt shade units based on the reflection of light.

runner. A strip of wood mounted on a roof or wall to which one end of a solar collector, or a rack holding a solar collector, is securely attached.

runoff. Water from rain, snow melt, or irrigation that flows over the ground surface and returns to streams. It can collect pollutants from air or land and carry them to the receiving waters.

S

sabin. A measure of the sound absorption of a square-foot of a perfectly absorbing surface.

Saint Elmo's fire. A more or less continuous, luminous, electrical discharge of weak or moderate intensity in the atmosphere, emanating from elevated objects at the earth's surface (lightning conductors, wind vanes) or from aircraft in flight (wingtips, propellers, etc.).

saline water conversion process classification.
A. Processes that separate water from the solution
  1. Distillation or evaporation
    a. Multiple-effect long-tube vertical
    b. Multistage flash
    c. Vapor compression
    d. Humidification (solar)
  2. Crystallization or freezing
    a. Direct freezing
    b. Indirect freezing
    c. Hydrates
  3. Reverse osmosis
  4. Solvent extraction
B. Processes that separate salt from the solution
  1. Electrodialysis
  2. Osmosis
  3. Adsorption
  4. Liquid extraction
  5. Ion exchange
  6. Controlled diffusion
  7. Biological systems

salinity. (1) The amount or concentration of salt in a solution such as seawater. When salt brines from mines or oil wells are released into normally fresh water, some organisms can tolerate a certain range of salt concentration; many disappear as fresh water becomes brackish. If the salt level remains constant, salt-tolerant species often appear and reproduce successfully. Because of the constantly fluctuating levels of most pollutants, organisms that might otherwise adjust to the new conditions rarely have the opportunity to become naturalized, so neither old nor new organisms can survive. (2) For seawater, the total amount of solid material in grams contained in one kilogram of seawater when all the carbonate has been converted to oxide, the bromine and iodine replaced by chlorine, and all organic material completely oxidized. Desalinated freshwater is usually defined as "water containing less than 1000 parts per million of dissolved salts." In the open ocean the total salinity varies between 33,000 and 38,000 ppm, depending upon the geologic location. Seawater carries great quantities of mineral matter in solution. One thousand parts of seawater contains 34.4 parts by weight of mineral matter, or 3.44 percent. For one cubic mile of seawater this amounts to 151,025,000 tons.

Salts Present in the Ocean

Salt	Percentage
Sodium chloride, NaCl	77.758
Magnesium chloride, $MgCl_2$	10.878
Magnesium sulfate, $MgSO_4$	4.737
Calcium sulfate, $CaSO_4$	3.600
Potassium sulfate, K_2SO_4	2.465
Calcium carbonate, $CaCO_3$	0.345
Magnesium bromide, $MgBr_2$	0.217
	100.000

salt. Any substance which yields ions, other than hydrogen or hydroxyl ions. A salt is obtained by displacing the hydrogen of an acid by a metal. When fatty acids contact metal, they form a salt called soap. Salt occurs in nature as rock salt or halite. It also occurs in solution in salt springs, in salt lakes, and in the ocean. Halite is sodium chloride, NaCl. It is colorless or white when pure, but impurities may color it gray, green, blue, or even red.

saltwater desalting. A process for removing salt from saltwater. The basic method is to heat water to convert it to water vapor, leaving the solids behind. In a usual type of still, the water vapor is led from the evaporator to a condenser, where it is cooled and again becomes liquid water, but without the salt it originally contained.

saltwater intrusion. The invasion of fresh surface or groundwater by saltwater. If the saltwater comes from the ocean, it's called seawater intrusion.

salvage. The utilization of waste materials.

sanitary landfill. (1) An engineered burial of refuse. Refuse is dumped into trenches and compacted by a bulldozer, and microorganisms decompose the organic matter to stable compounds. Moisture is essential for the biological degradation. Groundwater assists this process, except when it fills air voids and prevents aerobic metabolism. (2) A method of waste disposal by spreading the wastes over land and covering them with a seal of earth. Communities often use landfills to reuse wastes, to help fill low areas, or to build recreational facilities.

sanitary sewers. Underground pipes that carry only domestic or commercial waste, not stormwater.

sanitation. The control of physical factors in the human environment that can harm development, health, or survival.

Santa Ana wind. A hot, dry, desert wind, generally from the northeast or east, especially in the pass and river valley of Santa Ana, California, where it is further modified as a mountain gap wind. It blows, sometimes with great force, from the deserts to the east of the Sierra Nevada Mountains and may carry a large amount of dust. It is most frequent in winter.

saturant. A substance used to neutralize another substance or to saturate it so that the affinity between the two substances is completely satisfied.

saturated. Said of an organic compound that cannot incorporate additional hydrogen atoms into its structure; such compounds are not as reactive as unsaturated ones.

saturated air. Air that contains the maximum amount of water vapor it can hold at a given pressure and temperature (its relative humidity is 100 percent).

saturated compound. A compound to which no other element can be added for the reason that there are no free valences to which hydrogen atoms or their equivalent can be added.

saturated steam. Steam or water vapor carrying all of the water that can be evaporated at a given temperature and pressure.

saturation vapor pressure. For pure water vapor, the pressure of the vapor when in a state of neutral equilibrium with a plane surface of pure water at the same temperature. The saturation vapor pressure is then equal to the vapor tension of the water surface at this temperature. It varies with the temperature exactly as does the vapor tension of the water, so the higher the temperature, the greater the vapor pressure required for saturation. The dependence of the saturation vapor pressure on temperature is a specific characteristic; it is dependent on the pressure of the other gases at all atmospheric pressures.

scintillation counter. An instrument using a photomultiplier tube, used for counting the scintillations produced by subatomic particles hitting a fluorescent screen. In effect, the photomultiplier complements the spinthariscope principle; the spinthariscope changes particle energy to light energy and the photomultiplier tube changes light energy to electric energy. The electric current is then used to activate a counter.

scram. The sudden shutting down of a nuclear reactor, usually by the dropping of safety rods. This may be arranged to occur automatically at a predetermined neutron flux or under other dangerous conditions, the reaching of which causes the monitors and associated equipment to generate a scram signal.

scrap. Materials discarded from manufacturing operations that may be suitable for reprocessing.

screen. A plate, sheet, woven cloth, or other device with regularly spaced apertures of uniform size, mounted in a suitable frame or holder for use in separating materials according to size.

screening. The removal of relatively coarse floating and suspended solids by straining through rocks or screens.

scrubber. An apparatus used in sampling and gas cleaning in which gas is passed through a space containing wetted "packing" or spray; a device which utilizes a liquid to achieve or assist the removal of solid or liquid particles from a carrier gas stream. In general, particles are collected in scrubbers by one mechanism or a combination of mechanisms: the impingement of particles on a liquid medium; the diffusion of particles onto a liquid medium; the condensation of liquid-medium vapors on the particles; partitioning of gas into extremely small elements to allow the collection of particles by Brownian diffusion and gravitational settling on the gas—liquid interface. Designs include spray towers, jet scrubbers, Venturi scrubbers, cyclonic scrubbers, inertial scrubbers, mechanical scrubbers, and packed scrubbers. Scrubbers in which liquid contact of the particles is obtained as a result of gas velocity are classified as inertial-type scrubbers; these in turn consist of the impaction and deflection types. Mechanical scrubbers operate by liquid particle contact achieved by the simultaneous introduction of a liquid medium and a gas stream onto rotating disks, blades, or perforated plates. The packed scrubber is a conventional packed tower employing Raschig rings, Berl saddles, fiberglass, or other packing. Normally, gas flow is counter to the liquid flow. Simple types of scrubbers with low-energy inputs are effective in collecting 5- to 10-μm particles; more efficient scrubbers will collect particles as small as 1 to 2 μm.

seasonal periods. A prevernal period (early spring) begins in early March and terminates in mid-April, an interval during which some of the hardier vegetation begins its growth and blooms. Late spring is called the vernal period and usually lasts a month and a half (from late April through May), during which period the soil temperature rises with atomspheric temperature. The next two periods comprise the summer interval. Early summer (June through early July) is called the estival period; late summer (late July through early September) is called the serofinal period. During these periods the conditions for the growth of plants and animals are optimal. The autumnal period beginning in mid or late September and terminating in early November, is often an interval during which the temperature lowers and frost occurs. The hiemal period, or winter interval, begins some time in mid or late November and lasts until the end of February. In many areas this is a period during which precipitation in the form of snow, frozen rain, or rain is plentiful.

seawater. A complex solution of organic and inorganic salts derived over the course of geological time from the solution of rocks, gaseous effusion of volcanoes, biological activity, and, to a much lesser extent, meteoritic material in the earth's atmosphere. Seawater is about 2700 times more abundant on earth than impounded freshwater. Its physical characteristics are quite different from those of freshwater because as a solution, its characteristics differ not only with temperature and pressure, but with the concentration of the salt in solution, commonly called salinity.

second law of thermodynamics. An inequality asserting that it is impossible to transfer heat from a colder to a warmer system without the occurrence of other simultaneous changes in the two systems or in the environment. It follows from this law that during an adiabatic process, entropy cannot decrease. For reversible adiabatic processes, entropy remains constant, and for irreversible adiabatic processes, it increases. Another equivalent formulation of the law is that it is impossible to convert the heat of a system into work without the occurrence of other simultaneous changes in the system or its environment. This version, which requires an engine to have a cold source as well as a heat source, is particularly useful in engineering applications.

secondary emission. A product of a polluted air reactant, such as those which occur in atmospheric, photochemical reactions. Secondary pollutants include ozone, formaldehyde, organic hydroperoxide, free radicals, NO, O, etc.

secondary emission. (1) The result of a primary emission, such as x rays falling on a solid body and giving rise to another emission of a similar or different character. (2) The emission of subatomic particles or photons, stimulated by primary radiation; for example, cosmic rays impinging on other particles and causing them, by the disruption of their electron configuration or even their nuclei, to emit particles and photons or both.

secondary radiation. The electromagnetic or particulate radiation resulting from absorption of other radiation in matter.

secondary treatment. A process used to reduce the amount of dissolved organic matter and further reduce the amount of suspended solids in waste water. The effluent from the primary treatment process is given this additional treatment along with activated sludge or trickling filter processes.

sedimentation. A process which utilizes the natural separation tendencies of an insoluble material in water. It is applicable to the removal of insoluble oils and/or particulate solids. Separation is effected by holding the water in a quiescent or controlled low-velocity condition long enough for the oil to rise to the surface and the solids to settle to the bottom.

sedimentation, final. The settling of partly settled, flocculated, or oxidized sewage in a final tank.

sedimentation, plain. The sedimentation of suspended matter in a liquid unaided by chemicals or other special means and without provision for the decomposition of deposited solids in contact with the sewage.

sedimentation tank. A holding area for waste water, where floating wastes are skimmed off and settled solids are pumped out for disposal.

seeding. The introduction of atoms, such as sodium, with a low ionization potential into a hot gas for the purpose of increasing the electrical conductivity.

seepage. Water that flows through the soil.

selective black paint. A paint formulated to be more absorbent of the long infrared wavelengths of sunlight than commercial black paint, and therefore more efficient in absorbing solar heat. There is some doubt as to the value of selective black paints or coatings.

selective pesticide. A chemical designed to affect only certain types of pests, leaving other plants and animals unharmed.

selenium (Se). A gray, nonmetallic element of the sulfur group. Selenium, when present with heavy metals in a waste stream, will be removed in much the same way as arsenic is removed in heavy metal precipitation. It has been recommended that the value 0.01 mg/liter be established as the permissible criterion for the presence of selenium in public water supplies. Selenium is toxic to animals when it accumulates in tissue at the level of 5 mg/kg. It is also an essential trace mineral and is commonly fed in quantities sufficient to cause accumulations of 1 to 2 mg/kg.

semiconductor. A substance that has a high electrical resistance, but not high enough to be classed as an insulator; a material like silicon, which is neither as conductive of electricity and heat as metals nor as insulative as glass or ceramic. Most useful photovoltaic materials used in solar cells are semiconductors.

senescence. The aging process. It can refer to lakes in advanced stages of eutrophication.

sensible atmosphere. The part of the atmosphere that offers resistance to a body passing through it.

sensible temperature. The temperature at which average indoor air of moderate humidity would induce in a lightly clothed person the same sensation of comfort as that induced by the actual environment.

septic tank. An enclosure that stores and (processes) wastes where no sewer system exists, as in rural areas or on boats. Bacteria decompose the organic matter into sludge, which is pumped off periodically.

serpentine and asbestos. A fairly soft, greasy-looking green, yellow, brown, or black hydrous magnesium silicate. Massive serpentine forms serpentine rock. Verd antique is an ornamental variety. Chrysotile, a silky variety that can be separated into strong flexible, infusible, threadlike fibers, is high-grade asbestos. "Asbestos" is not the name of one specific mineral, but a commercial term applied to any mineral separable into more or less flexible fibers. It was given originally to amphiboles, such as tremolite and actinolite, which may occur as long, silky fibers, but are usually brittle and have little tensile strength. The main uses for asbestos are in asbestos yarns and papers, asbestos shingles and sidings, and in heat-insulating materials. Asbestos is also used in automobile brake-band linings and various types of gaskets.

settleable solid. A material heavy enough to sink to the bottom of waste water.

settling chamber. A chamber designed to reduce the velocity of gases in order to permit the settling out of fly ash. It may be either part of, adjacent to, or external to an incinerator.

settling tank. A holding area for waste water, where heavier particles sink to the bottom and can be siphoned off.

settling velocity. The terminal rate of fall of a particle through a fluid as induced by gravity or other external forces.

sewage. The organic waste and waste water produced by residential and commercial establishments.

sewage, combined. A sewage containing both sanitary sewage and surface or storm water with or without industrial waste.

sewage, dilute. Sewage containing less than 150 ppm of suspended solids and BOD (weak sewage).

sewage, industrial. Sewage in which industrial waste predominates.

sewage lagoon. See **lagoon.**

sewage, raw. Sewage prior to treatment.

sewage, settled. Sewage from which most of the settleable solids have been removed by sedimentation.

sewage, storm. The liquid flowing in sewers during or after a period of heavy rainfall and resulting therefrom.

sewer. A pipe or conduit, generally closed, but normally not flowing full, for carrying sewage and other waste material.

sewer, intercepting. A sewer which receives dry-weather flow from a number of transverse sewers or outlets, and, frequently, additional, predetermined quantities of storm water (if from a combined system), and conducts such water to a point for treatment or disposal.

sewerage. The entire system of sewage collection, treatment, and disposal; also applies to all effluent carried by sewers.

shell. A series of concentric spheres formed by the orbits of electrons around a nucleus. According to Pauli's exclusion principle, the extranuclear electrons do not circle around the nucleus all in orbits of the same radius, but are arranged in orbits at various distances from the nucleus. Shells are designated in the order of increasing distance from the nucleus, such as *K, L, M, N, O, P,* and *Q* shells. The number of electrons which each of these shells can contain is limited. All electrons arranged in the same shell have the same principal quantum number. The electrons in the same shell are grouped into various subshells and all the electrons in the same subshell have the same orbital angular momentum.

shield. A material placed around a nuclear reactor core or radioactive material to prevent radiation from escaping.

shortwave. The "short" radio wave band covering the range of 10–55 m, corresponding to the "high-frequency band" of 3000–30,000 kHz.

sierozem soil. A gray desert soil that develops in semidesertic areas with temperate to cool temperatures and dry conditions. Surface soils are brownish gray, grading into lighter-colored subsoils, with a calcium carbonate stratum layer only a foot or so from the soil surface. Sierozem soils support desert shrubs and grasses.

significant deterioration. Pollution from a new source in previously "clean" areas.

silicon (Si). A nonmetallic element closely related to carbon in its atomic structure and having similar chemical properties. It is never found pure in nature but is always combined with oxygen and other elements. These compounds are called silicates. Of every 100 atoms in the earth's crust, 47 are oxygen and 28 are silicon. Many common building materials such as bricks, cement, and mortar are made of silicates. Sand, which is the compound silicon dioxide, is melted and used in making glass. Silicon combines with carbon to form silicon carbide, or carborundum.

silicon tetrachloride. $SiCl_4$, a colorless, fuming liquid with a suffocating odor. It reacts with water or steam to produce heat and toxic and corrosive fumes; it reacts violently with sodium or potassium. When heated, it may decompose to emit hydrochloric acid fumes. The inhalation LC_{50} for rats is 8000 ppm for four hours. In a 1975 accident in Chicago, a storage tank containing 850,000 gallons of silicon tetrachloride ruptured, forming a thick cloud of hydrochloric acid fumes over an area within a mile of the tank. Approximately 8000 people living in the area were evacuated and 300 were hospitalized with burns of the skin and mucous membranes, including the eyes and nose, throat irritation, breathing difficulty, nausea and vomiting, and dizziness.

silicone. A polymeric organic compound containing a —Si—O—Si—O group, available as oil, grease, or resin. Silicone is highly stable to heat and chemicals; it is used in high-temperature electrical insulations.

silt. Fine particles of soil or rock that can be picked up by air or water and deposited as sediment.

silviculture. The management of forest land for timber, sometimes contributing to water pollution, as in clear cutting.

simple sound source. A source that radiates sound uniformly in all directions under free-field conditions.

simple tone. (1) A sound wave, the instantaneous sound pressure of which is a simple sinusoidal function of time. (2) A sound sensation characterized by its singleness of pitch.

sinking. Controlling oil spills by using an agent to trap the oil. Both sink to the bottom of the body of water and biodegrade there.

sintering. Baking at a very high temperature. Fly ash accumulated in smokestacks is a salable item for use in the manufacturing of cinder blocks and other ceramic products. However, it must first be sintered.

skimming. Using a machine to remove oil or scum from the surface of the water.

sleet. Generally transparent, globular, solid grains of ice formed by the freezing of raindrops, or the freezing of largely melted snowflakes as they fall through a below-freezing layer of air near the earth's surface.

slow neutron. A neutron that has been slowed down sufficiently by collisions with the atoms in a moderator to be captured by a ^{235}U nucleus and cause fission.

sludge. The concentration of solids removed from sewage during waste water treatment.

sludge, activated. Sludge floc produced in raw or settled sewage by the growth of zoogleal bacteria and other organisms in the presence of dissolved oxygen, and accumulated in sufficient concentrations by returning floc previously formed.

sludge, cake. The material resulting from air drying or dewatering sludge (usually forkable or spadable).

sludge dewatering. The process of removing a part of the water in sludge by any method, such as draining, evaporation, pressing, centrifuging, exhausting, passing between rollers, or acid flotation, with or without heat. It involves the reduction from a liquid to a spadable condition, rather than merely changing the density of the liquid (concentration), on the one hand, or kiln drying, on the other hand.

sludge, digested. Sludge digested under anaerobic conditions until the volatile content has been reduced, usually by approximately 50 percent or more.

sludge, excess activated. Excess activated sludge removed from the activated sludge system for ultimate disposal.

sludge lagoon. A relatively shallow basin, or natural depression, used for the storage or digestion of sludge, sometimes for its ultimate detention or dewatering.

sludge reaeration. The continuous aeration of sludge after its initial aeration in the activated sludge process.

sludge seeding. The inoculation of undigested sewage solids with sludge that has undergone decomposition for the purpose of introducing favorable organisms, thereby accelerating the initial stages of digestion.

slurry. A watery mixture of insoluble matter that results from some pollution control techniques.

small ion. An atmospheric ion, apparently a single-charge atmospheric molecule (or rarely, an atom) about which a few other neutral molecules are held by the electrical attraction of the central ionized molecule. Estimates of the number of satellite molecules range as high as 12. Small ions may disappear either by direct recombination with oppositely charged small ions or by combination with neutral Aitken nuclei to form new large ions, or by combination with large ions of opposite sign. The small ion, collectively, is the principal agent of atmospheric conduction.

smog. A term to describe a combination of smoke and fog, but widely used in the United States to describe objectionable air pollution; a natural fog contaminated by automobile exhausts and industrial pollutants; a mixture of smoke and fog. This term coined in 1905 by Des Voeux has experienced rapid acceptance, but so far it has not been given a precise definition. It is also used in the term "Los Angeles smog" to denote a haze produced in the atmosphere by a large-scale, photochemical oxidation process accompanied by eye irritation, plant damage, ozone formation, and characteristic odor. Smog is frequently laden with noxious gases and other irritants produced in the smoke of industrial plants and motor vehicles. Smog causes smarting eyes, fits of coughing, and even death in some instances. What intensifies man-made contributions to smog is a meteorological condition called "inversion" or "temperature inversion," which causes smog to hang motionless over a city. The smog is locked in place by a warm air mass, blanketing a lower, cool layer of air. The situation is further aggravated by the absence of wind to move the atmosphere horizontally.

smog alert. A set of conditions indicated by meteorological forecasts, air sampling, etc., where authorities responsible for air pollution control in a given area recommend or impose a reduction of pollutant output by industry and other sources which contribute to smog development. Alert levels of gaseous pollutants (CO, nitrogen oxides, sulfur oxides, O_3) are considered to be concentrations indicative of an approach to danger to public health. There are three such levels promulgated by the Los Angeles County Air Pollution Control District, the first being an initial warning level, the second a signal for the curtailment of certain significant sources, and the third the level at which emergency action must be taken.

smog index. A mathematical correlation between smog and meteorological conditions associated with it which has resulted in a formula, such as the Stanford Research Institute (SRI) smog index, which can ascertain with accuracy whether eye-irritating smog is or is not present in the area at the moment for which the index is calculated. Calculation of the SRI smog index S is $S = 10(T + 10)/RW = (I/V)^{1/2}$, where T is the °F deviation of the 24-hr mean temperature from the mean temperature for that particular day of the year, R is the relative humidity at noon, W is the total 24-hr wind movement in miles, and I is the inversion intensity. The five separate aspects of meteorological conditions originally used in the calculations were the wind movements for 24 hr, the noon visibility in miles, the relative humidity at noon, the difference between the day's 24-hr mean temperature and the normal mean temperature, and the height of the temperature inversion level.

smoke. *Principal sources*: photochemical reaction products of combusted gases. *Effects*: a variety of symptoms typical of smog components; observed effects include "silver leaf," chlorosis, and leaf drop. *Example*: Small grapefruit trees exposed to natural concentrations of smoke in the Los Angeles area dropped 36 percent of their leaves, compared with 6 percent for trees exposed to filtered air.

smoke air pollution. Small gas-borne particles resulting from incomplete combustion, consisting predominantly of carbon and other combustible materials, and present in sufficient quantities as to be observable independently of the presence of other solids. It is used in English in three ways. (a) An aerosol consisting of all the dispersible particulate products from the incomplete combustion of carbonaceous materials entrained in flue gas as a gaseous medium. (b) The persistent aerosol of solid particles and tarry droplets arising from dilution of the above with air. (c) The particulate material separated from the above aerosol. Black smoke, as dark as or darker than shade 4 on the Ringelmann chart. The smoke trail from a chimney ultimately becomes so well mixed with the air that it ceases to be visible, except possibly as a bluish haze. After such an attenuation, smoke remains a potential cause of dirt and damage. Coal smoke contains a high proportion of carbon and tarry hydrocarbons, which add to the smoke's sticking power and its tendency to form sooty deposits in chimneys and elsewhere. Smoke has the important property of, because of the small size of its particles, behaving in many ways like a gas and it has the same powers of penetration. The average diameter of a smoke particle is about 0.75 μm, or three millionths of an inch. Smoke particles do not fall to the ground by their weight, but are swept this way and that by every current of air. Smoke also sticks to the outside walls of buildings; rain will not wash it away. Smoke does not remain permanently in the atmosphere, the average time for which a smoke particle remains in suspension has been estimated as one to two days.

smoke alarm. An instrument which can provide an objective method of continuous measurement and recording smoke density by measuring the amount of light obscured by smoke when a beam of light is shone through the smoke in a flue. Most of the instruments have a scale, graded according to the Ringelmann shades. They can be fitted with an alarm which operates when the smoke is above a preset density. The Ringelmann scale is not applicable to white or colored smokes.

smoke candle. An apparatus used in collecting acid mists; tubes or candles made from glass or plastic fibers which are pressed into pads with thicknesses up to 2 in. and are mounted in banks. The efficiency is much increased when the glass is treated with silicone oil to repel water, or when normally water-repellent plastic is used.

smoke density. The amount of solid matter contained in smoke and often measured by systems that relate the grayness of the smoke to an established standard.

smoke horizon. The top of a smoke layer which is confined by a low-level temperature inversion in such a way as to give the appearance of the horizon when viewed from above against the sky. The true horizon is usually obscured by the smoke in such instances.

smoke plume. A stack effluent, the geometrical form and concentration of which depend on meteorological factors such as atmospheric turbulence and wind. The five types of smoke plumes due to distinctly different atmospheric conditions are looping, coning, fanning, fumigating, and lofting.

smoke shade. A method for measuring the particulate matter in the air. A clean white piece of filter paper is placed in an apparatus which forces a known quality of air through it. The particulates which adhere to the paper darken it. A "smoke shade" reading may be obtained by comparing the discolored paper with known standard shades.

smoking. Smoking took a sharp jump worldwide in 1978 with its greatest rise recorded in the Third World nations. According to the U.S. Department of Agriculture, world tobacco consumption reached a record 4.96 billion tons in 1978, an annual increase of 3.5 percent. Although smoking was on the increase worldwide, it continued to decline overall in the United States and other industrially advanced countries. At the same time, residents of the world's poorer countries in Asia, Africa, and Latin America were reportedly smoking more. In 1977, the United States exported nearly 67 billion cigarettes, twice the number it was sending overseas in the early 1970s. Other industrial nations, including West Germany, Great Britain, the Netherlands, and Switzerland, also showed significant gains in tobacco exports in the 1970s. The per capita use of tobacco by Americans, according to the U.S. Department of Agriculture, declined for the sixth consecutive year. It is estimated that the per capita cigarette consumption in 1979 was 3000 cigarettes, or about 195 packs. A report released by the Federal Trade Commission on September 21, 1979, found that Americans who had stopped smoking gained an average of two extra years of life. The Federal Trade Commission report also said that smokers lived an average of three months longer than those smoking before the warnings, because of the reduced levels of tar and nicotine in newer cigarettes. Smoking among teenagers had declined by 25 percent since 1974, but for the first time more girls in that age group were smoking than boys.

Snell's law of refraction. If i is the angle of incidence, r is the angle of refraction, v is the velocity of light in the first medium, v' is the velocity in the second medium, the index of refraction is $n = \sin i/\sin r = v/v'$.

snow. The precipitation of ice crystals, most of which are branched (sometimes star shaped). Snow pellets are precipitation of white and opaque grains of ice. These grains are spherical or sometimes conical; their diameter is from about 2 to 5 mm. Snow grains are precipitation of very small white and opaque grains of ice. These grains are fairly flat or elongated; their diameter is generally less than 1 mm.

sodium (Na). A silvery white, soft metallic element of atomic weight 22.991; specific gravity, 0.971; melting point, 97.8°C; boiling point, 883°C; soluble in water with decomposition to $NaOH + H_2$. More than 10 million pounds of sodium are produced every year in the United States for use in the manufacture of tetraethyl lead, a material in gasoline sold under the name "ethyl gas." Vast quantities are used in the manufacture of dyestuffs, cyanides for electroplating and metal refining, plastics, and sodium peroxide. Sodium is an excellent conductor of heat, and is sometimes used in machines to transfer heat where heat must be applied or removed quickly. Sodium is one of several metals which will emit electrons under the influence of light.

sodium azide. N_3Na, a common preservative found in many *in vitro* diagnostic products used in some 15,000 hospitals and clinical laboratories in the United States. In 1976 NIOSH advised that violent sodium azide-related explosions had occurred at a number of hospitals in the United States and Canada. Most of the accidents were associated with the use of the azide in automatic blood cell counters. The explosions were powerful enough to propel metal fragments over a wide area, posing a potentially serious hazard for workers in the vicinity. In many instances, the problem has been traced to discharging the azide waste into the plumbing system where the sodium azide reacts with copper, lead, brass, or solder in the pipes, forming an accumulation of lead and/or copper azide. Lead azide is a more sensitive primary explosive than nitroglycerine and a more effective detonating agent than mercury fulminate, according to NIOSH, and copper azide is even more explosive and too sensitive to be used commercially. Serious explosions have occurred when plumbers have attempted to penetrate blocked azide-contaminated drainage systems with a flexible metal "snake" probe or when cutting or sawing through drain pipes coated internally with an accumulation of azide deposits. As a preventive measure, NIOSH has recommended that plumbing systems into which sodium azide waste may be discharged be manufactured of materials that are free of lead and copper.

sodium hydroxide. HNaO, a white, waxy-looking solid which dissolves easily in water; it is commonly called caustic soda. When concentrated, sodium hydroxide is highly corrosive, destroying the skin and attacking a great many substances. It dissolves animal fibers, like wool, silk, and hair. Large amounts are used by the rubber, petroleum, paper, rayon, explosive, dye, drug, and general chemical industries. When sodium hydroxide is allowed to act on cotton cloth under proper control, the cotton fibers undergo a change and acquire considerable luster, a process called mercerizing. Practically all of the soap presently manufactured is made by treating fats and oils with caustic soda. A byproduct of soap is glycerol, commonly known as glycerin.

soft detergent. A cleaning agent that breaks down in nature.

soft radiation. The radiation absorbable by an absorber equivalent to 10 cm of lead or less. Radiation which can penetrate more than 10 cm of lead is termed "hard radiation."

soil acidity. As the clay-humus particles lose their mineral nutrients, the soil becomes less fertile. When these minerals are replaced, however, with hydrogen ions, the soil becomes more acidic. Acidity is defined as the concentration of hydrogen ions, which is usually expressed by the symbol pH. A pH of 1 means that one in 10,000 parts by weight of the soil is composed of hydrogen ions. If most of the minerals normally held by the clay-humus particles are replaced by hydrogen ions, the pH drops to about 4. On the other hand, if the clay-humus particles are saturated with minerals, the pH rises to about 7, close to neutral. If the soil is supplied with organic acids or lime, the pH may be lower or higher than these two values. For most soils the pH is between 5 and 7. A low soil pH may affect the solubility of various minerals, and hence adversely affect the activities of vital microorganisms. The fertility of the soil cannot be inferred from its pH alone. A more accurate way to measure the fertility of soils is to find the number of clay-humus particles present, determine from that the nutrient exchange capacity of the soil, and from that its fertility.

soil classification.   Soils are classified according to the soil series and soil class to which they belong. A soil series is based on the type of soil profile and named for the locality in which the soil was identified. All soils of the same series have developed from the same parent rock. Soil class is based on the size of soil particles in the surface layer, such as gravel, sand, silt, or clay. A particular class of soil in a given series is called a soil type.

soil conditioner.   An organic material like humus or compost that helps soil absorb water, build a bacterial community, and distribute nutrients and minerals.

soil profile.   The main division of a soil profile consists of the *A* horizon (the topmost layer), the *B* horizon (the layer beneath the *A* horizon), the *C* horizon (the layer beneath the *B* horizon), and the *D* horizon (the lowest layer). Each horizon in turn may be subdivided into several layers.

soil profile (*A* horizon).   In the *A* horizon, an upper A_{00} region is well developed, consisting of fresh leaf material and other organic detritus that has not undergone any chemical or mechanical breakdown. Below this is an A_0 level containing organic matter in a variable state of decomposition. This is commonly referred to as the *F* layer or fermentation level. The lower portion of the A_0 horizon is called the *H*, or humus, level; here annelids, insects, fungi, and other soil organisms have chemically and mechanically acted upon the material to such an extent that the source from which the material was derived is no longer identifiable. Below these upper layers is the mineral soil, or solum, the genetic soil developed from the various soil-forming processes.

soil profile (*B* horizon).   The *B* horizon can be divided into three layers: B_1, B_2, and B_3. B_1, if present, is transitional between the leached *A*-type layers and the *B* horizon, though darker in color, indicating the presence of leached chemicals. B_2 is the layer containing the maximum amount of leached material, the zone of maximum illuviation, the collection of materials that have been transported downward by gravitational water. The B_3 layer will contain some leached materials, but not as much as in the B_2 layer, and it may contain rather large chunks of parent rock material that are in a gradual process of weathering and chemical breakdown. In general, the *B* horizon is characterized by deeper colors because of illuviation and coarser textured mineral matter.

soil profile (*C* horizon).   The *C* horizon may be present in the normal profile, though it is occasionally absent because soil building may follow the weathering process so rapidly that there is no stratum located between the *B* horizon and the parent rock material (horizon *D*). The *C* horizon is often thick, consisting of large masses of weathered mineral material. The upper portion in some areas may become gleyed, meaning that an accumulation of water and lack of oxygen have caused a uniform, gray, soggy layer. Below the gleyed levels there may be thin layers of calcium carbonate (hard pan) or calcium sulfate.

soil profile (*D* horizon).   The *D* horizon is normally comprised of unweathered rock or possibly a clay or sand.

soiling. The visible damage to materials by deposition of air pollutants. Soiling may be used as an index for collector efficiency by relating the quantities of gas before and after passage through a separator which are necessary to produce equal soiling of a standard filter paper.

soiling particulate. A product of combustion emitted into the atmosphere, such as fly ash, charred paper, cinders, dust, soot, or other partially incinerated matter.

solar. Of or pertaining to the sun; by the sun, as in solar radiation and solar atmospheric tide; relative to the sun as a datum or reference, as in solar time.

solar absorber. A surface which converts solar radiation into thermal energy.

solar air mass. The optical air mass penetrated by light from the sun for any given position of the sun in the sky.

solar battery. A large photosensitive surface that generates useful amounts of electric energy when sunlight falls upon it.

solar cell. A semiconductor device which when subjected to varying levels of illumination will vary its resistance to electric current flow and is able to produce electric power; also known as a photovoltaic cell. If a small amount of phosphorus is introduced into a layer of pure crystalline silicon, this doped crystal is called an N-type semiconductor, because it has an excess of electrons which carry negative electrical charges. If a silicon crystal is doped with boron, a P-type semiconductor is created, so named because it has an excess of holes, a hole being defined as the absence of an electron and therefore a positively charged spot. Between the two layers is an intermediate surface called a P–N junction, where there is considerable mobility of electrons and holes. In a silicon solar cell, the N-type layer is extremely thin, about 0.5 µm, half of one-millionth of a meter. Photons penetrate this layer far enough to create pairs of electrons and holes near the P–N junction. Thus electrons collect in the N layer and holes in the P layer until there is a voltage built up in the silicon crystal, which pushes any additional electrons back into the thicker P layer. This voltage is about 0.65 V in a typical silicon solar cell. Electric current is drawn from the cell by making a conductive circuit from the front surface to the back. The front carries a grid of wires connected to the back, which is completely plated with a coating of conductive material.

solar collector. A collector that accumulates the sun's energy and transfers it to water moving through coils on a flat plate. In a solar collector there are several factors that affect the rate of heat loss or gain. Most important are losses by convection to air or water currents, and conduction to cooler parts of the collector by infrared radiation and reflection. The various parts of a collector are designed to reduce such losses of energy to an efficient minimum. There are two different kinds of solar collectors, one that is composed of a receiving surface covered with sheets of glass or plastic, and one composed of mirrors or lenses which focus solar radiation into a smaller area for more intense heat. Solar collectors used in residential dwellings are mostly classed as flat-plate collectors which do not focus or bend light, but absorb it directly from the sun. There are many designs of flat solar collectors. The diagram is a simple illustration of one showing sunlight going through transparent cover plates or glazing, and striking the absorbing surface, painted black. The absorber may have pipes mounted on it or include the piping as an integral feature. Both the absorber plate and pipe tubing carrying the water are painted a flat black in order to absorb a maximum amount of insolation and reflect a minimum of the sun's energy. Attached on the upper side of the thick layer of insulating material such as fiberglass or polyurethane foam under the absorber is a sheet of aluminum foil, so that heat radiated from the underside of the absorber bounces back and is not wasted.

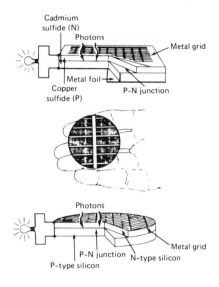

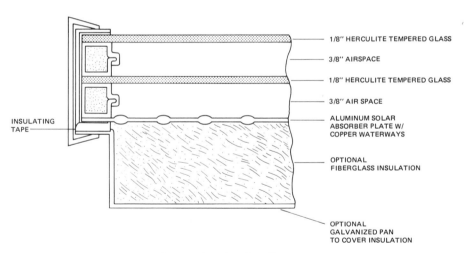

Operation of a solar cell.

1/8" HERCULITE TEMPERED GLASS

3/8" AIRSPACE

1/8" HERCULITE TEMPERED GLASS

3/8" AIR SPACE

INSULATING TAPE

ALUMINUM SOLAR ABSORBER PLATE W/ COPPER WATERWAYS

OPTIONAL FIBERGLASS INSULATION

OPTIONAL GALVANIZED PAN TO COVER INSULATION

Cross section of a solar collector.

solar collector, air-medium. In the air-medium collector, heated air can be trapped in the space between the double panes of glass or plastic and the absorber plate, or underneath the absorber plate. The heated air is ducted out to a storage system. The air-medium collector is normally composed of a thick slab of insulation on the bottom, an absorber plate painted black resting on that, and double panes of glass or plastic on the top acting as a cover. Air from the interior of the house circulates through the collector between the black-painted surface of the absorber and the enclosing cover. As the air is heated by the sun, it rises and is drawn in through ducts at the top to a storage or distribution zone inside the house. The diagram shows details of an air-medium solar collector developed by the Solaron Corporation. Air channels are underneath the absorber plate. The glass cover on top is double paned. A diagram showing an air-medium solar collector system used with domestic hot-water systems is also given. Hot air in the collectors (1) moves to the heat exchanger unit (2), where it imparts heat to water circulating from the storage tank (3) and moved by the pump (6). When hot water is drawn from regular hot-water tank (4), the storage tank water replenishes its supply. The mixing valve (5) adds cold water if the temperature is over 120°F.

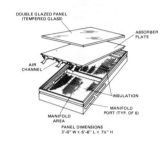

Air-medium solar collector (Solaron Corp.).

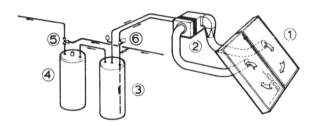

Air-medium solar collector for domestic hot-water systems.

solar collector, fluid-medium. The fluid-medium collector, from bottom up, is normally composed of a thick slab of insulation to keep the heat inside the collector box. Directly above the insulation backing is the absorber plate. Embedded in the absorber plate is a circuit of pipes running either vertically up and down the collector box, or in S shapes and other configurations. The circuit carries the liquid through the absorber plate for maximum heat transfer. As the sun shines through the cover, the black absorber plate inside the box warms up. The fluid passing through the conduit in the absorber plate collects the heat and carries it through pipes to the storage zone.

solar concentrator. A device used for increasing the intensity of solar energy by optical means.

solar constant. (1) The rate at which solar radiation is received outside the earth's atmosphere on a surface normal to the incident radiation and at the earth's mean distance from the sun. (2) The total radiation energy (including all wavelengths from the whole disk of the sun) received vertically on 1 cm^2 in 1 min, the receiver being theoretically just outside the earth's atmosphere and the earth being at its average distance from the sun. The average measured value is about 1.94 cal/cm^2 min (1.39 kW/m^2) or 1.8 hp falling upon each square meter of the earth, neglecting the effects of atmospheric absorption.

solar day. The duration of one rotation of the earth on its axis with respect to the sun.

solar domestic hot-water system.   In the conversion of solar radiation into usable energy in a solar domestic hot-water system, solar radiation is absorbed by a collector, placed in storage as required, with or without the use of a transport medium, and distributed to the point of use. The performance of each operation is maintained by automatic or manual controls. An auxiliary energy system is usually available both to supplement the output provided by the solar system and to provide for the total energy demand should the solar system become inoperable. The parts of a solar system—collector, storage, distribution, controls, and auxiliary energy—may vary widely in design and performance. They may be arranged in numerous combinations dependent upon the function, component compatability, climatic conditions, required performance, site characteristics, and architectural requirements. At present, the simple flat-plate collector has the widest application. Most solar systems can be characterized as either active or passive in their operation. An active solar system is generally classified as one in which an energy resource, in addition to solar energy, is used for the transfer of thermal energy. A passive solar system, on the other hand, is generally classified as one where solar energy alone is used for the transfer of thermal energy.

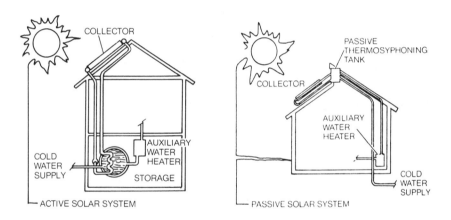

solar energy.   At a distance of 93 million miles from the sun, the earth intercepts only a small fraction of the sun's energy. With temperatures ranging from 10,000°F at its surface to an estimated 30 million °F at its center, the sun converts 4 million tons of hydrogen to helium per second. This conversion releases a constant energy flow of about 380 billion trillion kilowatts. Being electromagnetic radiation, this energy travels at the speed of light and reaches the earth in around eight minutes. Of the total energy emitted by the sun, the earth receives about 173 trillion kilowatts at its outer atmosphere. Of that amount, 30 percent is immediately reflected back into space by the atmosphere. Another 47 percent is absorbed by the atmosphere, land surfaces, and oceans and converted into heat, and 23 percent is involved in the evaporation, convection, precipitation, and surface runoff water in the earth's hydrological cycle. A small fraction, 0.2 percent, drives the atmosphere and oceanic convections and circulations, and an even smaller fraction, 0.02 percent, is utilized in the photosynthesis production of chlorophyll in green plants, which feeds the life forms on this planet.

solar energy (desalination).   Solar energy can be converted to heat at low or high temperatures (by direct absorption or by absorption after focusing), and the resulting heat can be used directly or converted to electrical energy by use of heat engines;

electricity can also be generated by thermoelectric or photovoltaic devices. The economics of solar energy use may be appraised by comparing the cost of providing a certain quantity of heat at a given temperature by conventional methods. Solar energy may be used to generate heat or power, which can then be used for operating any desalination process, or it may be employed directly for distillation as saline water in equipment which serves both as a solar energy absorber and as a distiller. Processes employing heat can be supplied with steam from boilers heated by concentrated solar energy, and those operating at low temperatures, up to about 200°F, can be heated by nonconcentrating solar collectors. Solar distillation refers to several variations of a process which involves the direct absorption of solar energy in saline water or on an adjacent surface and the evaporation of a portion of the water into an enclosed air space. In the most common forms of experimental solar distillers, condensation of pure water from the vapor–air mixture occurs in the same enclosure on a cooled surface. In other forms, the air–vapor mixture is removed from the solar evaporator, and water is condensed in a separate heat exchanger.

solar energy system. Every solar energy system has the following parts: (a) a collector for gathering and directing the heat from the sun; (b) a circulation system for moving the energy to the point of use or to storage (for later use); (c) a storage system which holds the heat, releasing it upon demand or at a controlled rate; (d) the point(s) of use which, in a residential application, would be the living space or some other place; and (e) an envelope which surrounds individual components and/or the entire closed system. In addition, most solar heating systems must have auxiliary heating plants to carry the load when the immediately available or stored energy is not sufficient to meet the demand. The diagram is a typical solar system in which the fluid medium from the collector flows into the exchanger heating storage tank for domestic hot water and space heating.

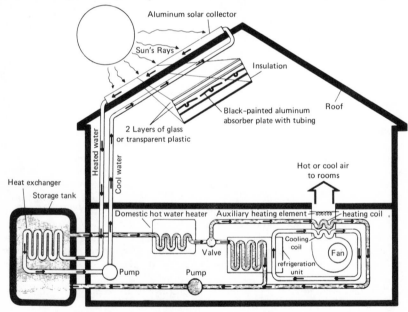

Solar energy system with fluid-medium collector.

solar energy trapping and utilization. There are essentially two ways available for trapping and utilizing solar energy. One is to let the sun's rays fall on some type of flat plate and heat either water or air which, in turn, can be used to heat, or possibly cool, a building. The other is to allow the sun's rays to fall on an array of cells which directly convert energy to electricity. Such cells are photovoltaic cells, which convert the absorbed photons of sunlight into electrons.

solar flare. A sudden and temporary outburst of light and matter from an extended region of the sun's surface; a very bright spotlike outburst on the sun, caused by an eruption of hydrogen gas, generally observed in the vicinity of large, irregular sunspots and lasting from a few minutes to an hour or even longer. Solar flares very often cause radio fading and interfere with telephonic and telegraphic communication.

solar heat. The internal temperature of the sun is of the order of $15–20 \times 10^6$ °C and that of the photosphere, 6000°C. In a normal day the solar radiation on the earth's surface is about 120 Btu/ft^2 hr and the maximum that could be anticipated is about three times this amount.

solar-heated system. A system depending on solar radiation focused by a collector onto a working fluid for the production of energy. The working temperatures of 700–815°C are less than those for an arc-heating system and no dissociation occurs in the former.

solar heater collector, batch-type. These solar heater collectors are called batch heaters because they are filled in the morning, left to heat during the day, and drained around sunset. They can be used to heat water for a large bathing tub or drained into a storage tank and linked to a pressure system by using a shut-off valve. The mass of water develops convection currents which aid heat loss, making batch-type solar heater collectors have a low efficiency. Temperatures attained can range from 100 to 125°F during the summer months to 75 to 90°F during the winter. Batch-type solar heater collectors are best used in connection with a large tub or small storage system, since they will lose most of their heat when exposed to cooler night air.

solar heater collector, cylindrical-type. A solar heater collector similar to batch and flat-plate heaters. They have a large mass which heats up slowly during the day and can be used in the evening. They suffer some heat loss through convection currents, but can be tied in with a line-pressure or forced circulation system. Commonly called preheaters, they are generally used in connection with a standard gas/electric heater tank. In terms of efficiency, they can produce a reasonable amount of hot water if a proper thermal gradient is maintained (obtained by mounting at an angle between 45–90° from the horizontal). The ideal use for cylindrical heaters is with a standard water heater. As a preheater, a solar-heated cylinder raises water temperatures from 65 to 110–120°F and can cut down on fuel bills.

solar heating component. Most solar heating systems have some or all of these basic components in common: (a) the collector, which is the part of the system that comes in contact with solar radiation; (b) a transfer medium, either air or water in virtually all systems which remove heat from the collector surfaces; (c) a storage vessel which holds the heat for use at night or on sunless days, and (d) a heat-circulating system.

solar heating panel. A panel used for solar heating, usually a shallow, rectangular box covered by one or more panes of glass or clear plastic with a special coating and/or internal properties to maximize transmission and reflection. Under the panes is a black heating surface, usually of copper or aluminum painted or coated with a compound. A desirable black surface absorbs sunlight with little reflectance. But when hot, it does not radiate much heat, which can then be lost through the panes to the atmosphere. The panes may also have a special optical coating so that they do not readily transmit the radiation from the black surface. If water is to be heated, tubing is embedded in the black surface. If air is to be heated, the black surface probably has fins on the backside to promote the transfer of heat to air as it is blown along the back of the heating surface. In appearance these panels are similar to the accompanying diagram.

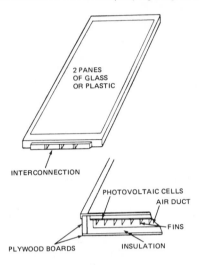

INTERCONNECTION

PHOTOVOLTAIC CELLS

AIR DUCT

FINS

PLYWOOD BOARDS

INSULATION

Solar heating panels.

solar heating system, liquid-type. A solar heating system used for space heating and domestic hot water. Sun heats the water in the collector panels, the warm water then flowing into the water storage tank. Hot water from this tank goes into an auxiliary heater and then into a loop through the house, with coils of pipe carrying hot water to each point needing space heating. Room thermostats turn on blowers to circulate air past the heated coils of water. Additional sun-heated water flows into a second storage tank, and from it into the conventional water heater. When the outlet temperature falls a preset number of degrees below the storage temperature, the pump is turned off with appropriate gate valves, and the solar collector system is shut down.

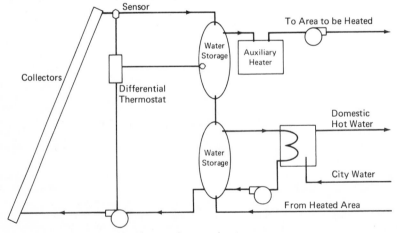

Liquid-type solar water-heating system.

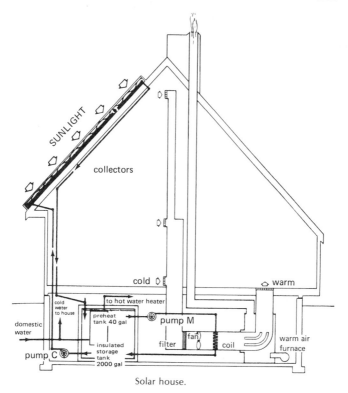

Solar house.

solar house. A house where solar collectors bring heat down from the roof and impart it to a storage tank through a heat exchanger. Air blows across hot coils carrying warmth up into the house through ducts.

solar pond. A pond with a blackened base containing an appropriate salt solution to suppress convection in order to store solar radiation effectively.

solar radiation. The total electromagnetic radiation emitted by the sun. To a first approximation, the sun radiates as a black body at a temperature of about 5700°K, 99.9 percent of its energy output falling within the wavelength interval from 0.15 to 4.0 μm, with peak intensities near 0.47 μm. About one-half of the total energy in the solar beam is contained within the visible spectrum from 0.4 to 0.7 μm, and most of the other half lies in the near infrared, a small additional portion lying in the ultraviolet spectrum. Solar radiation is an intermittent, low-intensity, abundant source of energy. In the United States, a daily average of about 1500 Btu/ft² (406 cal/cm²) is received in the ground. In the summer, daily averages in the southwest are above 2000, whereas winter in the northern parts of the world experiences average daily radiation levels in the range of a few hundred to about 2500 Btu/ft².

solar spicule. A glowing tongue of gas, like a tiny prominence, which shoots out of the sun's chromosphere near its poles and lasts only a few minutes.

solar time. Time based upon the rotation of the earth relative to the sun. Solar time may be designated as mean or astronomical if the mean sun is the reference, or apparent if the apparent sun is the reference. The difference between mean and apparent time is called the equation of time. Solar time may be further designated according to the reference meridian, either the local or Greenwich meridian, or additionally in the case of mean time, a designated zone meridian. Standard or daylight-saving times are variations of zone time.

solar window. A window oriented to trap the sun's winter heat directly in the building's interior; the collector. The temperature of the interior surfaces and the inside air is raised so that the mass of the building materials themselves serve as the heat storage. The heat is distributed by natural air movement inside the building.

solarate. A coined word meaning "to equip a house with a solar heating system, or at least a solar hot-water heater."

solenoid. An electromechanical device so designed that when electric current is applied to a coil of wire wound around a cylinder, the electromotive force causes a bar or plunger inside the cylinder to move.

solid. A state of matter in which the relative motion of the molecules is restricted and they tend to retain a definite fixed position relative to each other, giving rise to crystal structure. A solid may be said to have a definite shape and volume.

solid angle. The angle formed by three or more planes meeting in a common point, or formed at the vertex of a cone, measured by the ratio of the surface of the portion of a sphere enclosed by the conical surface forming the angle to the square of the radius of the sphere. The unit of solid angle is the steradian, the solid angle which encloses a surface on the sphere equivalent to the square of the radius.

solid state. A state of matter in which a substance possesses definite volume, definite shape, and elasticity of shape and bulk, resisting any force that tends to alter its volume or form; solids are characterized by very stable surfaces of distinct outline on all sides.

solid waste. The United States is now producing over five billion metric tons of solid waste each year. Of concern are the hazardous materials in industrial solid wastes, which amount to 10 to 30 million metric tons (dry weight) annually. Public health is directly affected by the disposal of solid waste. Traditional dumps and open landfills have served as breeding grounds for rats, flies, and other disease carriers. Leachate-form landfills often contain high concentrations of organic and inorganic materials that can seriously pollute surface and groundwater. The use of landfills for the disposal of industrial wastes containing hazardous materials heightens immensely the concern for adequate environmental protection measures. Recent studies revealed that some of these hazardous wastes have been inadequately disposed of and are leaking in high concentrations into areas where people may be exposed. Resource conservation and recovery through in-plant process changes, recycling, and waste exchange offer alternatives to the traditional patterns of waste management. Waste from agricultural and forestry activities, especially when concentrated and improperly managed, can result in the runoff of large quantities of dissolved and solid materials into surface waters, thereby seriously degrading water quality. In some water sheds, nutrients in such runoff can be the major cause of eutrophication. The runoff of pesticide contaminants can also pose hazards to animals and humans. Solid wastes are commonly disposed of by one of three methods: (a) incineration, or the burning of solid matter, now considered undesirable because of its major contribution to air pollution; (b) land-filling or sanitary landfill, the disposal of solid wastes by dumping them into a pit, natural canyon, marsh, or coastal estuary (as the depression is filled, the waste material is covered with soil), (c) reclamation, the environmentally beneficial way to avoid solid waste pollution.

solid waste disposal. The final placement of refuse that cannot be salvaged or recycled.

solid waste management. The supervised handling of waste materials from their source through recovery processes to disposal.

solifluction. The term applied to a type of creep that takes place in regions where the ground freezes to a considerable depth. As the ground thaws during warm seasons, the upper thawed portion creeps downhill over the frozen material at greater depths. As thawing continues from the surface downward, eventually the upper unfrozen layers of soil become saturated. In this condition, soil will move as a viscous liquid down slopes of as little as two to three degrees.

soluble. Capable of being dissolved in a fluid. Such a solution is generally regarded as a purely mechanical process in which no chemical change takes place in the matter dissolved or the solution.

solute. That constituent of a solution which is considered to be dissolved in the other, the solvent. The solvent is usually present in larger amounts than the solute. A solution is saturated if it contains at a given temperature as much of a solute as it can retain in the presence of an excess of that solute. A true solution is a mixture—liquid, solid, or gaseous—in which the components are uniformly distributed throughout the mixture. The proportion of the constituents may be varied within certain limits.

solvent. That constituent of a solution which is present in the larger amount; the constituent which is liquid in the pure state, in the case of solutions of solids or gases in liquids; a substance, usually a liquid, capable of absorbing another liquid, gas, or solid to form a homogeneous mixture.

solvent-refined coal. *Process:* The solvent-refined coal process produces a low-sulfur, ash-free material which can be handled in either a liquid or solid form. Coal is ground (minus 200 mesh), dried, and mixed with a coal-derived solvent having a boiling range of 500 to 800°F. Slurry, together with hydrogen, is heated to 815°F at 1000 psi, causing complete solution of the organic matter. The hydrogen requirement varies from 15,000 to 80,000 scf/ton of coal. Gas from the dissolver is separated in a high-pressure flash vessel operating at 995 psia and 625°F. Liquid and solids from the flash vessel go to a rotary filter where undissolved coal solids are removed. Filtrate is sent to a vacuum distillation. The overhead fraction is a solvent for recycle and product light oils. The bottom fraction is a hot liquid with a solidification point of about 300°F (solvent-refined coal), having a heating value of about 16,000 Btu/lb. Gas from the high-pressure flash tank is treated for acid gas removal before recycling it to the dissolver. In this process all the pyritic sulfur and over 60 percent of the organic sulfur and most of the ash are removed from the coal. Solids from the filtration system are dried to recover solvent. The dried solids containing about 35–55 percent by weight undissolved carbon and about 5–8 percent sulfur are burned in a fluidized combustor to generate steam. Stack gas containing SO_2 reacts with H_2S from the gas scrubbing system in a Claus plant to produce sulfur. Liquid product can be transported hot as a liquid fuel or allowed to cool below 300°F to solidify and produce a relatively clean solid fuel (Pittsburg and Midway Coal Mining Co.).

sone. A unit of loudness. By definition, a simple tone of frequency 1000 Hz, 40 dB above a listener's threshold, produces a loudness of one sone. The loudness of any sound that is judged by the listener to be *n* times that of the one-sone tone is *n* sones.

sonic boom. Loud noise associated with supersonic speeds. Bodies moving through the air at speeds less than the local speed of sound produce spherical sound waves propagated at the prevailing velocity of sound. If the velocity of the body exceeds that of sound, the compression of the air as a result of its motion cannot be transmitted forward, only laterally. The result of all the disturbances set up is that at speeds appreciably exceeding the local speed of sound, a very thin conical shell of a much increased pressure is shed from the tip of the projectile, and the higher the projectile air speed (indicated by the Mach number), the smaller the cone angle. At the nose of the projectile and in its immediate proximity the difference in pressure is so great that the noise wave is propagated at a velocity greater than that of sound. Subsequently, normal sonic-speed propagation ensues. At the rear of the moving body a similar cone of reduced pressure is left behind. The exact pattern of shock waves of compression and rarefaction depends on the shape of the body. In some commercial jet aircraft, cruising at subsonic speeds, the local airflow at the wingtips is apparently sufficiently rapid to create a shock wave.

soot.   Carbon particulate matter, the product of incomplete combustion. Soot is made of very finely divided carbon particles clustered together into long chains. Because they are so fine, the particles have an exceptionally broad surface in proportion to their weight, and attract a great variety of chemicals from the air around them. The ability to attract other substances makes soot not only prevalent, but dangerous.

soot fall.   During combustion, high-velocity exhaust gases will draw particulate matter up into the atmosphere. Many of these particles, however, are too large to remain suspended and will be deposited as "soot fall" onto the surrounding terrain.

sorption.   A term including both adsorption and absorption. Sorption is basic to many processes used to measure, analyze, and remove both gaseous and particulate pollutants.

sound.   An oscillation in pressure, stress, particle displacement, particle velocity, etc., which is propagated in an elastic material or a medium with internal forces (e.g., elastic, viscous), or the superposition of such propagated oscillations.

sound absorption.   The change of sound energy into some other form, usually heat, in passing through a medium or upon striking a surface; the property possessed by materials and objects, including air, of absorbing sound energy.

sound absorption coefficient.   The incident sound energy absorbed by a surface or medium, expressed as a fraction.

sound analyzer.   A device for measuring the band-pressure level or pressure-spectrum level of a sound as a function of frequency.

sound energy.   The energy which sound waves contribute to a particular medium.

sound energy density.   At a point in a sound field, the sound energy contained in a given infinitesimal part of the medium divided by the volume of that part of the medium. The terms "instantaneous energy density," "maximum energy density," and "peak energy density" have meanings analogous to the related terms used for sound pressure. In speaking of an average energy density in general, it is necessary to distinguish between the space average (at a given instant) and the time average (at a given point).

sound intensity.   The power flowing through one square meter of area, taken normal to the direction of the waves. A common method of specifying intensity is to compare the power in a given sound with the power in another. When the power in one sound is 10 times that in another, the ratio of intensity is said to be one bel.

sound intensity level.   In decibels, 10 times the base 10 logarithm of the ratio of the intensity of a sound to a reference intensity. The reference pressure must be exactly stipulated.

sound level.   A weighted sound-pressure level, obtained by the use of metering characteristics and weighting A, B, or C specified in *American Standard Sound Level Meters* for *Measurement of Noise and Other Sounds*, Z24.3−1944. The weighting employed must always be stated. The reference pressure is 0.0002 μbar.

sound level meter. An instrument consisting of a microphone, an amplifier, and an indicating meter. It responds to sound frequency components in the audible range between 20 and 20,000 Hz, depending upon the manufacturer, which meets the specifications of the American Standard Association (ASA). The sound level meter indicates the sound pressure level by a single number or a range of numbers, which is descriptive of sound pressure levels over the whole audible frequency range.

sound power. The amount of acoustic energy produced per unit time by a source. The sound power W is related to the sound intensity by $W = I_{avg} 4\pi r^2$, where I_{avg} is the average sound intensity at a distance r from a sound source whose acoustical power is W. The quantity $4 \pi r^2$ is the area of a sphere over the surface of which the intensity is averaged.

sound pressure. The instantaneous pressure at a point in a medium in the presence of a sound wave, minus the static pressure at that point.

sound-pressure level (SPL). Most sound-measuring instruments are calibrated to read in terms of the common logarithm of the ratio of the root-mean-square (rms) sound-pressure level (SPL), which is expressed in dB. The word "level" emphasizes the fact that the value is above a given pressure reference. For sound measurements in air, 0.0002 μbar commonly serves as the reference sound pressure. The sound-pressure level in decibels of a sound is 20 times the ratio of the pressure of this sound to the reference pressure. The reference pressure shall be explicitly stated.

sound wave. A disturbance whereby energy is transmitted in the medium by virtue of the inertial, elastic, and other dynamical properties of the medium. Usually the passage of a wave involves only a temporary departure of the state of the medium from its equilibrium state.

spallation. A term used to denote a nuclear reaction induced by high-energy bombardment and involving the ejection of more than two or three particles (neutrons, protons, deuterons, alpha particles, etc.).

spark chamber. An ionization chamber where the positive and negative electrodes are placed in a series so that the migration of ions can be made directly visible. When a particle or ray triggers a current in such a spark chamber, its path becomes visible as a series of light flashes. These flashes can be easily photographed to provide permanent records. An image amplifier may be used to increase the sensitivity of the spark chamber. Computers can also be linked with the chamber to analyze the resulting photographs or to make only the desired tracks visible.

specific acoustic impedance. The ratio of acoustic pressure in a medium to the particle velocity of the medium. The meter-kilogram-second unit of specific acoustic impedance is one kilogram per square meter-second or rayl.

specific gravity. (1) The ratio of the mass of a body to the mass of an equal volume of water at 4°C or another specified temperature. (2) The ratio between the density of a substance at a given temperature and the density of some substance assumed as standard. For liquids and solids, the standard assumed is either the density of distilled water at 4°C or the density of distilled water at 60°F. For gases the standards are air, hydrogen, or oxygen at 0°C and a pressure of 760 mm Hg, or distilled water at 4°C. Specific gravity is a relative property that varies with temperature.

specific heat. The amount of heat in kcal which is necessary to raise the temperature of 1 kg of material by 1°C.

specific heat capacity. That quantity of heat which will increase a mass of material by a given temperature. Materials differ widely in the quantity of heat required to raise a standard amount of mass a standard temperature (fahrenheit or centigrade) unit increment. Specific heat is normally expressed in Btu/lb°F or cal/g°C.

specific humidity. The weight of water vapor contained in a unit weight of air (dry air plus water vapor), expressed in grams per gram, or grams per kilogram.

spectrum. (1) The distribution in frequency of the magnitude (and sometimes phase) of the components of a wave. Spectrum is also used to signify a continuous range of frequencies, usually wide in extent, within which waves have some specified common characteristics, e.g., an audio frequency spectrum, etc. The term spectrum has also been applied to the distribution of sound-pressure level as a function of frequency or frequency band. (2) Light or radiant energy arranged or displayed in proper order of its component colors or wavelengths. A body emitting radiation shows a characteristic distribution of energy emitted over the various wavelengths under given temperature conditions of the body. With the proper apparatus, this distribution can be analyzed in detail and the results can then be presented as plots of emissive power against wavelength for different temperatures. Such a plot is often called the emission spectrum of a body at the temperature in question. The spectrum of the sun shows a peak of emissive power in the visible wavelengths, but about half its total emission is outside the visible range, in the infrared and ultraviolet wavelengths.

spectrum pressure level. The level of sound of a continuous spectrum and at a specified frequency; the band-pressure level for a bandwidth of one cycle per second centered at a specified frequency .

speech interference level (SIL). The average, in decibels, of the sound pressure levels of the noise in the three octave bands of frequency 600–1200, 1200–2400, and 2400–4800 Hz.

speed of light. The speed of propagation of electromagnetic radiation through a perfect vacuum; a universal dimensional constant equal to 299,792.5 ± 0.4 km/s.

spherical aberration. When large surfaces of spherical mirrors or lenses are used, the phenomenon of light divergent from a point source not being exactly focused at a point.

spin. In nuclear physics, a number used to describe the angular momentum of elementary particles or of nuclei; the unceasing rotation, with an angular momentum of $\frac{1}{2}h/2\pi$, which had to be attributed to the electron to account for spectroscopic and other facts. Dirac modified Schrödinger's wave equation to fit into Einstein's special theory of relativity and found that the electron spin was a direct consequence of that modification. Since then, nearly all fermions have been found to spin with the same angular momentum; exceptions are the omega-minus hyperon ($\frac{3}{2}h/2\pi$) and some of the very short-lived baryons.

spoil. The overburden removed and dumped to the side in gaining access to a coal bed in strip mining operations.

spot test. A means of detecting a chemical constituent, or class of constituents, by a color-producing chemical reaction. The function of spot tests is the rapid, simple qualitative detection of air pollutants.

spout. A phenomenon consisting of an often violent whirlwind, revealed by the presence of a cloud or inverted cloud cone (funnel cloud) protruding from the base of a cumulonimbus, and of a "bush" composed of water droplets raised from the surface of the sea, or of dust, sand, or litter raised from the ground; tornado, or waterspout.

sprawl. The unplanned development of open land.

squall line. A nonfrontal line of thunderstorms that generally precedes fast-moving cold fronts.

stable. Not likely to change.

stable air. A mass of air that is not moving normally, so that it holds, rather than disperses, pollutants.

stable isotope. An isotope of an element which is not radioactive; a mixture of isotopic, nonradioactive nuclides of compositions different from those occurring in nature.

stability curve. In a chart of the nuclides (different kinds of nuclei), a line that runs through the most stable isotope of each chemical element.

stabilization. The conversion of active organic matter in sludge into inert, harmless material.

stabilization pond. See **lagoon.**

Stablex. The tradename for a toxic waste treatment product developed by Stablex Canada Ltd., a subsidiary of the Stablex Corporation of London. The toxic waste materials are neutralized with chemicals, and the resulting compounds are mixed with wet cement and poured into pits to harden.

stack A vertical passage or chimney, whether of refractory, brick, tile, concrete, metal, or other material, for conducting cooled products of combustion to the atmosphere from a process.

stack effect. Used air, as in a chimney, that moves upward because it is warmer than the surrounding atmosphere.

stack effluent. Gaseous and particulate waste products discharged to the atmosphere through stacks of some form.

stack gas. Gaseous waste products discharged to the atmosphere through a stack.

stack height selection. Various limitations prevent the construction of stacks to be as high as might be desired from the polution viewpoint. Among these limitations are cost, air traffic considerations and hazards, esthetic considerations, aerodynamic factors, vibration, and materials. Stock stacks are now reaching the 700-ft level and the 1000-ft level is anticipated.

stack sampling. The collection of representative gaseous and particulate samples of matter flowing through a duct or stack. Acceptable performance should indicate a collection efficiency of 95 ± 5 percent. It may be necessary to use more than one sampling device if partition of the contaminant is desired. Three major requirements must be met in sampling particulate matter: The sample must be taken under isokinetic conditions; there must be a sufficient number of samples; and the long axis of the sampling should be parallel to and facing the gas flow direction. Filter paper holders or electrostatic sampling may be used for particulates, and scrubbers for gases; the standard impinger may be used for either or both together.

stagnant inversion. During cloudless evenings and nights, particularly in the fall and winter, the earth's surface loses heat by radiation and in consequence, the air adjacent to the ground is cooled and reduced to some temperature below that of the air at some height above. Thus air pollutants are trapped in this layer of air, which, being colder than the layer above it, will not rise. During extreme stagnation, pollution reaches such high concentrations that a major decrease in solar radiation results over a period of time, coincidentally ensuring the stability of the lower atmosphere.

stagnation. An atmospheric phenomenon responsible for air pollution inversion, where concentrations of air pollutant develop in air layers close to the ground. (1) *Stagnation point*. A point in a field of flow about a body where the fluid particles have zero velocity with respect to the body. (2) *Stagnation pressure*. The pressure at a stagnation point. (3) *Stagnation region*. Specifically, the region at the front of a body moving through a fluid where the fluid has a negligible relative velocity.

staining. The visible damage to materials caused by chemical reactions between air pollutants and substances contained in materials.

standard conditions for gases. Measured volumes of gases are quite generally recalculated to 0°C and 760 mmHg, which have been arbitrarily chosen as standard conditions.

standard temperature and pressure (STP). A pressure of one atmosphere (the average atmospheric pressure at sea level) and a temperature of 0°C. Since the volume of a gas is affected by temperature and pressure, these variables must be fixed when volumes are being determined.

standing wave. A periodic wave having a fixed distribution in space, the result of interference of progressive waves of the same frequency and kind. Such waves are characterized by the existence of nodes or partial nodes and antinodes that are fixed in space.

statcoulomb. (1) That quantity of electric charge which, when placed in a vacuum 1 cm distant from an equal and like charge, will repel it with a force of 1 dyn. (2) The unit of electric charge in the metric system; 3×10^9 statcoulomb = 1 C.

state. A particular position and shape in which an electron (or other particle) can come to rest; also, a collection of such positions and shapes that can be occupied by an equal number of particles.

static head. The pressure differential between two points in a fluid system in terms of the vertical distance between these two points, measured while the system is static.

static pressure. At a point in a medium, the pressure that would normally exist at that point in the absence of sound waves. The unit is the newton per square meter.

stationary source. A pollution location that is fixed, rather than moving; one point of pollution rather than a widespread one.

steady state. A state said to have been reached by a system when the relevant variables of the system no longer change as functions of time.

steam plume. A wet-type scrubber or wet-type electrostatic precipitator that may show a steam discharge plume upon becoming saturated with moisture. Steam plumes are the result of the rapid cooling of gases or air carrying moisture to below the saturation temperature. In addition to their appearance, steam plumes may have other side effects such as SO_2 (or other corrosive gases) possibly becoming aggravated through steam plume condensation when absorbed by the newly formed droplets into sulfurous acid and then falling on homes and industrial sites. In some cases odoriferous constituents may be entrapped by falling droplets to increase odor at ground elevation.

Stefan–Boltzmann constant. A universal constant of proportionality between the radiant emittance of a black body and the fourth power of the body's absolute temperature; 5.6697×10^{-5} erg/cm² s °K.

Stefan–Boltzmann law of radiation. The energy radiated in a unit of time by a black body is given by $E = K(T^4 - T_0^4)$, where T is the absolute temperature of the body, T_0 is the absolute temperature of the surroundings, and K is a constant.

steradian. The unit solid angle which cuts a unit area from the surface of a sphere of unit radius centered at the vertex of the solid angle. There are 4 π steradians in a sphere.

Stoke's law. The empirical law stating that the wavelength of light emitted by a fluorescent material is longer than that of the radiation used to excite the fluorescence. In modern language, the emitted photons carry off less energy than is brought in by the exciting photons; the details are in accordance with the energy conservation principle.

stoker. A machine for feeding coal into a furnace and supporting it there during the period of combustion; any mechanical device that feeds fuel uniformly onto a grate or hearth within a furnace. It may also perform other functions, such as supplying air, controlling combustion, and distilling volatile matter. Modern stokers may be classified as overfeed, underfeed, and conveyor types.

stomata. Tiny openings on the underside of a leaf, through which a plant takes in carbon dioxide and some polluting gases.

storage cell. An electrochemical cell in which the reacting materials can be renewed by the use of a reverse current from an external source.

storm sewer. A system that collects and carries rain and snow runoff to a point where it can soak back into the groundwater or flow into surface waters.

strain. The deformation resulting from a stress measured by the ratio of the change to the total value of the dimension in which the change occurred.

stratification. (1) A process used in solar thermal storage systems whereby hot water, being less dense, floats on the top of denser cold water. (2) In ecological terminology, a vertical layering of organisms or environmental conditions within a biotic community, considered as one of the integral properties of nearly every natural community.

stratification, animal. Animal life is stratified within plant communities with some mobility. Often animals will move from one stratum to another in search of food or in response to numerous abiotic factors. Arboreal species that may move from the seedling stratum to higher strata are, among many, insects, snails, birds, squirrels, opossums, etc. Animals that are most commonly situated above the soil include turtles, snakes, some lizards, certain insects, some birds, and a variety of mammals such as rabbits, deer, wolves, and foxes. A faunal stratum inhabits the litter and humus microstrata of the soil, which includes springtails, beetles, fly larvae, millipedes, centipedes, mites, spiders, annelids, snails, and sow bugs. At a still lower level in the mineral soil are protozoans, nematodes, annelids, and some springtails.

stratification, aquatic. Aquatic communities exhibit vegetative stratification. Emergent hydrophytes located around the periphery of ponds and lakes are partially submerged in water, with a greater portion extending above the water's surface; cat tails, bulrushes, reeds, and sedges are examples. Floating leaf aquatics, including water lilies and water hyacinths, will often shade out a submergent vegetation in shallow waters. Submerged vascular vegetation comprises a stratum relatively close to the shoreline, the vegetation rooted, but unable to carry out sufficient photosynthesis to compensate for metabolic losses at increased depths. These include the water weed, tape grass, pickerel weed, or parrot's feather, and water crowfoot.

stratification, deep freshwater and oceanic. Deep freshwater communities and oceanic areas may be stratified into a photic and an aphotic (devoid of all light) layer. Marine ecologists recognize basically three zones: the epipelagic layer extending down to about

200 m in clear ocean waters, supporting a large photosynthetic phytoplanktonic population; the mesopelagic region, a twilight area extending from 200 to 5000 m, often with very little light below 1000 m, preventing efficient photosynthetic activity; and the lower bathypelagic stratum which in the open ocean will have the greatest depth (5000 to 10,000 m), a region of permanent darkness.

stratification forest. A typical forest will have an overstory stratum comprised of trees that are 40 or more feet in height in mature stands. These trees will form the canopy, under which there is often an understory stratum that extends from 20 feet in height to a short distance below the overstory. A transgressive stratum extending from about 4 to 20 feet or more, is comprised of a shorter, shade-tolerant species that, in the course of germination and growth, are passing through the transgressive stratum. The transgressive stratum is dependent in part upon a seedling stratum that begins at the soil level and extends to the lower limit of the transgressive stratum.

stratification, marine animal. Some animals move from one stratum to another, but most spend more time in one particular stratum. Various groups may be distinguished, such as the neuston, the faunal group closely associated with surface film covering a body of water, usually quiet waters such as ponds, lakes, shallow pools, and backwater areas along streams. Aquatic birds, water striders, whirligig bettles, spiders, and egg rafts of mosquitoes are commonly associated with surface film. Such creatures are referred to as supraneustons. The infraneustrons, which come in contact with the submerged surface of the film, are mosquito larvae, mosquito pupae, aquatic snails, and some cladocerans. Plankton is also found at this level, often comprised of many species of algae and bacteria. Zooplankton contains such animals as protozoa, jellyfish, shrimp larvae, rotifers, and certain annelids, and may be found at different times and in different stages of their life cycles from the surface film to the bottom. Nannoplanktons exceed in total volume the larger planktonic forms in many areas. The nektons, which occupy the same stratum, include the more advanced phyla such as the arthropods (mature shrimp, some crabs, diving beetles, giant water bugs, back swimmers, water boatmen), mollusks (squid, scallops, sea hares), and the chordates, including a great variety of bony and cartilaginous fish, amphibians, reptiles (crocodiles, alligators, turtles), and mammals (seals, walruses, whales, porpoises, muskrats, beavers). The lowest stratum of aquatic faunal life is the pedon, which includes all the bottom-dwelling types that live within the substrate; such protozoans as coelenterates, flatworms, annelids, arthropods, some chordates, and snakes make up the faunal representatives of this stratum.

stratopause. The upper limit of the stratosphere and the lower limit of the mesosphere in the earth's atmosphere.

stratosphere. A layer of the earth's atmosphere in which there is no convection and practically no temperature gradient. It extends between the troposphere and the ionosphere, roughly between the average heights of 9 and 50 km above the earth's surface. One of its chief characteristics is the scarcity of oxygen. Scientists discovered that the upper level of the stratosphere (ozonosphere) is quite different from its lower portion. This difference is attributed to the fact that the upper stratosphere soaks up large quantities of the sun's radiation and manufactures ozone (O_3), an unstable gas both formed and broken up by solar radiation.

stress. The force producing or tending to produce a deformation in a body, measured by the force applied per unit area.

strip mining. A type of surface mining whereby overburden and coal are removed from successive, long, parallel cuts with overburden from the second and successive cuts being placed in the previously mined cut. This system is practicable only where the coal lies close to the surface. Strip mines operate where the depth of the minable coal beds increases gradually, allowing mining to proceed at distances of up to a mile or more into the outcropping seam. This method is common in the midwest and west of the United States.

stripcropping. Growing crops in a systematic arrangement of strips or bands which serve as barriers to wind and water erosion.

subatomic particle. A particle smaller than an atom. Through the use of detection devices and particle accelerators physicists have observed and identified more than 200 subatomic particles. Almost all of these particles are unstable, many of them with extremely short lives, moving at very high speeds. The names and symbols of subatomic particles are taken from Greek words and letters. The names refer to masses: baryon means heavy, meson means medium, and lepton means small. Baryons and mesons are sometimes also called hadrons, which come from a Greek word meaning bulky. Each of the fundamental particles has a corresponding antiparticle. In some cases a particle is its own antiparticle. The masses of the particles are based on a rest mass of unity for the electron. Neutrinos and photons are given a rest mass of zero because they do not exist as stationary entities. There are now 82 particles listed in the Gell-Mann classification. This classification of subatomic particles is comparable to the classification of elements by Mendeleev in his periodic table of the elements. Subatomic particles are classified according to their special characteristics: their mass, charge, lifetime, spin, parity (left handedness or right handedness), and strangeness (how they interact with other particles). Most variations in subatomic particles are due to differences in energy levels.

subatomic particle symbols.

Symbol	Particle
$_{0}^{1}n$	Neutron
$_{1}^{1}H$	Proton (hydrogen nucleus)
$_{-1}^{0}e$	Electron (beta particle)
$_{+1}^{0}e$	Positron (positive electron)
$_{2}^{4}He$	Alpha particle (helium nucleus)
γ	Gamma ray

subatomic particles (effect of interaction). There are three main types of reactions which can occur when subatomic particles interact. One is the scattering of particles as they collide and bounce away in different directions like billiard balls. Another is the creation of new particles. This occurs when a particle is chipped off or created as particles collide or are bombarded. A third reaction is the annihilation of particles. This results when a particle is transformed, by either the addition or loss of energy, into a new and different particle.

sublimation. The process by which a gas is changed to a solid or a solid to a gas, without going through the liquid state.

subscript. A number written below and to the side of a symbol. If at the left, it represents the atomic number; if at the right, it represents the number of atoms of the element.

subshell. A different and specific energy level within the main electron shell of an atom. Electrons within the same shell (energy level) of an atom are characterized by the same principal quantum number (*n*), and are further divided into groups according to the value of their azimuthal quantum number (*l*). Electrons which possess the same azimuthal quantum number for the same principal quantum number are considered to occupy the same subshell (or sublevel). The individual subshells are designated with the letters *s*, *p*, *d*, *f*, *g*, and *h*, as follows:

l value	Designation of subshell
0	*s*
1	*p*
2	*d*
3	*f*
4	*g*
5	*h*

An electron assigned to the *s* subshell is called an *s* electron, one assigned to the *p* subshell is referred to as a *p* electron, etc. In formulas of electron structure, the value of the principal quantum number (*n*) is a prefixed to the letter indicating the azimuthal quantum number (*l*) of the electron; thus, e.g., a 4*f* electron is an electron which has principal quantum number 4 (i.e., assigned to the *N* shell) and orbital angular momentum 3 (*f* subshell).

succession.   The orderly sequence of different communities over a period of time in some particular area.

succession, hydrarch.   The biotic factor that has the most drastic effect on the types of communities inhabiting an area is the amount of moisture present. Water or its relative absence will govern the entire character of the successional pattern. In regions where water is plentiful, a hydrarch succession will ensue. This sequence of communities will occur in ponds, lakes, streams, swamps, or bogs. In these areas the conditions vary from one extreme, where animals and plants are completely submerged in water, to where the water table is high and soil moisture values are consequently great enough to produce a soggy, water-bound environment.

succession, mesarch.   An intermediate type of succession, with moisture present in adequate amounts. The successional series is much shorter because moisture conditions are more ideal, and the initial water problems that must be resolved in a xerach or hydrarch type of succession are nonexistent.

succession, primary.   A succession in which the biota become established on a particular substrate for the first time. The first group of plants and animals to become established is called the pioneer community. The aggregations that follow are secondary communities. All of the communities that will exist for an interval of time are collectively termed a sere; any one group of organisms or community is a seral stage.

succession, secondary.   In secondary succession the substrate has been occupied by aggregations of organisms in the past, but a catastrophic event, a change in climatic factors, or man's intervention has caused the community occupying that area to disappear. Very often the seral stages that become established during the latter stages of succession may be similar or identical to the seral communities that would have occupied the area if primary succession had been allowed to run its course.

succession, xerarch.   A type of succession that occurs when moisture is present in minimal amounts or is not available to plants and animals because of some intrinsic factor, such as in a desert community.

sugarcane residue. Situated on the island of Hawaii, the Pepeekeo Mill of the Hilo Coast Processing Company produces 90,000 tons of raw sugar annually plus 100 million kW hr of electricity which is sold to the Hawaiian Electric Light Company. The development of this dual product concept was stimulated by a need for improved environmental control techniques, particularly those pertaining to the ocean discharge of soil-bearing effluent from cane cleaning operations. Following a series of cleaning operations, sugarcane is crushed between large rollers to extract the juice. The remaining sugarcane fiber (bagasse) plus residual leaf trash is then stockpiled before fueling a large steam boiler producing 330,000 lbs of steam per hour, at 120 psi. High-temperature steam from this boiler passes sequentially through three distinct processes which extract energy from the steam at successively lower temperatures in a clear-cut example of "cogeneration." Initially, the steam enters a 35,000 hp turbogenerator capable of supplying 20 percent of the island's electrical demand. Exhaust steam from this turbine then drives three low-pressure turbines to furnish mechanical power to the mill. Finally, the exhaust from these three units supplies heat to a series of evaporators which convert cane juice to raw sugar. At this point the condensed steam has given up most of its energy and is returned to the boiler as feedwater.

sulfate and sulfides. Sulfides and their oxidation sulfates are found as the result of natural processes and as the byproduct of oil refinery, tannery, pulp and paper mill, textile mill, chemical plant, and gas manufacturing operations. The recommended permissible criterion is not more than 250 mg/liter of sulfate in public water supplies. Concentrations in the range of less than 1.0 to 25.0 mg/liter of sulfides may be lethal in one to three days to a variety of freshwater fishes.

sulfur dioxide. SO_2, corrosive and poisonous gas produced mainly from the burning of sulfur-containing fuel (82 percent) and from certain industrial processes (15 percent). Most sulfurous coal and oil is burned in urban areas, where population and industry are concentrated. Sulfur dioxide affects human breathing in a direct relation to the amount of gas in the air breathed. Many types of respiratory diseases, such as coughs and colds, asthma, and bronchitis are associated with sulfur pollution. Sulfur dioxide and particulates often occur together. At least four, and perhaps as many as six, different oxides of sulfur are known, all existing as gases under normal conditions: monoxide (SO), sesquioxide (S_2O_3), dioxide (SO_2), trioxide (SO_3), heptoxide (S_2O_7), and tetroxide (SO_4). For air pollution, the most important oxides are the dioxide (SO_2) and trioxide (SO_3). These gases are readily absorbed by vegetation, soil, and water surfaces, causing corrosion and other damages to wires, metals, textiles, and building materials. Many green plants are particularly sensitive to dioxide (SO_2) and are injured by exposure of a few hours to concentrations as low as 0.3 ppm. Sulfur dioxide is formed in considerable quantities when coal, coke, or certain fuel oils are burnt. Though it is not as chemically active as sulfur trioxide, hydrochloric acid, and fluorine compounds, which are also liberated during the combustion of coal, it is emitted in much greater quantities and is thus capable of doing more harm. From a consideration of atomic weight, a molecule of sulfur dioxide weighs twice as much as an atom of sulfur. So for each ton of sulfur in a fuel there are nearly two tons of sulfur dioxide in the flue gases, and it is estimated that an average of nearly three tons of sulfur dioxide is emitted from every hundred tons of coal or coke burnt. Sulfur dioxide is a gas with a choking smell. It can be smelled in the flue gases from coke fines and gas fires, but its smell is often obscured by smoke in the flue gases from coal fires. Sulfur dioxide is soluble in water, and is particularly liable to attack paint, metals, stone work, and slate when water is present. Some sulfur dioxide is removed from the air by solution in cloud droplets, in falling rain, and in surface water, but less than one-fifth of the sulfur dioxide emitted to the atmosphere is brought down with rain. It is believed that much of the remainder is dissolved in the water on buildings, soil, and vegetation, both after the rain and at other times. The average time during which sulfur dioxide remains in the air has been estimated at less than 12 hours.

Sulfur dioxide enters a growing plant through the stomata, the tiny openings on the underside of the leaf, as carbon dioxide does. The injury it causes may show up as markings along the edges or between the veins of the leaf. The damaged area usually appears dried and bleached, white or ivory in color. Sulfur dioxide seems to undergo a swift chemical change in the leaf, and if exposure is low and brief, plant development may be only temporarily inhibited. On the other hand, a long period of sublethal concentrations may result in chronically injured areas that never recover; and at higher concentrations, the plant cells die, the tissues between the veins collapse, and the leaf slowly takes on the typical scars of sulfur dioxide. Different species and varieties within a species may vary considerably in their susceptibility, primarily because of their different rates of absorption. Plants with thin leaves such as alfalfa, barley, cotton, and grapes usually suffer the most. Plants with fleshy leaves or needles, such as citrus and pine, tend to be resistant, except when the leaves are newly formed.

sulfur dioxide smog. When coal is burned, sulfur compounds, especially sulfur dioxide, are released in the smoke. Sulfur dioxide is found in greatest concentrations in the air surrounding the major industrial cities of the world. Sulfur dioxide is not an especially toxic gas, but in a humid atmosphere it is converted into sulfur trioxide or sulfuric acid, and adsorbed onto the fine particles or fly ash that result from combustion. The major point of vulnerability in the human body lies in the delicate membrane lining of the eyes, nose, and respiratory tract. A polluted atmosphere containing sulfur dioxide, ozone, or nitric acid can inhibit cilial action, allowing particles to remain in the respiratory system and cause damage.

sulfur oxides. Compounds containing sulfur and oxygen. When fuels containing sulfur are burned, the sulfur joins with the oxygen of the air, and gaseous oxides of sulfur are the byproducts. Fuel combustion is the major source of the polluting sulfur oxides, although they are also produced in chemical plants and, to a lesser extent, by processing metals and burning trash. The major oxide of sulfur that is produced in combustion is sulfur dioxide (SO_2), a heavy, pungent, colorless gas that combines easily with water vapor to become sulfurous acid (H_2SO_3), a colorless liquid. Sulfurous acid, which is mildly corrosive, is used as a bleaching agent in industry. It joins easily with oxygen in the air to become even more corrosive.

sump. In liquid-type gravity solar heating systems, a small tank into which fluid from the collectors may be drained. The sump may perform the functions of an accumulator.

sun. The central body of the solar system, a gigantic thermonuclear reactor, the gravitational pull of which keeps the bodies of the solar system tightly together. It has a diameter of 870,000 miles, mass of 2×10^{27} tons, a specific gravity of 1.41, a surface gravity of 27.9 g; and a period of rotation of about 27 days. The sun is about 8000 parsecs from the center of our galaxy and makes one revolution at 225 km/s in 200×10^6 years. The sun rotates on an axis inclined at $83°$ to the ecliptic in about a month. The sun emits many spectral lines from about 1800 down to 80 Å. It emits radio waves of varying duration and different wavelengths, some of which are circular polarized and others unpolarized; most are related to optical phenomena.

sunlight to electricity, conversion of. Methods using sunpower to generate electricity are defined as being photovoltaic, involving the direct conversion of sunlight to electricity. There are presently two types of solar photovoltaic cells of commercial significance. One uses very thin, single-crystal wafers of pure silicon about the size of the palm of a hand. The other is basically a thin film of polycrystalline cadmium sulfide, which can theoretically be almost any size or shape. To employ these cells, a large expanse of panel on a roof is needed, with the topmost layer being a clear glazing material, such as plastic or glass. Below this layer will be an array of small photovoltaic cells, tightly packed to capture the maximum amount of sunshine. The individual cells will be wired to each other on the panels, the panels interconnected, and, finally, a wire will lead into the home to deliver the electricity. A photovoltaic system will not only provide power for heating and air conditioning, but can also generate electricity for lights, appliances, or anything that uses electricity.

supersonic transport (SST).   A jet airplane that flies faster than the speed of sound; it may be extremely noisy upon takeoff and landing. See also **sonic boom.**

surfactant.   A surface-active chemical agent, usually made up of phosphates, used in detergents to cause lathering. The phosphates may contribute to water pollution.

surveillance system.   A series of monitoring devices designed to determine environmental quality.

suspended solids (SS).   Tiny pieces of pollutants floating in sewage that cloud the water and require special treatment to remove.

symbiosis.   A long-term, interspecific relationship in which two species live together in more or less intimate association. This is not a social system, but an ecological association involving a transfer of energy or adaptive benefit. Symbioses are of three general types: commensalism, occurring when one species benefits from the association but the other species is not significantly affected; mutualism, where both species in the relationship benefit; and parasitism, where one species, the parasite, is benefited, while the other species, the host, is harmed. All three types are very important interactions in biological communities. Without the mutualistic association between nitrogen-fixing bacteria and legume plants, as an example, the nitrogen biogeochemical cycle would be far slower and would probably limit the expansion and evolution of life on earth to a lower level than exists today.

synchrocyclotron.   An improved cyclotron, its electrical frequency being synchronized with the particles and keeping in step with their increased mass. This allows the synchrocyclotron to accelerate particles to higher energies. Both the cyclotron and synchrocyclotron use electric fields to accelerate the charged particles and magnetic fields to guide them along their prescribed paths. The moving charged particles generate their own magnetic fields which interact with the magnetic fields guiding them.

synergism.   The cooperative action of separate substances such that the total effect is greater than the sum of the effects of the substances acting independently.

synoptic chart.   A weather map showing the weather conditions over a large area at a given time.

Synthane process.   *Process:* Crushed and dried coal is fed to the fluid-bed pretreater (if necessary) through lock hoppers. About 12 percent of the total steam and oxygen required in the process is fed to the pretreater. The operating temperature of this stage is about 800°F. Coal is partially devolatilized and its caking tendency is destroyed. Coal along with any separated volatile matter and excess steam is fed to the top of the fluid-bed gasifier. A mixture of steam and oxygen is introduced at the bottom. The gasifier operates at about 1800°F and 500–1000 psi. Char, containing roughly 30 percent of the carbon from the original coal, is removed, combined with the char from the cyclone separator, and sent to the power plant. Dust particles are removed from the raw product gas in cyclones. Tar is removed by a water wash. Cleaned gas goes to the shift converter. The overall H_2/CO ratio is raised from 1.7 to 3.0. Gas from the shift converter is treated to remove CO_2 and H_2S. The purified gas is methanated and dehydrated to produce pipeline-quality gas. The substitution of air for oxygen to the gasifier will produce a low-Btu raw gas (U.S. Bureau of Mines).

Synthoil process.   *Process:* The Synthoil process has been developed to convert coal into fuel oil. Coal is crushed, dried, and slurried in a recycled portion of its own product oil.

Slurry is fed to a fixed-bed catalytic reactor with turbulently flowing hydrogen. The combined effect of hydrogen, turbulence, and catalyst is to liquefy and desulfurize the coal. A commercially available catalyst is used, consisting of cobalt molybdate on silica-activated alumina. The conditions of operations are 850°F and 2000–14,000 psig. Violently moving slurry prevents plugging of the packed bed as the coal passes through its sticky plastic phase before becoming liquid. This also increases the mass transfer of hydrogen into slurry and keeps the catalyst surface clean by controlled attrition. The slurry residence time is less than 14 min. The reactor pressure drop is 150 psi. Product passes through a high-pressure receiver, where the gas is separated and recycled after H_2S and NH_3 removal. Product-slurry oil pressure is reduced in passing to a low-pressure receiver. It is centrifuged to remove ash and organic coal residues. Part of the centrifuged oil is recycled as slurry oil and the remainder is the product oil.

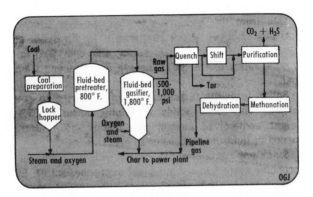

Synthane process (U.S. Bureau of Mines).

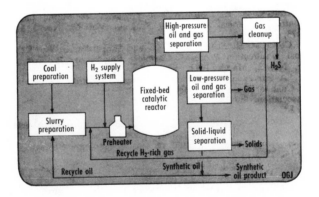

Synthoil processing (U.S. Bureau of Mines).

systemic pesticide. A chemical that is taken up from the ground or absorbed through the surface and carried through the systems of the organism being protected, making it toxic to pests.

T

taiga. The northern coniferous forest biome found particularly in Canada, northern Europe, and Siberia.

tailing. The residue of raw materials or waste separated out during the processing of crops or mineral ores.

tape sampler. A device used in the measurement of both gases and fine particulates. It allows air sampling to be made automatically at predetermined times.

Teflon. Polymerized tetrafluoroethylene; a synthetic plastic material.

telluric line. An absorption line in a solar spectrum produced by constituents of the atmosphere of the earth itself, rather than by gases in the outer solar atmosphere such as those responsible for the Fraunhofer lines. The terrestrial nature of the absorption processes responsible for telluric lines is revealed by their intensity variations with the solar zenith angle and by their freedom from any Doppler broadening due to solar rotation. Water vapor produces the strongest of the telluric lines in the visible spectrum.

temperature. The condition of a body which determines the transfer of heat to or from other bodies; particularly, a manifestation of the average translational kinetic energy of the molecules of a substance due to heat agitation. The customary unit of temperature is the centigrade degree, 1/100 the difference between the temperature of melting ice and that of water boiling under standard atmospheric pressure. The Celsius temperature scale is another designation for the centigrade scale. The degree Fahrenheit is 1/180 the temperature difference between melting ice and boiling water at STP, and the degree Reaumur is 1/80 this same difference of temperature. The fundamental temperature scale is the absolute, thermodynamic, or Kelvin scale, where the temperature measure is based on the average kinetic energy per molecule of perfect gas. The zero of the Kelvin scale is -273.16°C. The temperatue scale adopted by the International Bureau of Weights and Measures is that of the constant-volume hydrogen gas thermometer. The magnitude of the degree in both these scales is defined as 1/100 the difference between the temperature of melting ice and that of boiling water at 760 mm Hg pressure. Frequently, the Kelvin scale is defined as degrees C + 273.16°C and the Rankine scale as degrees F + 459.69°F.

temperature inversion. A layer in which temperature increases with altitude. The principal characteristic of an inversion is its marked static stability, so that very little turbulent exchange can occur within it.

temperature of the dew point. For every value of the water-vapor content, the temperature at which this quantity will produce saturation. Since supersaturation does not ordinarily occur, this temperature would be lower than the actual temperature or equal to it in the case of saturation from the beginning. The temperature of the dew point may be achieved by cooling the air at constant pressure and constant total water vapor content.

tempering valve. A mixing valve set by the user and placed at the outlet of the domestic hot-water system (input to the home) so that some cold water is mixed with the hot water, heated either by the sun or conventional fuel. This keeps scalding water out of faucets.

teratogenic. A substance that is suspected of causing malformations or serious deviations (which can't be inherited) from the normal type in or on animal embryos or fetuses.

terracing. Dikes built along the contour of agricultural land to hold runoff and sediment, thus reducing erosion.

terrigenous sediment. A sediment derived from the erosion of land. Large quantities of gravel, sand, and silt have their sources on the continents. In addition, land-derived solutions yield various chemical precipitates, including sodium chloride; calcium and magnesium carbonates; and iron, manganese, phosphatic and baritic concentrations and nodules. Some of the inorganic precipitates are difficult to separate from those of biochemical origin. These terrigenous materials, both solids and solutions, are formed by the disintegration and decomposition of rocks of all kinds and represent the end products of erosion.

tertiary treatment. A process used to remove practically all solids and organic matter from waste water. Granular activated carbon filtration is a tertiary treatment process. Phosphate removal by chemical coagulation is also regarded as a step in tertiary treatment.

tetrachlorodibenzodioxin. $C_{12}H_4Cl_4O_2$, a compound formed from other halogenated hydrocarbons in the presence of oxygen at high temperatures. A common source of the chemical is the burning or heating of polychlorinated biphenyls (PCB's) through trash incineration or during the welding of transformers or capacitors. Dibenzodioxin, called simply dioxin, is also released in the process of converting trichlorophenol into hexachlorophene, the disinfectant. When the process is not kept under control, the reaction has a tendency to form increasing percentages of dioxin as the mixture increases in temperature. The ultimate hazard is an explosion that would release large quantities of dioxin into the environment, as occurred on July 10, 1976, in a chemical factory near Seveso, Italy. Evacuation of the town was ordered after domestic and wild animals in the area began to die and children developed a distinctive skin rash. Other residents developed signs of damage to the eyes, nerves, blood, and liver, and about 100 women underwent abortions because of the fear of teratogenic effects. The dermal exposure effects were of an acne form that did not appear in some cases until two years after the accident. Animal carcasses, vegetation, and even the topsoil from the contaminated area were removed in plastic bags and stored, awaiting a decision that was to come several years later concerning the safest method of disposal. More than 200 residents were unable to return to their homes because of the danger of contamination and the Italian government filed a $135 million damage suit against the chemical company (Givaudan, a subsidiary of Hoffman-La Roche) to reimburse the government for its clean-up and health monitoring expenses.

tetraethyl lead. $Pb(C_2H_5)_4$, a gasoline additive introduced in 1923 to reduce the "knock" in high-compression internal combustion engines. The addition of lead compounds permits the synthesis of highly branched hydrocarbon molecules which are less likely to detonate rapidly at the end of a combustion cycle, the factor responsible for the knock effect. Before the introduction of unleaded gasolines, the petroleum industry consumed about one-fifth of the total lead production, of which about 300,000 tons per year entered the atmosphere, mainly in the form of aerosol particles less than 10 μm in size. Studies of lead particulate emissions from motor vehicles indicate that the lead content of exhaust averages slightly more than 30 μg per mile of travel, with amounts increasing with the speed of the vehicle. Lead particulates tend to accumulate in plants and soil closest to heavily traveled roads. Soils rich in organic matter and particularly suitable for food production also have a high affinity for lead and bind the element strongly. In addition to lead particulates entering plants through the soil, samples of the aerosol particles have been recovered from the bark of trees, indicating that lead can be absorbed directly into plant tissues. Evidence regarding the absorption of lead by humans consuming food grown in soil contaminated by the combustion of leaded gasoline is inconclusive.

tetranitromethane. $CH_3(NO_2)_4$, an energetic liquid oxidant, sensitive and toxic and having a high flame temperature, a molecular weight of 196, and a boiling point of 125.7°C.

textile mill residue conversion.   The Russel Corporation of Alexander City, Alabama, operates a large textile mill which employs large quantities of steam in its bleaching and dyeing plant. Following the curtailment of natural gas supplies in 1975 and facing the prospect of large price increases for fuel oil, the company investigated several alternatives and finally developed a plan designed by Russel Lands Inc., a local timber management firm, for a wood-fired boiler plant which started operation in 1976. The boiler plant itself consists of two 60,000-lb/hr steam boilers fired with sawmill residues made up largely of wood chips. These residues are consumed at a rate of 20 tons per hour, replacing six million gallons of oil on an annual basis.

theodolite.   An optical instrument used in weather forecasting to observe a pilot balloon for the purpose of determining wind speed and direction. It has a sighting telescope and a scale.

therm.   A small calorie; the amount of heat required to raise one gram of water by one degree centigrade.

thermal.   Pertaining to all forms of thermoelectric thermometers, including a series of couples, thermopiles, and single thermocouples; of or pertaining to heat or temperature.

thermal capacity.   The capacity for absorbing heat in British thermal units, referring especially to the heat required to raise a given mass of substance to a given temperature; the total quantity of heat necessary to raise any body or system by a unit of temperature, measured in calories per degree centigrade in the cgs system.

thermal conductivity, specific.   For a substance, the quantity of heat that flows in one second across a unit area of unit thickness when the temperatures of the two faces differ by one degree.

thermal efficiency.   The efficiency with which a heat engine transforms the potential heat of its fuel into work or output, expressed as the ratio of useful work done by the engine in a given time interval to the total heat energy contained in the fuel burned during the same time interval, both work and heat being expressed in the same units.

thermal emission.   The process by which a body emits electromagnetic radiation as a consequence of its temperature only.

thermal emissive power.   The rate of thermal emission of radiant energy per unit area of emitting surface.

thermal energy. For a material, the total potential and kinetic energies associated with the random motion of its particles. The quantity of thermal energy possessed by a body determines its temperature. Thermal energy which is absorbed, given up, or transferred from one material to another is heat. The temperature of a body is a measure of its ability to give up heat to or absorb heat from another body. Thus the temperature of a body determines whether or not that body will be in thermal equilibrium with any nearby body.

thermal expansion. The property of a substance which causes it to change its length, area, or volume with changes in temperature. The coefficient of linear expansion, or expansivity, is the ratio of the change in length per degree centigrade to the length at 0°C. The coefficient of volume expansion (for solids) is approximately three times the linear coefficient. The coefficient of volume expansion for liquids is the ratio of the change in volume per degree centigrade to the volume at 0°C. The value of the coefficient varies with temperature. The coefficient of volume expansion for a gas under constant pressure is nearly the same for all gases and temperatures and is equal to 0.00367 for 1°C. With few exceptions, solids expand when heated and contract when cooled; they not only increase in length, but also in width and thickness. Solids expand when their temperature is raised because the increase in thermal energy increases the amplitudes of vibration of the atoms and molecules composing the solid. As a result, the average distance between the atoms and molecules is greater than before, and the solid expands in all directions.

thermal neutron. A neutron slowed down by a moderator to an energy of a fraction of an electron volt, about 0.025 eV at 15°C.

thermal pollution. The discharge of heated water from industrial processes that can affect the life processes of aquatic plants and animals.

thermal precipitator. A precipitator operating on the principle that a hot body situated in a cloud of dust develops a particle-free zone around it. In a thermal precipitator the particles in the sampled stream move past a very hot wire and are repulsed to a nearby cold plate, where they are collected. The instrument is highly efficient, but operates at an extremely low sampling rate, approximately 50 milliliters per minute, and is thus largely a research tool.

thermal radiation. The electromagnetic radiation emitted by any substance as the result of the thermal excitation of its molecules. Thermal radiation ranges in wavelength from the longest infrared to the shortest ultraviolet radiation.

thermal transpiration. The passage of gas through a connection between two vessels at different temperatures, resulting in a pressure gradient when equilibrium is reached.

thermal unit. A unit, chosen for the comparison or calculation of quantities of heat such as the calorie or the British thermal unit.

thermic. Pertaining to heat.

thermie. A unit of energy; the quantity of heat required to raise the temperature of one metric ton (1000 kilograms) of water by one degree centigrade.

thermionic. Of or pertaining to the emission of electrons by heat.

thermionic conversion. The process whereby electrons are released by thermionic emission and collected and utilized as electric current.

thermionic emission. Electron or ion emission dependent upon the temperature of the emitter; the emission of electrons from a hot surface. The rate of emission increases rapidly with the increase of temperature. Thermionic emission is also very sensitive to the state of the surface. At a high temperature, the average kinetic energy of free electrons is large and a relatively large number of electrons will have enough energy to escape through the surface of the material. This is analogous to the evaporation of molecules from the surface of liquid water.

thermistor. A resistor composed of a synthetic material with a resistance that decreases rapidly with the rise of temperature. Thermistors are employed as safety devices, regulators in electric circuits, and nonlinear circuit elements.

thermochemical production. Direct hydrogen production by photochemical means.

thermocouple. A combination of two dissimilar metals with a junction across which an emf is developed when the junction is heated. Thermocouples form the basis of many devices for measuring and controlling temperatures.

thermodynamic equilibrium. A state of a system where all processes which can exchange energy are exactly balanced by reverse processes so that there is no net exchange of energy. For instance, ionization must be balanced by recombination, bremsstrahlung by absorption, etc. If a plasma is in thermodynamic equilibrium, the distribution function of particle energies and excited energy levels of the atoms can be obtained from the Maxwell–Botlzmann distribution, which is a function of temperature only.

thermodynamics. The science of heat energy and its conversion into mechanical energy; that branch of the theory of heat that treats the relations between heat and mechanical work.

thermodynamics, laws of. (1) When mechanical work is transformed into heat or heat into work, the amount of work is always equivalent to the quantity of heat. (2) It is impossible by any continuous self-sustaining process for heat to be transferred from a colder to a hotter body.

thermoelectric power. The power measured by the electromotive force produced by a thermocouple for a unit difference of temperature between the two junctions. Thermo-electric power varies with the average temperature and is usually expressed in microvolts per degree centigrade.

thermoelectricity. The electricity generated by heat, as in the unequal heating of a circuit of two dissimilar metals.

thermonuclear. Pertaining to a nuclear reaction which is triggered by particles of high thermal energy.

thermonuclear reaction. A fusion reaction initiated by extremely high temperatures, such as those generated by a fission reaction; a nuclear reaction in which the energy necessary for the reaction is provided by colliding particles that have kinetic energy by virtue of their thermal agitation. Such reactions occur at appreciable rates only for temperatures of millions of degrees and higher, the rate increasing enormously with the temperature. The energy of most stars is believed to be derived from exothermic thermonuclear reactions.

thermopile. A transducer for converting thermal energy directly into electrical energy, composed of pairs of thermocouples which are connected either in series or in parallel.

thermosiphoning. The upward movement of heated fluid through a pipe. This phenom-enon is very useful in making a solar water heater if a pump is not desired.

thermosphere. An atmospheric shell extending from the top of the mesosphere to outer space. It is a region where the temperature more or less steadily increases with height, starting at 70 to 80 km. The thermosphere includes the exosphere and most or all of the ionosphere.

thermostat. A control which delivers an electrical signal when the temperature reaches a certain level; this signal usually turns on a pump, fan, or both in solar heating systems.

third law of thermodynamics. Every substance has a finite positive entropy, and the entropy of a crystalline substance is zero at the temperature of absolute zero. Modern quantum theory has shown that the entropy of crystals at 0° absolute is not necessarily zero. If the crystal has any asymmetry it may exist in more than one state, and there is, in addition, an entropy residue derived from nuclear spin.

thorium (Th). A radioactive element used as a source of fuel for atomic furnaces. Thorium atoms do not split, but they can be converted inside an atomic furnace into uranium atoms that do split. Its atomic number is 99 and its atomic weight is 232.038.

Three Mile Island. A series of breakdowns in the cooling system of the Three Mile Island nuclear power plant No. 2 reactor led to a major accident on March 28, 1979. The Three Mile Island facility, owned by the Metropolitan Edison Company and two other utilities, was located 10 miles south of Harrisburg, Pennsylvania in the Susquehanna River Valley. By Friday, March 30, 1979, the Nuclear Regulatory Commission (NRC) warned of a possible core meltdown, a catastrophic event that could involve a major loss of life, and the possible threat of explosion of a hydrogen gas bubble that had formed in the overheated reactor vessel of the crippled plant. Over the weekend, nuclear experts worked to cool down the overheated uranium fuel core and to reduce the size of the hydrogen bubble. On the fifth day after the accident an announcement was made that the reactor was stable and that radiation levels near Three Mile Island were "safe." On April 2, the NRC announced the reduction in size of the potentially dangerous gas bubble and a further cooling of the reactor core. On April 3, seven days after the accident, it was announced that the hydrogen bubble was eliminated. The accident threatened the future of nuclear power in the United States and called into question the safety systems regulated by the NRC and used by the nuclear power industry. At the time of the accident, 72 nuclear reactors in the United States provided 13 percent of the nation's electrical power. The fate of 125 projected nuclear plants was left in the air after the Three Mile Island accident. Three Mile Island No. 2 had only been operating for a short period of time, since December 1978. Three Mile Island No. 1, in operation since 1974, had been closed for routine maintenance when the accident occurred.

The NRC estimated that it might take as long as four years before Three Mile Island No. 2 could be decontaminated, overhauled, and placed back on line. The containment building had the highest level of contamination recorded in commercial nuclear operations history. Waste water in sump tanks in an auxiliary building was also so highly radioactive that it could not be inspected. The Food and Drug Administration (FDA) indicated from their studies that the highest iodine level discovered was about 31 picocuries per liter, and the level in most of the samples was from 10 to 20 picocuries. The FDA considered 12,000 picocuries per liter dangerous. Federal and state officials projected that the cumulative radiation dose level for local residents would be less than 100 millirems.

The Three Mile Island nuclear power plant, a pressurized water system, generated electricity on the same principles as plants powered by fossil fuels: Heated water creates steam, the steam drives a turbine that turns a generator, and the generator produces electricity. The heat source was the energy released from the fission, or splitting, of the nuclei of fissionable materials, principally ^{235}U. Energy from the chain reaction of fissioning uranium in the reactor core (neutrons bombarding ^{235}U nuclei) heated the surrounding

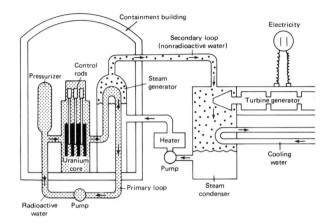

Three Mile Island nuclear power plant.

water. That water was pumped under pressure into the tubes of a power generator. Heat from the tubes converted water in the generator to steam, which turned the rotors of a high-pressure turbine. An electric generator converted the energy from the turbine shaft into power for transmission to consumers through high-voltage lines. Steam from the turbine passed over the cooling coils of a condenser and was converted to water, which returned to the steam generator to be heated again.

threshold. The value at which a stimulus just produces a sensation, or comes just within the limits of perception; absolutely the lowest possible limit of stimulation capable of producing sensation.

threshold dose. The minimum dose of a given substance necessary to produce an effect.

threshold limit value (TLV). From AGGIH (documentation of threshold limit values), the concentration of an airborne contaminant to which workers may be exposed repeatedly, day after day, without adverse affect. The TLV recognizes that there are individual variations among workers and that maintaining exposures within the prescribed limits may not prevent discomfort or the aggravation of a preexisting condition in certain individuals. Most TLV's represent "average" exposures, and some fluctuation during the working day is possible without causing adverse effects.

threshold of audibility. For a specified signal, the minimum effective sound-pressure level of the signal that is capable of evoking an auditory sensation in a specified fraction of the trials; also called the threshold of detectability. The characteristics of the signal, the manner in which it is presented to the listener, and the point at which the sound pressure is measured must be specified.

threshold of feeling. For a specified signal, the minimum effective sound-pressure level at the entrance to the external auditory canal which, in a specified fraction of the trials, will stimulate the ear to a point at which there is a sensation of feeling that is different from the sensation of hearing; also called tickle.

threshold shift. The deviation in decibels of a measured hearing level from a previously established one.

thunderstorm. One or more sudden electrical discharges, manifested by a flash of light (lightning) and a sharp or rumbling sound (thunder). The life cycle of a thunderstorm divides itself into three stages determined by the magnitude and direction (upward and downward) of the predominantly vertical motions. These stages are: the cumulus stage, a cell formed by a collection of cumulus clouds, characterized by an updraft throughout the cell; the mature state, characterized by the existence of both updrafts and downdrafts, at least in the lower half of the cell; the dissipating stage, characterized by weak downdrafts throughout.

tidal marsh. A low, flat marshland traversed by interlaced channels and tidal sloughs and subject to tidal inundation. Normally, the only vegetation present is salt-tolerant bushes and grasses.

timbre. The attribute of auditory sensation by which a listener discriminates between two sounds of similar loudness and pitch, but of different tonal quality. Timbre depends primarily upon the spectrum of the stimulus, but it also depends upon the waveform, the sound pressure, the frequency location of the spectrum, and the temporal characteristics of the stimulus.

tinnitus. A subjective sense of "noises in the head" or "ringing in the ears" of which there is no observable cause.

titration. A test method for determining the strength of a solution; also a test method for determining the concentration of a substance in a solution. Briefly, it consists of the addition of a liquid reagent to a known amount of another liquid until a change of color, or some other change or effect, takes place. This test is also referred to as volumetric analysis.

tolerance. The ability of an organism to cope with changes in its environment; also, the safe level of any chemical applied to crops that will be used as food or feed.

toluene. $C_6H_5CH_3$, a colorless liquid with a benzol-like odor, used as a solvent in rubber and plastic cements and in the manufacture of some paints. Toluene has been identified as the central nervous system depressant involved in glue sniffing. It is derived from coal tar, and commercial grades are often contaminated with small amounts of benzene. Psychotropic effects from inhalation by humans can occur in concentrations of 100 ppm, with central nervous system effects observed at twice that concentration. The lowest lethal concentration reported in animal studies was 4000 ppm for the rat; an LC_{50} of 5300 ppm has been reported for mice. For humans, exposures of 200 ppm for eight hours result in the impairment of coordination and slowed reaction times. At higher inhaled concentrations, subjects experience headaches, nausea, anorexia, and lassitude. In general, acute effects are similar at both high and low concentrations, although the onset of the symptoms is more rapid at higher concentrations. Chronic effects include anemia; the destruction of bone marrow; and fatty degeneration of the liver, heart, and adrenals, accompanied by small hemorrhages. The presence of toluene in drinking water in Lekkerkerk, Netherlands, in 1980 forced the evacuation of nearly 900 residents from a six-year-old housing project and the closing of a 29-acre Utrecht city park. The toluene was traced to the burial of drums of the chemical used to fill land on which the homes were later erected. The chemical was also used as a landfill to build the park in Utrecht.

ton of cooling. 12,000 British thermal units per hour. The term is derived from the amount of heat energy required to convert a ton of water into ice at 32°F during a 24-hour period.

tone. (1) A sound wave capable of exciting an auditory sensation having pitch. (2) A sound sensation having pitch.

topography. The physical features of a surface area, including relative elevations and the position of natural and man-made features.

tornado. A violently rotating column of air extending toward the earth's surface from a thunderstorm cloud, nearly always shaped like a funnel or tube. A tornado dips earthward out of a cumulonimbus formation or an associated cloud type, and as it nears the earth's surface it sways with an undulating motion. The funnel's circular mouth sweeps over the land; it is usually quite small, from about 150 to some 600 feet in diameter. The tornado cloud ranges in color from gray to black. Sometimes the long, gray tornado funnel never touches the ground, but slowly rises and falls between the cloud cover and the earth as it swings from side to side. Two distinguishing characteristics of the tornado are the extraordinarily low barometric pressure that exists at its center and the fantastic speed of the winds whirling around within the columnar vortex.

torr. A unit of pressure equal to the pressure required to support a column of mercury 1 mm high at 0°C and standard gravity.

TOSCA. An acronym for the Toxic Substances Control Act of 1976. The law is the first legislation designed to control potentially hazardous substances before they reach the environment. Provisions include a requirement that manufacturers must notify the EPA in advance of any plans to produce a new chemical or make a new application of an existing chemical compound. The EPA in turn has the authority to require tests of the chemical or product and may, if deemed necessary, prohibit the manufacture of the substance or limit its production or application. In addition, the law authorizes the EPA to compile a list of chemicals in use and conduct tests of those considered likely to be carcinogenic, mutagenic, teratogenic, or otherwise unhealthy or a threat to the quality of the environment. During the first years since enactment of the law, TOSCA efforts were directed toward investigation of the health and environmental consequences of the use of five classes of halogenated aromatic hydrocarbons.

TOSCOAL process. *Process:* This pyrolysis process provides the required heat by using hot ceramic balls. Coal is crushed, dried, and preheated with hot flue gas. Preheated coal is transferred to a rotating pyrolysis drum. It is then heated to carbonization temperatures

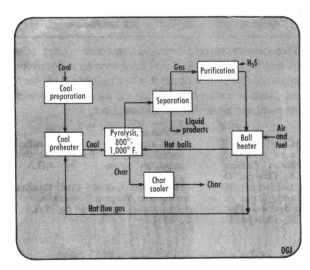

TOSCOAL Process (Oil Shale Corp).

by contact with hot ceramic balls. The balls are separated from the char and conveyed to a ball heater. The char produced is about 50 percent of the weight of the raw coal feed and has about 80 percent of the raw-coal heating value. An economical use should be found for this material. Pyrolysis vapors are cooled to condense oil and water from the gaseous products. The liquid products are fractionated into gas oil, naphtha, and residuum. Gas is used in the ball heater as a fuel (Oil Shale Corp.)

total hydrocarbon analyzer. Based on the principle of hydrogen flame ionization, an analyzer where the sample is burned in a hydrogen flame and the ions produced are collected and measured as a current. The instrument is very sensitive. Since it responds essentially only to carbon atoms, reasonable specificity for selected carbon compounds can be attained through the use of appropriate adsorption columns.

total organic carbon (TOC). A measure of the amount of carbon in a sample originating from organic matter only. The test is run by burning the sample and measuring the CO_2 produced.

total solids. The total amount of solids in a waste water in both solution and suspension.

toxic substance. A substance that acts like a poison. Toxic substances cause adverse environmental effects ranging from mild temporary dysfunction of organisms and ecosystems to acute symptoms, disorders, or death. Some toxic substances cause cancer. Some estimates are that up to 90 percent of all malignancies may be induced by, or related to, environmental factors. In this context, environmental factors include all stresses within the human environment, with the exception of an inherited susceptibility to cancer. Environmental factors include both voluntary and involuntary exposure to carcinogenic substances, smoke, x rays, occupational and household toxicants, water pollutants, naturally occurring toxins, and "life-style" traits such as delayed childbirth. Many potentially toxic chemicals are components of consumer products. These include tars from cigarettes, chloroform in cough medicines, benzene in paint remover. Substances such as asbestos and vinyl chloride have been linked to the incidence of cancer among exposed workers. Potentially toxic substances have been identified in several drinking water supplies.

toxicant. A chemical that controls pests by killing rather than repelling them.

toxicity. The quality of being poisonous, especially the degree of virulence of a toxic microbe or a poison. It is expressed by a fraction indicating the ratio between the smallest amount that will cause an animal's death and the weight of that animal. Heavy metals such as lead, mercury, copper, arsenic, and chromium which are used in tin plating, galvanizing, and chromium plating solutions can act as metabolic or respiratory blocks in many organisms, inhibiting or destroying enzymes that are essential to life processes. Since metals may affect the decomposer organisms as well as the higher forms of life, their excessive addition to an aquatic system can be quite destructive. These pollutants are actual poisons which may be absorbed by the bottom mud and be released whenever bottom deposits are disturbed.

tracer. A foreign substance mixed with or attached to a given substance to enable the distribution or location of the latter to be determined subsequently. A physical tracer is one that is attached by physical means to the object being traced. A chemical tracer has chemical properties similar to those of the substance being traced, with which it is homogeneously mixed. A radioactive tracer is a physical or chemical tracer having radioactivity as its distinctive property. An isotopic tracer is a radionuclide or an allobar used as a chemical tracer for the element with which it is isotopic.

trade-wind cumulus. A characteristic type of convection cloud occurring where no large-scale convergence has taken place in the low level. It has a base at about 2000 feet and a top at 5000 to 8000 feet. Characteristically, about one-third to one-half of the ocean area is covered by these clouds. The convection currents in them are weak, but they are capable of producing showers of fine rain. In the daytime, in areas of convergence and islands, they mass together to produce large towering clouds and copious downpours of rain.

transducer. A device capable of being actuated by waves from one or more transmission systems or media and of supplying related waves to one or more other transmission systems or media. The waves in either input or output may be similar or different types (e.g., electrical, mechanical, or acoustic).

transistors. A semiconductor replacing a thermionic valve, its most attractive feature being its high reliability and small size, ruggedness, and instant operation when switched on. Germanium and silicon are the two semiconductors of major practical importance, the former being more efficient and usable up to 70°C and the latter usable up to 200°C. Minute additions of other elements modify the semiconducting properties.

transmission. The process by which radiant flux is propagated through a medium or body; the passage of energy through a given medium.

transmission loss. In decibels, 10 times the base 10 logarithm of the ratio of the acoustic energy transmitted through a barrier to the acoustic energy incident upon its opposite side. The airborne sound attenuation provided by a barrier is described in terms of its sound transmission loss. Transmission loss is a physical property of the barrier.

transmissivity. In a partly transparent medium such as the atmosphere or the oceans, the fractional part of the radiation transmitted through the medium per unit of thickness along the path of the radiant beam. The thickness is often called the optical thickness and is usually expressed in units of mass of the absorbing and backscattering medium. The transmissivity is a dimensionless number between zero and one. It can be applied monochromatically or to the total spectrum. The fraction depleted per unit optical thickness defines the extinction coefficient, one minus the transmissivity.

transmissometer. An instrument used for measuring the extinction coefficient of the atmosphere and for the determination of visual range; also called telephotometer, transmittance meter, and hazemeter.

transpiration. The passage of gas or liquid through a porous solid (usually under the condition of molecular flow).

transpiration cooling. A process by which a body having a porous surface is cooled by a forced flow of coolant fluid through the surface from the interior.

transuranium. Any element with an atomic number greater than 92, the atomic number of uranium. Neptunium and plutonium were the first two man-made transuranium elements. At present there are 13 artificial transuraniums confirmed. All of them are radioactive, and some of them exist only for a small fraction of a second.

trichloroethylene. $CHClCCl_2$, a heavy colorless liquid with a powerful solvent action for fats, greases, and waxes. About 90 percent of the annual production is consumed by the metal degreasing and drycleaning industries. It is also used as an ingredient in printing inks, paints, lacquers, varnishes, and adhesives. Small amounts have been used for miscellaneous purposes, such as coffee decaffeination until 1975, and as a local anesthetic. NIOSH has estimated that as many as 5000 doctors, dentists, nurses, and other health professionals are exposed to trichloroethylene regularly because of its use as an anesthetic in the treatment of trigeminal neuralgia, the extraction of teeth, and the incision of furuncles. Much information about the toxic effects of trichloroethylene was derived from the use of the substance as an anesthetic in veterinary treatment of animals. Clinical animal studies showed that the chemical could cause death by respiratory failure or cardiac arrest, as well as growth retardation, liver and kidney tissue changes, residual brain damage, and myelotoxic anemia. Human inhalation of trichloroethylene can cause visual disturbances, mental confusion, fatigue, nausea, and vomiting. Skin contact may result in irritation and blister formation, and under industrial conditions with continuous or intermittent exposure, paralysis of the fingers. Cardiac rhythm disturbances are common effects of the use of trichloroethylene as an anesthetic for humans and, in rare cases, convulsions may result. Trichloroethylene diffuses rapidly across the placenta when inhaled by pregnant women. Trichloroethylene has been associated in tests by the National Cancer Institute with the development of cancer in animals, but no cases of human cancer resulting from exposure to the chemical have been observed.

trickling filter. A biological treatment device: Waste water is trickled over a bed of stones covered with bacterial growth, the bacteria break down the organic wastes in the sewage and produce cleaner water.

Tris. $C_9H_{15}O_4PBr_6$, tris (2,3-dibromopropyl) phosphate, is an experimental carcinogen, mutagen, and teratogen, and can cause testicular atrophy and sterility. In the rat the LD_{50} dose is 1010 mg/kg. Tris has been used to control the flammability of cloth, especially in children's sleepwear, where it can be absorbed via the skin, or chewed or sucked off by infants. The use of Tris as a flame resistant has been banned by the U.S. government since 1977.

tritium. The heavy isotope of hydrogen, having an atomic weight of 2 and a half-life of 12.5 years; an isotope of hydrogen with a mass of 3, with two neutrons and one proton in its nucleus.

trophic level. The transfer of food energy from the source in plants through a series of organisms with repeated stages of eating and being eaten is known as the food chain. In complex natural communities, organisms whose food is obtained from plants by the same number of steps are said to belong to the same trophic level. Thus green plants occupy the first trophic level (the producer level) and plant eaters (herbivores, etc.), the second level (the primary consumer level). Trophic classificaton is one of function, and not of species as such; a given species population may occupy one or more trophic levels according to the source of energy actually assimilated. At each transfer of energy from one organism to another, or from one trophic level to another, a large part of the energy is degraded into heat, as required by the second law of thermodynamics. The shorter the food chain, or the nearer the organism to the beginning of the food chain, the greater the available food energy.

tropic and polar circles. The latitude circles of 23 ½ °N and S are called the Tropic of Cancer and Tropic of Capricorn. They are the highest latitudes from the equator, where the sun can be observed directly overhead, and then only one day each during the year. As a consequence of the inclination of the earth's axis, when the sun shines directly on latitude 23 ½ ° in one hemisphere, the portion of the earth poleward from 90 − 23 ½ ° = 66 ½ ° in the other hemisphere is without sunlight. For the Northern and Southern Hemispheres these latitude circles are called the Arctic and Antarctic Circles. Every point poleward from these circles has at least one day of darkness during the year. At the poles there are six months without sun and six months with continuous sunlight between the equinoxes.

tropical air. Warm air having its source in the low latitudes, chiefly in the regions of the subtropical anticyclones.

tropical cyclone. A storm of great power originating over tropical oceans. Tropical cyclones are classified according to their force; tropical depressions, hurricanes, typhoons, etc.

tropopause. The boundary between the troposphere and stratosphere, usually characterized by an abrupt change of lapse rate. The tropopause is a thin layer of air separating the troposphere from the third level, the stratosphere. The troposphere varies in thickness, ranging from more than 50,000 feet in depth at the Equator to about 25,000 feet at the earth's polar extremities. The tropopause is noted particularly for its steady, powerful wind currents, which have come to be known as jet streams.

troposphere. The lower layer of the atmosphere where most vertical currents, much water vapor, and weather phenomena exist. The troposphere is the atmospheric level that lies closest to the earth's surface. It's thickness changes with the seasons and the altitude, but on the average it extends about 54,000 feet in height over the Equator to about 28,000 feet over the North and South Poles. In the temperate zones its thickness also fluctuates with the seasons, being thicker in the summer and thinner in the winter. This variation is believed to be caused by temperature changes in the earth's surface and atmosphere. The troposphere is the heaviest of the atmospheric levels, and contains three-quarters of all the atmosphere's weight and a major part of its water vapor and carbon dioxide.

trough. An elongated area of relatively low pressure, usually extending from the center of a low-pressure system. It is the opposite of a ridge.

tundra. A treeless plain between taiga in the south and polar ice caps in the north. Tundra is characterized by low temperatures, a short growing season, and ground that is frozen most of the year.

tundra soil. Tundra soils occur in northern Alaska and Canada, skirting the southern shores of the Hudson Bay and the northern edge of Newfoundland. This same soil group skirts the southern edge of Greenland, the southwestern tip of South America, and northern Russia and Siberia. The overlying vegetation consists of soil lichens, mosses, a variety of herbs, and dwarf shrubs. These soils have poor drainage, low temperatures (permafrost permeates the subsurface layers), and a peaty surface layer. Mineral soil below the peat layer is gray, mottled with brown, gradually grading into deeper soils of a slate-blue or green-blue coloration.

tungsten. A malleable and ductile metallic element of high tensile strength with atomic weight 183.86; specific gravity, 19.3; melting point, 3380°C. It is one of the five more abundant refractory metals in the earth's crust.

turbidimeter. A device that measures the amount of suspended solids in a liquid.

turbidity. (1) A condition of a liquid due to fine visible material in suspension, which may not be of sufficient size to be seen as individual particles by the naked eye, but which prevents the passage of light through the liquid. (2) A measure of fine suspended matter (usually colloidal) in liquid.

turbine. An engine that forces a high-pressure stream of gas or liquid through jets against the curved blades of a wheel, thus forcing the blades to turn.

turbulator. A coined word describing a small obstacle placed on the surface of an absorber in an air-type collector to increase air turbulence and thus get more heat from the solar collector.

turbulence. Atmospheric eddies classified as mechanical or thermal, since they are produced primarily by shearing stress or by convection. Turbulence or turbulent energy promotes the atmospheric dilution of pollutants. Turbulent fluctuations in stable air are mainly of high frequency with typical periods on the order of seconds (mechanical turbulence). The fluctuations often constitute major deformations of flow and are capable of transporting momentum, energy, and suspended matter at rates far in excess of the rate of transport by diffusion and conduction in a nonturbulent or laminar flow.

two-outlet heater. A water heater having one outlet going to the pipes for a domestic hot-water system, and the other through a separate pipe to an insulated storage tank.

U

233**U.** The symbol for one type (isotope) of uranium that splits, making it usable for atomic fuel. 233U is not found in natural uranium minerals, but must be manufactured from a different mineral, thorium, in an atomic furnace. The U stands for uranium, and the 233 identifies the number of protons and neutrons in the central core of each atom.

235**U.** The symbol for one type (isotope) of uranium that splits. 235U is a naturally occurring atomic fuel, but it is found mixed with another nonsplittable type of uranium. 235U makes up only 0.7 of one percent of the total uranium. The mixture called natural uranium can be used to operate an atomic furnace, although only its small percentage of 235U atoms splits.

238**U.** The symbol for the type (isotope) of uranium that makes up more than 99 percent of natural uranium. 238U does not split, but it can be transformed inside an atomic furnace into plutonium, which does split.

U-GAS process. *Process:* Crushed coal is fed through lock hoppers to a pretreater (necessary for caking coals only) which operates at 350 psi and 800°F. Air introduced into the pretreater partially oxidizes the coal to destroy its caking tendencies. From the pretreater, coal is fed directly to the fluid-bed gasifier that operates at 350 psi and 1900°F. Air and steam are introduced to the bottom of the gasifier. Dry ash is removed from the gasifier through lock hoppers. Gases produced in the pretreater and gasifier are combined

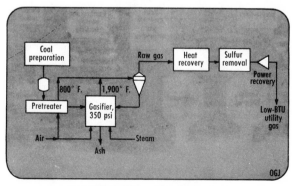

U-GAS™ (Institute of Gas Technology).

and pass through a cyclone for removal of coal dust. Raw gas passes through a heat-recovery system and then through a sulfur-removal system. Finally, the gas passes through a power-recovery turbine to recover energy available from the 350 psi pressure, down to whatever pressure the gas is to be used (Institute of Gas Technology).

ultrapure water. Water which, as far as is possible by present methods, is free of nearly all impurities, to meet the most rigid specifications. Some users specify a limit of 50 parts per billion (ppb) of total impurities and a minimum specific resistance of 10 MΩ cm. Ultrapure water, like many things in nature, exists only for short periods. Water in contact with container walls is susceptible to growths, leakage, corrosion, extraction, etc. All natural waters contain dissolved matter.

ultrasonic. A term used for sounds whose frequencies are too high to be audible to the human ear. Ultrasonics deals with mechanical vibrations and radiation in solids, liquids, and gases at frequencies of about 16,000 Hz.

ultraviolet radiation. An electromagnetic radiation of wavelengths ranging from 4000 down to 40 Å, with frequencies greater and wavelengths less than those of visible light. The region 3200–4000 Å is known as near ultraviolet or black light. Ultraviolet radiation affects photographic plates, can produce fluorescence, and is able to ionize gases. Ultraviolet radiation is emitted by the sun at times of increased activity and is absorbed by the ozone in the earth's atmosphere.

Union Carbide coal gasification. *Process:* A basic feature of this process is to provide heat for the reaction of carbon and steam by circulation of hot coal-ash agglomerates. These are produced by the combustion of char with air. Crushed (minus 35 mesh), dried coal is injected near the bottom of a fluidized-bed zone in the gasifier. Hot ash agglomerates (2000°F) from the combustor enter the gasifier near the top. The lighter coal particles (35 mesh, 1.3 specific gravity) migrate upward. Heavier ash agglomerates (6 mesh, 2.3 specific gravity) descend down through the bed. The bed is fluidized by steam and the products of gasification. Heat requirements of the coal–steam gasification reactions cool the agglomerates to about 1600°F. Coal entrained by the downflowing agglomerates is removed by elutriation in the bottom section of the gasifier. Part of the ash agglomerates are recycled to the combustor. The remainder is rejected from the system. Char from the top of the gasifier is removed continuously and burned with air in the combustor. The char burns at 2000°F under slagging conditions and amidst agglomerates. Hot agglomerates are recycled to the gasifier. Flue gas from the combustor is passed through waste-heat recovery and purification and compression energy-recovery systems. Raw product gas from the gasifier (1600–1800°F) passes through a heat-recovery section, scrubbing, and acid-gas removal. Purified gas is compressed, shifted, and methanated to produce pipeline-quality gas after compression and purification (Chemical Construction Corp.).

units of temperature. The British thermal unit (Btu) is the amount of heat required to raise 1 lb of water at 60°F by 1°F. This unit is defined for various temperatures, but the general usage seems to be to take the Btu as equal to 252.15°F gram-calories or 1055 joules. 1 Btu = 252 cal = 778.3 ft lb = 1055 joules. When greater precision is needed, the temperature rise is specified as from 39 to 40°F.

uraninite. A complex uranium compound and an ore of uranium and radium, moderately hard (3 to 6 on the Mohs' scale), very heavy (specific gravity 5 to 10), generally a brown to black mineral with a greasy, pitchlike, or dull luster and a brownish-black streak. Pitchblende, in which radium and helium were first found, is an amorphous or dense form of uraninite.

uranium (U).   A naturally occurring radioactive element with atomic number 92. Its two principal isotopes are fissionable ^{235}U (0.7 percent of natural uranium) and ^{238}U (99.3 percent), from which fissionable materials can be derived by nuclear bombardment.

urban runoff.   The storm water from city streets, usually carrying litter and organic wastes.

V

vacuum. A region in which the gas pressure is considerably lower than atmospheric pressure at sea level. At 45 km it is approximately 1 Torr; at 100 km, 10^{-4} Torr; at 1000 km, 10^{-10} Torr; at 10,000 km, with 10 atoms/cm³, it is less than 10^{-13} Torr; and it declines to about 10^{-16} Torr in higher regions. Thermal conduction and convection are negligible at 10^{-4} Torr, electrical discharge phenomena at 10^{-5} Torr, outgassing of metals, surface chemical effects only, at 10^{-7} to 10^{-6} Torr.

vacuum-evaporated film. A film formed on a sheet or plate by electrical evaporation of a metal or alloy in an evacuated chamber.

valence electrons. Electrons of an atom which are gained, lost, or shared in chemical reactions.

value of the solar constant. The amount of solar radiant energy received per minute outside the atmosphere on a surface of area 1 cm² normal to the incident radiation at the earth's mean distance from the sun.

Van de Graaff generator. A relatively low-energy particle accelerator producing streams of charged particles with energies up to 10 MeV. These particles moving at moderate energies generate x rays and produce radioactive isotopes.

vanadium (V). A chemical element of atomic number 23 and atomic weight 50.94. Absorption of vanadium compounds through the lungs results in chronic toxicity, irritation of the respiratory system, pneumonitis, and anemia. Although relatively rare in nature, vanadium concentrations in the atmosphere of some American cities is high because it is present in fuel oils imported from Venezuela. Vanadium-rich oil is used mainly by utility companies in the production of electric power. Concentrations of vanadium pollutants in the New York City area have tended to range from a low of 5 micrograms per 10 cubic meters during the summer to nearly double that level during the winter months. Studies have shown high concentrations of vanadium compounds in the atmospheres of Boston and Cincinnati also, while levels in the air of cities of the southwest and Pacific Coast regions are negligible. Although vanadium in the atmosphere may have toxic effects, the element is one of the few that is essential in trace amounts for normal animal life. Marine life in particular requires vanadium, which becomes concentrated in muscle, gonad, and shell tissue structures.

Van't Hoff's principle. If the temperature of interacting substances in equilibrium is raised, the equilibrium concentrations of the reaction are changed so that the products of that reaction which absorb heat are increased in quantity; or if the temperature for such an equilibrium is lowered, the products which evolve heat in their formation are increased in their amounts.

vapor. A substance which, though present in the gaseous phase, generally exists as a liquid or solid at room temperature. The words vapor and gas are often used interchangeably. Gas more frequently means a substance that generally exists in the gaseous phase at room temperature. Thus one would speak of iodine or carbon tetrachloride vapors and of oxygen gas.

vapor barrier. A covering of aluminum foil or plastic sheet on one side of a batt of fiberglass used for insulation. This covering prevents water vapor from permeating the insulation and thereby reducing its efficiency.

vapor plume. A flue gas that is visible because it contains water droplets.

vapor pressure deficit. The amount of moisture necessary, under existing conditions of temperature and pressure, to saturate a particular volume of air or to increase the relative humidity to 100 percent. Whether this vaporous water is leaving, entering, or moving through the soil, it always obeys the physical law of diffusion. It always moves from an area of high vapor pressure to a region of lower vapor pressure.

vapor recovery system. A system used in petroleum refining for separating a mixed charge of miscellaneous gases and gasolines into desired intermediates for further processing.

vapor tension. All measures of water vapor or atmospheric humidity are based upon quantities related to evaporation and condensation over a flat surface of pure water. For a water surface that is evaporating, the excess of molecules of water leaving the surface over those coming back in is expressed and measured as a pressure. This pressure, called the vapor tension, depends only on the temperature of the water surface. This concept is based on measurements made in a closed space.

vaporization. The change of a substance from the liquid to the gaseous state; one of the three basic contributing processes of air pollution, the others being attrition and combustion.

vaporization, heat of. The quantity of heat required at a given temperature to convert a unit mass of liquid into vapor.

variance. A government permission for a delay or exception in the application of a given law, ordinance, or regulation.

vector. An organism, often an insect, that carries disease.

velocity. A vector quantity that specifies the time rate of change of displacement with respect to a reference time or frame.

Venturi. A tube with a constricted middle, at which point a stream of air increases in speed and decreases in pressure. It is a feature of many scrubbers, among other kinds of equipment.

Venturi scrubber. A scrubber employed in installations requiring high-energy collisions of submicron particles. The Venturi is one of the most accurate of fluid meters, since it contains no moving parts to impede air flow. Atomization and impaction occur as the injected liquid is shattered by the unscrubbed gas into minute droplets which collide with and carry away minute particles. Atomization of the scrubbing liquid takes place in the throat of the Venturi. Here the liquid is introduced at relatively low pressures and is shattered into minute droplets by the onrushing gas flow.

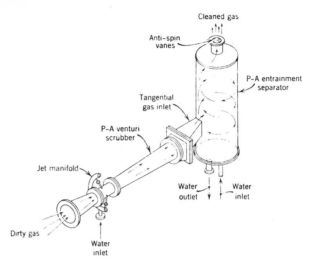

Venturi scrubber.

vermiculite. A generic name for a group of hydrous micas which, upon heating, expand like an accordion to 10 to 20 times their original volume. Vermiculite has approximately the composition of phlogopite, a magnesia mica. The expanded product can be used as an insulator against heat, cold, and sound in house walls, plaster, and various containers.

vibration. An oscillation wherein the quantity is a parameter that defines the motion of a mechanical system; the oscillatory motion of a system. The system might be gaseous, liquid, or solid. Of the flexural, torsional, compressional, and gross oscillations of solids which may produce sound waves in the adjacent air, flexural oscillations are the most important because they usually create the greatest disturbances. In the course of vibration one or more parts of the system oscillate about an equilibrium position. At any instant, the motion of a part may be characterized by its displacement from equilibrium or rest position, by its velocity (rate of change of displacement), or by its acceleration (rate of change of velocity). These quantities are usually described in the same way as sound pressure, i.e., by root-mean-square averages of instantaneous values.

vibration meter (vibrometer). An apparatus used for the measurement of displacement, velocity, or acceleration of a vibrating body.

vinyl chloride. A chemical compound used in producing some plastics. Excessive exposure to this substance may cause cancer.

viscosity. All fluids possess a definite resistance to change of form and many solids show a gradual yielding to forces tending to change their form. This property, a sort of internal friction, is called viscosity; it is expressed in dyne-seconds per square centimeter, or poises. If the tangential force per unit area exerted by a layer of fluid upon another adjacent one is one dyne for a unit space rate of variation of the tangential velocity, the viscosity is one poise. Kinematic viscosity is the ratio of viscosity to density. The cgs unit of kinematic viscosity is the stoke.

visibility. The ability to be seen, measured by the ratio of the luminous flux in lumens to the total radiant energy in ergs per second or in watts.

visibility meter. An instrument used to make direct measurements of visual range in the atmosphere or to determine the physical characteristics of the atmosphere. Visibility meters may be classified according to the quantities they measure. Telephotometers and transmissometers measure the transmissivity or, alternatively, the extinction coefficient of the atmosphere. Nephelometers measure the scattering function of atmospheric suspensoids. A third category of visibility meters makes use of an artificial "haze" of variable density which is used to obscure a marker at a fixed distance from the meter.

visibility reduction. A decrease in visibility produced by the scattering of light from the surfaces of airborne particles. The degree of light obstruction is a complex function of particle size, aerosol density, thickness of the affected air mass, and certain more subtle physical factors, e.g., coal smoke or secondary pollution.

visible light spectrum. Higher in frequency than radio waves are the electromagnetic radiations called the visible light spectrum. Within this range are all the colors of the rainbow. Starting with red, which has the lowest frequency, light progresses through orange, yellow, green, blue, indigo, and violet, which has the highest frequency. Frequencies just below visible red light cannot be seen. These are the infrared or heat radiations. At the other end of the light spectrum are frequencies just above visible violet light. These are the ultraviolet radiations, whose natural source is the sun.

visual threshold. The value of a visual stimulus that just produces a sensation, is just appreciable, or comes just within the limits of perception.

volatile. Any substance that evaporates at a low temperature.

volcanic sediment. A sediment of volcanic origin including the fragmental materials ejected from volcanoes and deposited in beds on land or in water. Volcanic sediments consist of fine volcanic dust, ash, sand, and possibly steam-borne cinders and other coarse particles.

volt. The unit of electromotive force. It is the difference in potential required to make a current of one ampere flow through a resistance of one ohm.

W

warm front. The boundary between two air masses which move so that warm air replaces cold air. The warm air is advancing and the cold air is retreating.

wash plant. A plant in which coal is freed from some of its ash and other impurities, such as inorganic sulfur, in order to produce a higher-quality fuel. The separation may be accompanied by a number of processes that generally take advantage of differences in density between pure coal and any contaminating ingredients. A waste product consisting of various amounts of ash, sulfur, water, and coal is rejected from a wash plant.

waste. Unwanted materials left over from manufacturing processes; refuse from places of human or animal habitation.

waste water. Water carrying dissolved or suspended solids from homes, farms, businesses, and industries.

water. A compound formed by two atoms of hydrogen bonded to one atom of oxygen, symbolized as H_2O. The bonds between the hydrogen and oxygen are primarily polar covalent. The water molecule itself is polar and in the liquid and solid states many water molecules are bound together by hydrogen bonds.

water color. The color of water, almost always coming from vegetation or other organic matter the water comes in contact with. Water color can be measured either by a colorimeter or by comparing it with standard solutions of known color. Good distilled water has a color of zero, as many well waters. The U. S. Public Health standards recommend the color be no more than 15 in public drinking water supplies.

water cooling systems. The principal methods used for cooling water in industrial processes, and the environmental effects of these methods, are related to meteorological conditions. Water is cooled either for recirculation or to meet temperature-rise environmental standards in the receiving body in once-through systems. The two types of wet cooling towers in common industrial use are the natural-draft and mechanical-draft systems, relying on the cooling of water by direct contact with air. The method of contact between water and air can be countercurrent or crossflow. During warm, dry weather, exhaust from a wet mechanical-draft cooling tower is often cooler than the ambient dry-bulb temperature, although the wet-bulb temperature of the exhaust will be higher than that in ambient air. This means that the exit air from the cooling tower will be denser than the ambient air. Although cooling towers were believed to be a solution to environmental problems, they produce their own set of environmental problems due to the release of large amounts of heat and water vapor. Fog from large cooling towers is frequent and can severely reduce visibility. Cloudiness can be increased and sunshading can result. In subfreezing temperatures, ice can occur. Another major environmental problem is to determine the drift deposition (where the droplets will settle) from cooling-tower plumes.

water cycle. The great reservoir of water is the ocean. The sun's heat vaporizes water and forms clouds. These, moved by winds, may pass over land, where they are cooled enough to precipitate the water as rain or snow. Some of the precipitated water soaks into the ground, some runs off the surface into streams and goes directly back to the sea. Groundwater is returned to the surface by springs and pumps and by the activities of plants (transpiration). Water inevitably ends up back in the sea, but it may become incorporated into the bodies of several different organisms, one after another, en route. The energy to run the cycle—the heat needed to evaporate the water—comes from sunlight.

water gas. If steam is blown through red hot coke at about 1100°C (2000°F), a simple chemical reaction occurs—$C + H_2O = CO + H_2$—and two combustible gases, carbon monoxide and hydrogen, are formed. The emerging mixtures of gases are called water gas. At gas works where continuous vertical retorts are used, the red-hot coke is cooled by steam blown upward through it as it nears the bottom of the retorts; water gas is thus formed, mixing with distillation gas.

water pollution. The release of materials into water sources or supplies which are damaging to life because of their toxicity, because of their reduction of the normal oxygen level of the water, or because they are esthetically unpalatable.

water pollution, deoxygenation. The most common pollutants are organic materials which are not poisonous to stream life, nor do they affect pH necessarily. Their effect is more subtle. Most organic materials are attacked by bacteria and broken down into simpler compounds. To do this, bacteria require oxygen. The greater the supply of organic food, the larger the population of bacteria that can be supported, and the greater the demand on the oxygen supply in the water. This demand for oxygen by bacteria is called the biological oxygen demand, the BOD. The BOD is an index of pollution related to the organic load of water. Because all stream animals are dependent upon the oxygen supply in water, the BOD is of particular importance in determining which forms of life a polluted stream is capable of supporting. Fish have the highest need, invertebrates can tolerate still lower concentrations of oxygen, and bacteria can tolerate still lower concentrations.

water pollution, toxicity. Heavy metals, such as lead, mercury, copper, arsenic, and chromium, which are used in tin plating, galvanizing, and chromium plating solutions, can act as metabolic or respiratory blocks in many organisms, inhibiting or destroying enzymes that are essential to life processes. Since metals may affect the decomposer organisms as well as the higher forms of life, their excessive addition to an aquatic system can be quite destructive. These pollutants are actual poisons which may be absorbed by the bottom mud and released whenever bottom deposits are disturbed.

water pollution, turbidity. Relatively inert and finely divided materials are easily suspended in water, cutting down light transmittance enough to inhibit photosynthesis of both macro- and microscopic water plants. The china clay industry, which washes its raw material to remove grit, is an example, as is the steel industry whose hot-strip mills produce very fine particles that give a red or black color to the effluent stream. Ultimately, the suspended particles settle out, smothering all life at the bottom of the body of water with a fine ooze.

water pressure. The pressure increase with increased depth of water at the rate of one atmosphere (760 mm Hg) for every 10 meters of descent. Organisms normally inhabiting the floor of deep sea areas at depths of 10,500 meters are exposed to pressures of about one ton per square centimeter. The swim bladder of a fish decreases to one-half of its surface volume at a depth of 10 meters. Most organisms living at great depths lack air-filled cavities, or spaces that may exist are filled with fluids.

water quality criteria. The levels of pollutants that affect use of water for drinking, swimming, raising fish, farming, or industrial use.

water quality standard. A management plan that considers (1) what water will be used for, (2) setting levels to protect those uses, (3) implementing and enforcing the water treatment plans, and (4) protecting existing high-quality waters.

water softening, lime–soda process. In the precipitation process known as the lime–soda process, the softening (precipitating) chemicals used are hydrated lime, calcium hydroxide, $Ca(OH)_2$, and soda ash (sodium carbonate, Na_2CO_3). The basic chemical principle is that no matter what the compounds of calcium and magnesium were that dissolved in the water originally, or what hypothetical combinations are shown in a water analysis, in the lime–soda process the calcium is always precipitated as calcium carbonate, $CaCO_3$, and the magnesium always as magnesium hydroxide, $Mg(OH)_2$. The hydrated lime, calcium hydroxide, $Ca(OH)_2$, used in the lime–soda process is made by adding water under controlled conditions to quicklime, calcium oxide, CaO. The reaction is called "lime-slaking."

water softening, lime–soda process plant. This process is much the same in its essential features as the filtration for treating turbid or colored water. The three basic steps are the same: forming a precipitate, settling the precipitate, and filtering out the particles of precipitate that didn't settle out. In lime–soda softening, a coagulant may be used to promote precipitate formation and settling. Lime–soda softening plants, like filtration plants, may be either the conventional or the suspended solids contact type.

water softening, lime–soda process (recarbonation). The addition of carbon dioxide gas to softened water to convert most of the residual calcium carbonate and magnesium hydroxide to their respective bicarbonates.

water supply system. The collection, treatment, storage, and distribution of potable water from source to consumer.

water table. The level of groundwater.

water vapor. The basic atmospheric ingredient from which come such forms of precipitation as rain, snow, hail, and sleet. These all originate when water vapor is condensed by the cooling process that normally occurs with the expansion of upward-flowing currents of air. Clouds consisting of myriads of tiny water droplets are formed. Before the droplets can fall as precipitation of one kind or another, they must grow to far larger sizes. The water droplets do not freeze at first, but they do become supercooled. As the clouds rise to still higher levels, some of the droplets are transformed into ice particles or ice crystals. These are quite minute in size, but gradually become larger by taking moisture from the supercooled water droplets, which condenses and freezes on them. The growing action continues until the ice particles become so heavy that they begin to fall.

water vapor mixing ratio. The weight of water vapor contained in a mixture with a unit of weight of dry air, expressed in grams per gram or grams per kilogram. It differs from specific humidity only in that it is related to dry air instead of to the total of dry air and vapor.

watershed. The land area that drains into a stream.

waterspout. A tornado over water. Waterspouts are of two kinds, those that build downward from heavy clouds and that are simply tornadoes over water and those that build upward from the surface of the water and are not directly related to the clouds. They are both called waterspouts because they draw water spray upward, just as tornadoes stir up dust. The noncloud type of waterspout is more of the nature of "dust devils," commonly seen over strongly heated deserts and normally less violent than the tornado cloud variety.

watt. The unit of power in the mks system; that power which produces energy at the rate of one joule per second.

wave. A disturbance which is propagated in a medium in such a manner that at any point in the medium the quantity serving as a measure of disturbance is a function of the time, while at any instant the displacement at a point is a function of the position of the point. Any physical quantity that has the same relationship to some independent variable (usually time) that a propagated disturbance has, at a particular instant with respect to space, may be called a wave.

wave filter. A transducer for separating waves on the basis of their frequencies. It introduces a relatively small insertion loss to waves in one or more frequency bands and a relatively large insertion loss to waves of other frequencies.

wave front. In water pollution applications, the capacity gradient that exists in the critical bed depth. It outlines the gradual transition of carbon from "fresh" to "spent."

wave mechanics. The theory of wave mechanics states that units of matter (the atom and its parts) and units of electromagnetic energy (the photon) both travel in packages and move through space in a wavelike pattern. Waves and particles are interdependent, with the waves directing the movement of the particles and the particles composing the waves.

wave motion. A progressive disturbance propagated in a medium by the periodic vibration of the particles of the medium. Transverse wave motion is that in which the vibration of the particles is perpendicular to the direction of propagation. Longitudinal wave motion is that in which the vibration of the particles is parallel to the direction of propagation.

waveform. The shape of the graph representing the successive values of a varying quantity, usually plotted in a rectangular coordinate system.

wavelength. For a sinusoidal plane progressive wave, the perpendicular distance between two wave fronts in which the phases differ by one complete period. Symbolized by λ, the wavelength is equal to the phase velocity divided by the frequency.

weight. The force with which a body is attracted toward the earth. The cgs unit is the dyne. Although the weight of a body varies with its location, the weights of various standards of mass are often used as units of force, such as pound weight, pound force, gram weight, etc. The weight W of mass m, where g is the acceleration due to gravity, is $W = mg$. The weight will be given in dynes when m is in grams and g in centimeters per second squared.

Wellman–Galusha gasification. *Process:* Crushed coal ($\frac{3}{16}-\frac{5}{16}$ in.) dried and fed by steam mixture is introduced through a revolving grate at the bottom. Gasifiers are available with and without an agitator. The agitator producer has a slowly revolving horizontal arm which spirals vertically below the surface of the fuel bed. The agitator reduces channeling and maintains a uniform bed. The temperature of the gas leaving the gasifier is in the range of 1000–1200°F, depending on the coal type. Pressure is about atmospheric. Ash is removed continuously through a slowly revolving eccentric grate at the reactor bottom. Substitution of air for oxygen to the gasifier will produce a low-Btu raw gas. Raw gas leaving the gasifier is passed through a waste-heat-recovery section. Ash, carried over by gas, and tar are removed by scrubbing. The gas is then compressed and shifted. Pipeline-quality gas is produced by purification, methanation, and dehydration (Wellman Engineering Co).

Westinghouse coal gasification. *Process:* A basic feature of this process is desulfurization of gases at high temperatures (1400–1800°F) using limestone or dolomite sorbent in the gasifier. Crushed (⅛ – ¼ in. × 0), dried coal is fed into a central draft tube of the devolatilizer–desulfurizer unit (gasifier). Coal and large quantities of internally recycled solids are carried upward in the draft tube by hot gases from a combustor flowing at a velocity greater than 15 ft/s. Recycled solids flow downward in a fluidized bed surrounding the draft tube at rates of up to 100 times the coal feed rate. They dilute the coal feed to prevent agglomeration as it devolatizes. Heat requirements of the coal–steam gasification reactions are provided by hot gases produced in the combustor. Lime sorbent is added to the devolatilizer–desulfurizer reactor to remove sulfur which is present as H_2S in the gas. Spent sorbent is withdrawn from the reactor after stripping out the char either in the transfer line or in a separator of special design. Spent sorbent is regenerated and recycled to the reactor. Char is withdrawn from the top section of the devolatilizer–sulfurizer and fed to the combustor. Char is gasified with air and steam at 2100°F. Ash agglomerates at the temperature of the combustor and is removed. Raw product gas from the devolatilizer–desulfurizer unit passes through a cyclone to remove fines and then through a heat-recovery system. Fines are recycled to the combuster (Westinghouse Research Laboratories).

wet air oxidation. A process which oxidizes sludge in the liquid phase without mechanical dewatering. High-pressure, high-temperature air is brought into contact with the waste material in a pressurized reactor. Oxidation occurs at temperatures from 300 to 500°F and pressures from several hundred to 3000 psig. Wet air oxidation has been used for pulp mill liquors, plating wastes, sewage, cyanide destruction, the treatment of sour waters containing sulfides and phenol, and the recovery of valuable products such as silver from x-ray film.

wet cooling tower. The most common practice is to run water in a thin sheet over baffles in hugh hyperbola-shaped towers relying on drafts of air entering at the base to remove heat by evaporation. A variant is to spray the hot water inside the tower in a fine mist. Cool air rising through the tower condenses the mist, releasing heat, which is carried by the air column out of the tower. In either case the cooled water is either discharged into the environment or recycled. Two problems associated with this wet tower technique are water loss and fog formation on cold days, which freezes on contact forming hoar frost, potentially damaging to vegetation and dangerous on highways.

wet scrubber. A scrubber most frequently chosen for industrial applications requiring the cleaning, cooling, and deodorizing of air and gas emissions. Except for the packed towers, all wet scrubbers operate on the basic aerodynamic principle of water droplets which are projected so they collide countercurrent with the particles carried by the gas stream. If the droplets and particles are of comparable size, collision will take place, with the result that the particles will adhere to the droplets and be easily collected. A surface film surrounding a water droplet has an approximate thickness of 1/200 of its diameter.

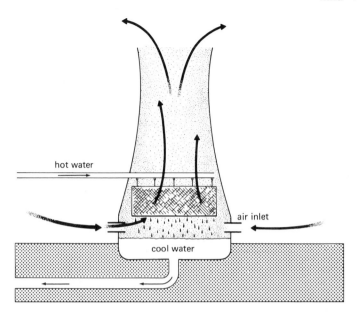

Wet cooling tower. In a wet cooling tower, the hot water is exposed to air circulating through the tower. As water evaporates, heat is lost. The cooled water is either recycled or released into the environment.

The probability of a droplet hitting the particles decreases in relation to dust concentration. To equalize this factor, scrubbers are regulated as to the volume of gas to be scrubbed (pressure drop) and water to be sprayed (pressure). (See p. 342.)

white noise. A noise whose spectrum density (or spectrum level) is substantially independent of frequency over a specified range.

wide band. Applies to a wide band of transmitted waves, with neither the critical nor cutoff frequencies of the filter being zero or infinite.

Wien law. One of the radiation laws which states that the wavelength of maximum radiation intensity for a black body is inversely proportional to the absolute temperature of the radiating black body. This law, established experimentally by Wien in 1896, describes the manner in which the wavelengths of maximum radiation shift toward lower values as the temperature of the radiator rises. It is to be distinguished from the Wien distribution law, which describes the variation of the intensity of emission at any wavelength with temperature. The Wien displacement law is used to compute the color temperature of a radiator.

wind. The horizontal movement of air. Its direction and speed are determined for the most part by the character of the major winds. However, other influences also play a part, such as the time of day. As a rule, winds blow most lightly during the early morning hours and the hardest in the early afternoon. So the morning's load of pollution is the heaviest of the day because it is especially likely to linger on for hours afterwards. Wind speed changes with seasons, and wind direction is also influenced by its environment. It can be affected by the heating ability of the terrain. Topography particularly affects the speed and direction of the wind. Even local obstructions change the character of a wind. As air blows over rough terrain or around buildings it is stirred into eddies and cross currents.

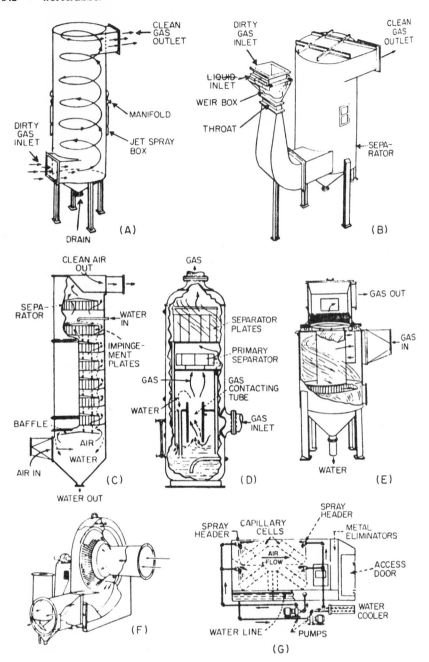

Scrubbers and washers. (a) Airetron/Mikro cyclonic scrubber. *(Pulverizing Machinery.)* (b) Airetron/ Mikro venturi scrubber. *(Pulverizing Machinery.)* (c) Multiwash scrubber. *(Claude B. Schneible Co.)* (d) Liquid vortex contractor. *(Blaw Knox Co.)* (e) Hydrocyclone type scrubber. *(Whiting Corp.)* (f) Centrimerge rotor scrubber. *(Schmieg Industries.)* (g) Wet filter. *(Air and Refrigeration Corp.)*

The obstruction of wind is common to cities, contributing to the dangers they engender. Because of the city's perpendicular buildings and its varve-like streets, the city absorbs more energy during the day and holds on to it longer at night. The many vertical surfaces have more opportunity to collect heat and reflect it to each other, rather than back to the sky. The resulting phenomenon is what meteorologists call the "heat island effect." Warm air rises, carrying with it its burden of pollution; then it expands, flows outward over the edges of the city, and, cooling, sinks. Cooler air from the edge of the city flows into the center to replace the rising air and is followed by the now-cooled dirty air. A self-contained circulatory system has been set up, one that is altered only by a strong wind.

wind rose. Any one of a class of diagrams designed to show the distribution of wind direction experienced at a given location over a considerable period. A wind rose thus shows the prevailing wind direction. The most common form consists of a circle from which 8 or 16 lines emanate, one for each compass point. The length of each line is proportional to the frequency of wind from that direction; the frequency of calm conditions is entered in the center. Many variations exist. Some indicate the range of wind speeds from each direction, some relate wind directions with other weather occurrences.

Winkler process. *Process:* Crushed coal (0–$\frac{3}{8}$ in.) fed to a fluid-bed gasifier through a variable-speed screw feeder. Coal reacts with oxygen and steam to produce offgas rich in carbon monoxide and hydrogen. The temperatures of the fluid bed are in the range of 1500 to 1850°F, depending on the coal type. The pressure is about atmospheric. Because of the high temperature, all tars and heavy hydrocarbons are reacted. About 70 percent of the ash is carried over by gas and 30 percent of it is removed from the bottom of the gasifier by the ash screw. Unreacted carbon carried over by gas is converted by secondary steam and oxygen in the space above the fluidized bed. As a result, maximum temperatures occur above the fluidized bed. To prevent ash particles from melting and forming deposits in the exit duct, gas is cooled by a radiant boiler section before it leaves the gasifier. Raw gas leaving the gasifier is passed through a further waste-heat-recovery section. Fly ash is removed by cyclones, followed by a wet scrubber, and, finally, an electrostatic precipitator. Gas is then compressed and shifted. Gas from the shift converter is purified, methanated, dehydrated, and compressed further, if necessary, to pipeline quality. Substitution of air for oxygen to the gasifier will produce a low-Btu raw gas (Davy Powergas Inc.).

wood fiber. A tiny strand of material that can be reclaimed from old paper to make new paper. Such strands make up the basic material in paper pulp.

wood residue conversion. The city of Eugene, Oregon, owns water, steam, and electric power utilities which are operated by the Eugene Electric Board. Electric power is supplied by three hydroelectric plants of 128-MW total capacity, plus a steam–electric plant of 33.8-MW capacity. This latter plant is fueled with surplus forest residues from the surrounding area to furnish electric power for the local grid. The plant also supplies steam heat service to an area of about two square miles, comprising a business district, a large hospital, a college campus, a large cannery, and a 16-acre greenhouse complex. Hogged wood residues consisting largely of bark are delivered by truck to an outdoor storage pile from nearby wood product mills. They are conveyed to the power house boilers by an elaborate handling system, featuring sizing screens, pulverizers, conveyor belts, metal detectors, and closed circuit television monitors.

work. When a force acts against resistance to produce motion in a body, the force is said to do work. Work is measured by the product of the force acting and the distance moved against the resistance. The cgs unit of work is the erg, a force of one dyne acting through a distance of one centimeter. The joule is 1×10^7 ergs. The foot-pound is the work required to raise a mass of one pound a vertical distance of one foot where $g = 32.174$ ft/s^2. The foot-poundal is the work done by a force of one poundal acting through a distance of one foot. The international joule, a unit of electrical energy, is the work expended per second by a current of one international ampere flowing through one international ohm. The kilowatt-hour is the total amount of energy developed in one hour by a power of one kilowatt.

X

x ray. An electromagnetic wave of short wave length (0.01–40 Å) produced when a cathode ray impinges on matter, when certain nuclear reactions take place, and from the sun. The last appear to be emitted intermittently from small "nozzles" on the sun and are significant in parts of the E region of the ionosphere in causing ionization. X rays of 3 Å from the sun indicate temperatures of 2.5×10^6 °K, higher than that of the chromosphere. X rays of wavelengths as short as 1 or 2 Å have been observed by instruments carried in rockets a few minutes after a large solar flare, indicating a flare temperature of 10^7 °K.

xenon (Xe). An inert gaseous element with atomic weight 131.3; density, 5.85 g/liter; liquid at -10°C; concentration in air, 1 part in 20×10^6.

xylene. $C_6H_4(CH_3)_2$, in commercial preparations a mixture of o, p, and m isomers; a clear to colorless liquid used as a solvent for rubber and plastic cements. The minimum lethal dose of xylene reported was 2000 mg/kg when administered orally and the LD_{50} oral dose has been established as 5000 mg/kg in the rat. Xylene is a common air contaminant, requiring adequate ventilation in industries that use the chemical. In large amounts, xylene can depress the central nervous system. Acute poisoning from inhalation or ingestion may be marked by dizziness, weakness, euphoria, headache, nausea, vomiting, breathing difficulty, and loss of coordination. In severe cases, there may be visual blurring, tremors, heart-beat irregularities, convulsions, paralysis, and loss of consciousness. Skin contact can result in irritation, scaling, and cracking. Contamination of land in a city park in Utrecht, Netherlands, and a housing development at Lekkerkerk was traced in 1980 to the dumping of drums of xylene, along with toluene, as part of the landfill beneath the projects. The incident was similar in cause and effect to the poisoned landfill situation in the Love Canal area of Niagara Falls, New York.

Z

Zeeman effect. The splitting of a spectrum line into several symmetrically disposed components, which occurs when a source of light is placed in a strong magnetic field. The components are polarized, the directions of polarization and the appearance of the effect depending on the direction from which the source is viewed relative to the lines of force.

zeolite softener. An ion exchange process used to soften water, consisting of replacing the calcium and magnesium ions with the sodium ion. Salt (sodium chloride) is used for regeneration. When operated on the hydrogen cycle, calcium and magnesium are replaced with the hydrogen, and a mineral acid (usually sulfuric acid) is used for regeneration.

zinc (Zn). A metallic element. Zinc-bearing effluents may be the result of primary metal and chemical process operations, among others. It has been recommended that zinc be present in public water supplies in concentrations no higher than 5 mg/liter. A complex relationship exists between zinc concentrations, dissolved oxygen pH, temperature, and calcium and magnesium concentrations. Zinc is normally found in seawater at a concentration of 0.01 mg/liter. Marine life may contain zinc in concentrations up to 1500 mg/liter.

zonation, freshwater. Lakes and ponds are zoned primarily on the basis of water depth and the types of vegetation that will appear over the course of time in freshwater areas. A zone just above the edge of standing water is designated as the supralittoral zone. Bulrushes, sedges, insects, and some annelids will live in this zone. From the water's edge to a depth of about six meters is the littoral zone; all of the rooted hydrophytes are limited to this zone. In shallow ponds, rooted vegetation may fill the entire body of water, in which case the pond is monozonal (containing one zone). The zoological representatives of this zone are directly or indirectly dependent on the vegetation present. Beyond this zone of rooted vegetation is the sublittoral zone, extending from the six-meter level out to the average upper limit of the hypolimnion, containing the zooplankton and phytoplankton, as well as the debris from adjacent rooted vegetation that may litter the bottom. The deepest water zone of ponds and lakes, the profundal zone, is not well developed unless the body of water is thermally stratified and it is usually characterized by a deep muddy substrate.

zonation, marine. The ocean community can be divided into two provinces: a neritic province, less than one percent of the entire marine biome, and an oceanic province of considerable magnitude. The neritic province extends from the high-tide mark to a depth of about 200 meters, which marks the outer limit of the continental shelf in most areas, containing illuminated water. The neritic province is comprised of a number of more restrictive zones such as the intertidal zone, which is delimited by high and low water marks of spring tides. There is a considerable variability in temperature, light, evaporation, and salinity in keeping with the movement of tidal water. Many marine organisms are sessile or burrow and depend on the type of substrate available. Along rocky shores, during low tide periods, inhabitants are subject to varying environmental conditions. The biota inhabiting the upper limits of loose rock areas are springtails and crustaceans. In the lower reaches of loose rock intertidal regions, annelids, hermit crabs, starfish, and sea urchins are common. Large stable rocks are inhabited by solid rock borers such as boring spongers, annelids, lamellibranches, sea urchins, and barnacles. The upper limits of stable rock coasts are populated by barnacles and lichens, and the lower part of a stable rock intertidal zone includes such organisms as sponges, tunicates, bryozoans, muscles, and scallops. Sandy areas, in general, are rather sterile with respect to biota; some lamellibranches such as burrowing copepods and mole crabs are characteristic inhabitants.

zonation, sublittoral. The sublittoral zone extends from the lowest low-tide mark out to a 200-meter depth. The substrate is usually of a rather soft consistency, a mixture of sand, silt, and clay, with small rocks or shells scattered over the surface. Zones of seaweed such as *Laminaria* and *Lithothamnion* are found in shallow waters out to a depth of 120 meters. Brown algae such as *Saccorhiza, Alaria,* and *Himanthalea* are found in certain areas. Red algae are common in deeper waters. In the Pacific many of the sublittoral regions are dominated by giant kelps such as *Macrocystis* and *Nereocystis,* which tend to wave action shoreward, stabilize the bottom by means of holdfasts, and provide points of attachment for small epiphytes and epizooids such as barnacles. Animals of different types are numerous in the area: protozoa (*Foraminifera*), sponges, worms including spinunculids, nemertines, polychaets, coelenterates (sea fans, sea pens), and arthropods such as spider crabs, hermit crabs, rock lobsters, and stone crabs. Mollusks of many types, whelks, boat shells, scaphopods, bivalves, cephalopods, echinoderms, including crinoids, brittle stars, and starfish, abound in this zone of the ocean.

zonation, terrestrial. In the ecological sense, zonation refers to the horizontal placement or arrangement of abiotical and living components. In nearly every environmental situation there are distinct zones where the ecological conditions vary from one place to another, as in stratification or vertical layering. A familiar type of vegetational zonation is found in mountainous areas. Montane zonation consists of a number of vegetational belts located at various altitudes along a mountain slope. The distribution of these vegetational zones is governed by temperature and rainfall. Montane zonation varies with regard to faunal and floral inhabitants throughout the world, in keeping with the climatology and geological history of the area. Latitudinal zonation of vegetation is influenced by thermal zonation and precipitation. Isotherms tend to run east and west, but are modified in coastal areas because warm or cold currents of water flowing along coastal regions will modify the terrestrial temperatures.

zooplankton. Tiny aquatic animals that fish feed on.